Applied Biocatalysis in Europe

Applied Biocatalysis in Europe: A Sustainable Tool for Improving Life Quality

Editors

Andres R. Alcantara
Francisco Plou

MDPI • Basel • Beijing • Wuhan • Barcelona • Belgrade • Manchester • Tokyo • Cluj • Tianjin

Editors
Andres R. Alcantara
Complutense University
Spain

Francisco Plou
Institute of Catalysis and Petrochemistry
Spain

Editorial Office
MDPI
St. Alban-Anlage 66
4052 Basel, Switzerland

This is a reprint of articles from the Special Issue published online in the open access journal *Catalysts* (ISSN 2073-4344) (available at: https://www.mdpi.com/journal/catalysts/special_issues/applied_biocatalysis).

For citation purposes, cite each article independently as indicated on the article page online and as indicated below:

LastName, A.A.; LastName, B.B.; LastName, C.C. Article Title. *Journal Name* **Year**, *Volume Number*, Page Range.

ISBN 978-3-0365-1126-9 (Hbk)
ISBN 978-3-0365-1127-6 (PDF)

Contents

About the Editors

Andres R. Alcantara (Ph.D. in Chemistry from the University of Cordoba, Spain) is Full Professor of Organic Chemistry at the Complutense University of Madrid (Spain), where he is co-leader of the TRANSBIOMAT Group (https://www.ucm.es/qffa/transbiomat). His research line is focused on Biotransformations and Applied Biocatalysis in Organic Chemistry, specifically, in the preparation of enantiomerically pure compounds as chemical drug precursors, using mainly proteases, lipases and alcohol dehydrogenases, either native or immobilized. He is a Member of the Scientific Board of the European Society of Applied Biocatalysis (ESAB), where he is the Leader of the Working Group on Sustainable Chemistry, and on the other hand, he is also a Member of the Board of SUSCHEM, the European Platform of Sustainable Chemistry.

Francisco Plou has a degree in Chemistry from the University of Zaragoza and a PhD in Chemistry from the Autonomous University of Madrid. He is a Research Professor at the Institute of Catalysis (Spanish Research Council, CSIC) in Madrid (Spain), where he leads the Applied Biocatalysis Group (http://www.franciscoploulab.eu). His research interests are the biotransformations of carbohydrates and polyphenols, using single or multi-step processes, enzyme immobilization and process optimization for the production of bioactive compounds. He is a representative of Biocatalysis at the Spanish Society of Biotechnology and at the European Section of Applied Biocatalysis.

 **catalysts**

Editorial

Special Issue on "Applied Biocatalysis in Europe: A Sustainable Tool for Improving Life Quality"

Andrés R. Alcántara [1],* and Francisco J. Plou [2],*

1   Department of Chemistry in Pharmaceutical Sciences, Pharmacy Faculty, Complutense University of Madrid (UCM), Ciudad Universitaria, Plaza de Ramon y Cajal, s/n., 28040 Madrid, Spain
2   Instituto de Catálisis y Petroleoquímica, CSIC, Marie Curie 2, 28049 Madrid, Spain
*   Correspondence: andalcan@ucm.es (A.R.A.); fplou@icp.csic.es (F.J.P.)

Citation: Alcántara, A.R.; Plou, F.J. Special Issue on "Applied Biocatalysis in Europe: A Sustainable Tool for Improving Life Quality". *Catalysts* **2021**, *11*, 339. https://doi.org/10.3390/catal11030339

Received: 1 March 2021
Accepted: 3 March 2021
Published: 6 March 2021

Publisher's Note: MDPI stays neutral with regard to jurisdictional claims in published maps and institutional affiliations.

Applied biocatalysis, i.e., the use of enzymes and whole-cell systems in manufacturing processes for synthetic purposes, has been experiencing a clear boom in recent years, which has led to the start of the so-called "fourth wave". In fact, the latest advances in bioinformatics, system biology, process intensification, and in particular, enzyme-directed evolution (encouraged by the 2018 Nobel Prize awarded to F. Arnold) are widening the range of the efficacy of biocatalysts and accelerating the rate at which new enzymes are becoming available, even for activities not previously discovered.

Biocatalysis is the preferred alternative compared with other methodologies of traditional chemistry (based on protection/deprotection steps of functional groups, resulting in multi-step processes with a low overall yield), due to the excellent regio-, chemo- and stereo-specificity of enzymes, as well as to the mild conditions required. Biocatalytic processes are mostly environmentally friendly and generate fewer amounts of waste compared with conventional organic synthesis.

Biocatalysis is a cost-effective and sustainable technology that fulfils most of the principles of green chemistry. This tool has been applied for the manufacture of a wide range of compounds for different areas, including the food, chemical, pharmaceutical and cosmetic industries. Figure 1 represents some of the most important substances currently synthesized with the use of enzymes.

Nine contributions dealing with different aspects of applied biocatalysis are gathered in this Special Issue. The main topics are summarized herein.

Zaccone et al. have reported on a transesterification strategy to synthesize the bioactive compound (*R*)-3-hydroxybutyl-(*R*)-3-hydroxybutyrate [1]. The reaction exploited the stereoselectivity of *Candida antarctica* lipase B, which was very efficient for the kinetic resolution of the racemate.

The issue of the production of omega-3 ethyl esters was investigated and thoroughly discussed by Aguilera-Oviedo et al. [2], who employed enzymes (*Candida antarctica* lipase B) or resting cells (*Aspergillus flavus* and *Rhizopus oryzae*) as biocatalysts.

Estévez-Gay et al. accomplished the computational study of halohydrin dehalogenases (HHDH), a family of enzymes that exhibit a promiscuous epoxide-ring opening activity [3]. Coupling the analysis of conformational landscapes with calculations of tunnels and channels by CAVER software, they assessed their impact on the active site tunnels and the potential ability of HHDH to catalyze the ring opening of bulky epoxides, providing key information for HHDH engineering.

Ramos-Martín et al. accomplished the challenge of obtaining high enantioselectivity at high temperatures [4]. For such purposes, they employed a thermophilic lipase for the synthesis of both enantiomers of mandelic acid by the ethanolysis of a racemate. The best performance was obtained in imidazolium-based ionic liquids at temperatures as high as 120 °C.

Coloma et al. developed an immobilization strategy on controlled-porosity glass to increase the stability in acidic media of hydroxynitrile lyase from *Arabidopsis thaliana* [5].

1

The enzyme was decorated with a His-tag to improve immobilization, and the resulting biocatalysts were efficient in the selective synthesis of (*R*)-cyanohydrins in batch and continuous flow systems.

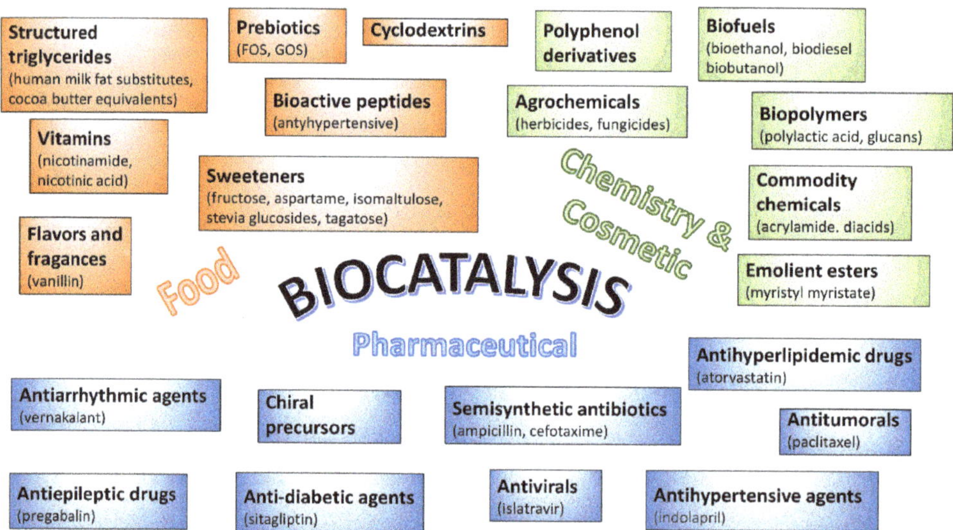

Figure 1. Examples of current applications of biocatalysis for the synthesis of compounds for the food, chemical, cosmetic and pharmaceutical industries. FOS: fructooligosaccharides; and GOS: galactooligosaccharides.

Arslan et al. developed a transesterification process to synthesize a group of fatty acid polyesters of great potential called estolides [6]. They used castor oil as raw material and *Candida antarctica* lipase A as biocatalyst. They obtained estolide trimers and tetramers and proposed multienzyme systems to control product selectivity.

Velasco-Bucheli et al. [7] studied in detail the hydrolysis of *N*-acyl-homoserine lactones using two enzymes: penicillin acylase from *Streptomyces lavendulae* and aculeacin A acylase from *Actinoplanes utahensis*. They proved the involvement of both enzymes in quorum quenching (QQ) processes by *Chromobacterium violaceum* CV026-based bioassays and the inhibition of biofilm formation by *Pseudomonas aeruginosa*, which suggests the application of these enzymes as quorum quenching agents.

Cervantes et al. presented a three-stage process for the valorization of cheese whey into the functional sweetener *D*-tagatose [8]. The *β*-galactosidase from *Bifidobacterium bifidum* hydrolyzed lactose, and the glucose was selectively removed by treatment with *Pichia pastoris* cells, and *L*-arabinose isomerase US100 from *Bacillus stearothermophilus* isomerized *D*-galactose into *D*-tagatose.

Ntana et al. reviewed the pivotal role of *Aspergillus* species in the field of industrial biotechnology [9], in particular as cell factories for the efficient production of recombinant proteins, including many commercial enzymes. The authors went over the advances in the *Aspergillus*-specific molecular toolkit and the use of engineering strategies to increase the expression levels of recombinant fungal proteins. The strategies to overcome the limitations in the production of non-fungal proteins was also covered.

In conclusion, the editors hope that the articles included in this Special Issue of *Catalysts* clearly illustrate the versatility of soluble and immobilized enzymes as efficient catalysts in their different areas of application, and the potential of European research laboratories devoted to this field. We really appreciate the authors for their excellent contributions and the reviewers for their comments, which highlighted certain shortcomings in the manuscripts and helped improve them. We thank the staff of the editorial office of *Catalysts*, and in particular the formidable work done by Caroline Zhan, assistant editor.

Funding: This work was supported by grants from the Spanish Ministry of Science and Innovation (Grants PID2019-105838RB-C31 and PID2019-105337RB-C22).

Conflicts of Interest: The authors declare no conflict of interest.

References

1. Zaccone, F.; Venturi, V.; Giovannini, P.P.; Trapella, C.; Narducci, M.; Fournier, H.; Fantinati, A. An Alternative Enzymatic Route to the Ergogenic Ketone Body Ester (R)-3-Hydroxybutyl (R)-3-Hydroxybutyrate. *Catalysts* **2021**, *11*, 140. [CrossRef]
2. Aguilera-Oviedo, J.; Yara-Varón, E.; Torres, M.; Canela-Garayoa, R.; Balcells, M. Sustainable Synthesis of Omega-3 Fatty Acid Ethyl Esters from Monkfish Liver Oil. *Catalysts* **2021**, *11*, 100. [CrossRef]
3. Estévez-Gay, M.; Iglesias-Fernández, J.; Osuna, S. Conformational Landscapes of Halohydrin Dehalogenases and Their Accessible Active Site Tunnels. *Catalysts* **2020**, *10*, 1403. [CrossRef]
4. Ramos-Martín, J.; Khiari, O.; Alcántara, A.R.; Sánchez-Montero, J.M. Biocatalysis at Extreme Temperatures: Enantioselective Synthesis of both Enantiomers of Mandelic Acid by Transesterification Catalyzed by a Thermophilic Lipase in Ionic Liquids at 120 °C. *Catalysts* **2020**, *10*, 1055. [CrossRef]
5. Coloma, J.; Lugtenburg, T.; Afendi, M.; Lazzarotto, M.; Bracco, P.; Hagedoorn, P.-L.; Gardossi, L.; Hanefeld, U. Immobilization of *Arabidopsis thaliana* Hydroxynitrile Lyase (AtHNL) on EziG Opal. *Catalysts* **2020**, *10*, 899. [CrossRef]
6. Arslan, A.; Rancke-Madsen, A.; Brask, J. Enzymatic Synthesis of Estolides from Castor Oil. *Catalysts* **2020**, *10*, 835. [CrossRef]
7. Velasco-Bucheli, R.; Hormigo, D.; Fernández-Lucas, J.; Torres-Ayuso, P.; Alfaro-Ureña, Y.; Saborido, A.I.; Serrano-Aguirre, L.; García, J.L.; Ramón, F.; Acebal, C.; et al. Penicillin Acylase from *Streptomyces lavendulae* and Aculeacin A Acylase from *Actinoplanes utahensis*: Two Versatile Enzymes as Useful Tools for Quorum Quenching Processes. *Catalysts* **2020**, *10*, 730. [CrossRef]
8. Cervantes, F.V.; Neifar, S.; Merdzo, Z.; Viña-Gonzalez, J.; Fernandez-Arrojo, L.; Ballesteros, A.O.; Fernandez-Lobato, M.; Bejar, S.; Plou, F.J. A Three-Step Process for the Bioconversion of Whey Permeate into a Glucose-Free D-Tagatose Syrup. *Catalysts* **2020**, *10*, 647. [CrossRef]
9. Ntana, F.; Mortensen, U.H.; Sarazin, C.; Figge, R. *Aspergillus*: A Powerful Protein Production Platform. *Catalysts* **2020**, *10*, 1064. [CrossRef]

catalysts

Article

An Alternative Enzymatic Route to the Ergogenic Ketone Body Ester *(R)*-3-Hydroxybutyl *(R)*-3-Hydroxybutyrate

Ferdinando Zaccone [1], Valentina Venturi [1], Pier Paolo Giovannini [1,*], Claudio Trapella [1], Marco Narducci [2], Hugues Fournier [2] and Anna Fantinati [1]

[1] Department of Chemistry and Pharmaceutical Sciences, University of Ferrara, Via Luigi Borsari, 46, 44121 Ferrara, Italy; zccfdn@unife.it (F.Z.); vntvnt@unife.it (V.V.); trpcld@unife.it (C.T.); fntnna1@unife.it (A.F.)

[2] Impact Science Co., LTD, 3F Toutosui Bldg., 6-22-4 Tsukiji, Chuo-ku, Tokyo 104-0045, Japan; mnardux@gmail.com (M.N.); hugtango@gmail.com (H.F.)

* Correspondence: pierpaolo.giovannini@unife.it; Tel.: +39-0532-974532

Abstract: Recent studies have highlighted the therapeutic and ergogenic potential of the ketone body ester, *(R)*-3-hydroxybutyl-*(R)*-3-hydroxybutyrate. In the present work, the enzymatic synthesis of this biological active compound is reported. The *(R)*-3-hydroxybutyl-*(R)*-3-hydroxybutyrate has been produced through the transesterification of racemic ethyl 3-hydroxybutyrate with *(R)*-1,3-butanediol by exploiting the selectivity of *Candida antarctica* lipase B (CAL-B). The needed *(R)*-1,3-butanediol was in turn obtained from the kinetic resolution of the racemate achieved by acetylation with vinyl acetate, also in this case, thanks to the enantioselectivity of the CAL-B used as catalyst. Finally, the stereochemical inversion of the unreacted *(S)* enantiomers of the ethyl 3-hydroxybutyate and 1,3-butanediol accomplished by known procedure allowed to increase the overall yield of the synthetic pathway by incorporating up to 70% of the starting racemic reagents into the final product.

Keywords: ketone body ester; lipase; kinetic resolution; asymmetric synthesis; configuration inversion

Citation: Zaccone, F.; Venturi, V.; Giovannini, P.P.; Trapella, C.; Narducci, M.; Fournier, H.; Fantinati, A. An Alternative Enzymatic Route to the Ergogenic Ketone Body Ester *(R)*-3-Hydroxybutyl *(R)*-3-Hydroxybutyrate. *Catalysts* **2021**, *11*, 140. https://doi.org/10.3390/catal11010140

Received: 24 December 2020
Accepted: 15 January 2021
Published: 19 January 2021

Publisher's Note: MDPI stays neutral with regard to jurisdictional claims in published maps and institutional affiliations.

1. Introduction

The ketone bodies *(R)*-3-hydroxybutyrate and acetoacetate, are short chain acids produced by the liver from the free fatty acids released from adipose tissue. The blood ketone bodies concentration normally ranges below 1 mM [1] increasing up to 5–7 mM during prolonged fasts [2]. Under this metabolic condition, known as ketosis, ketone bodies efficiently replace glucose as respiratory substrate, furnishing a higher adenosine triphosphate (ATP) yield with respect to pyruvate, the end-product of glycolysis [3]. This explain why a mild ketosis is beneficial for muscle and brain during prolonged physical exercise [4–6]. Furthermore, significant results in the treatment of patients affected by neurodegenerative diseases [1,7–9] and epilepsy [10] have been obtained through the increasing of blood ketone bodies induced by consumption of a ketogenic diet. However, a nutrition devoid of carbohydrate and rich of saturated fats is scarcely tolerated by most of the patients, and increases plasma cholesterol and free fatty acids, both known risk factors for several pathologies [11,12]. On the other hand, administration of therapeutically relevant amounts of the ketone bodies as free acids or sodium salts resulted in dangerous acidosis or sodium overload, respectively [13]. The oral assumption of ketone bodies esters has been demonstrated as a successful alternative to induce beneficial levels of ketosis avoiding the increase of blood levels of cholesterol and fatty acids as well as the risk of acidosis or sodium overloading [14]. The most employed ketone body ester is the palatable and nontoxic *(R)*-3-hydroxybutyl *(R)*-3-hydroxybutyrrate [15,16] which is cleaved in vivo to *(R)*-3-hydroxybutyrate and *(R)*-1,3-butanediol. The former is the most abundant ketone body of the entire circulating pool (about 70%) [4], the latter is converted to acetoacetate and *(R)*-3-hydroxybutyrate in the liver [17]. The *(R)*-3-hydroxybutyl *(R)*-3-hydroxybutyrate has been produced by enzymatic reduction of 3-oxobutyl acetoacetate (in turn obtained by transesterification of diketene with 4-hydroxybutan-2-one) [18]. The fermentative production by means of metabolically engineered anaerobic microorganisms has been also reported [19]. The simplest strategy for producing this ketone body ester is the lipase-catalyzed transesterification of ethyl *(R)*-3-hydroxybutyrate with *(R)*-1,3-butandiol [20]. This approach requires enantiopure reagents. The *(R)*-3-hydroxybutyrate can been obtained by enzymatic kinetic resolution of the racemate [21] as well as by alcoholysis of polyhydroxybutyrate, a polyester produced on large scale by bacterial fermentation [22]. Recently, *(R)*-3-hydroxyburate and *(R)*-1,3-butandiol have been respectively obtained by acid catalyzed ethanolysis or sodium borohydride reduction of *(R)*-β-butyrolactone deriving from enzymatic hydrolysis of the corresponding racemate (Scheme 1) [23,24]. Herein we report a new enzymatic approach which, starting from both racemic ethyl 3-hydroxybutyrate and 1,3-butandiol, affords the ketone body ester *(R)*-3-hydroxybuthyl *(R)*-3-hydroxybutyrate. The overall yield of the synthetic pathway was pushed up to 70% thanks to the recycling of the *(S)* reagents by stereochemical inversion.

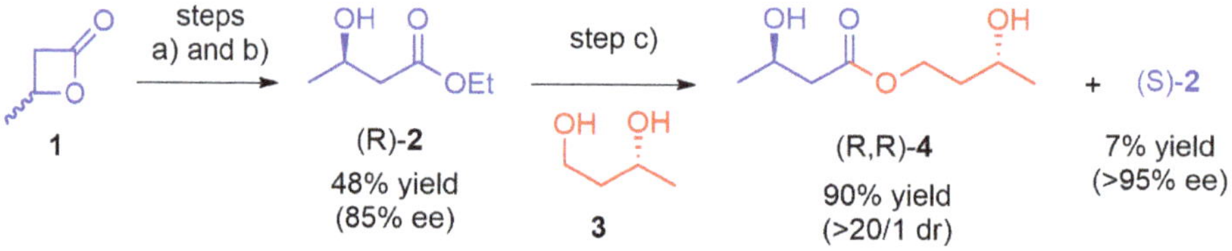

Scheme 1. Synthesis of *(R)*-hydroxybuthyl *(R)*-3-hydroxybutyrate **4** starting from racemic β-butyrolactone **1** following the methodology reported in reference 24. Reaction conditions: Step **(a)** racemic-**1** (50 mmol), H_2O (30 mmol), methyl tert-butyl ether (MTBE) (250 mL), *Candida antarctica* lipase B (CAL-B) (0.3 g), 25 °C, 2 h; Step **(b)** *(R)*-**1** (23 mmol), ethanol (50 mL), H_2SO_4 (0.2% *v/v*), 25 °C, 48 h; Step **(c)** *(R)*-**2** (20 mmol), *(R)*-**3** (20 mmol), CAL-B (0.2 g), 30 °C, 80 mm Hg, 6 h. Yields of *(R)*-**2** and *(R,R)*-**4** referred to the isolated products; the yield of compound *(S)*-**2** was deduced from the CG analysis of the crude reaction mixture.

2. Results and Discussion

2.1. Synthesis of (R)-3-Hydroxybutyl (R)-3-Hydroxybutyrate from Enantioenriched (R)-3-Hydroxybutyrate

To produce *(R)*-3-hydroxybutyl *(R)*-3-hydroxybutyrate on a gram scale, we followed the procedure recently proposed by Ulrich and coworkers [24]. In this procedure, the racemic β-butyrolactone (compound **1**, Scheme 1) was kinetically resolved by *Candida antarctica* lipase B (CAL-B) catalyzed hydrolysis. After aqueous workup to remove the *(S)*-3-hydroxybutanoic acid, the resulting *(R)*-β-butyrolactone was transesterified with ethanol to give ethyl *(R)*-3-hydroxybutyrate **2** (steps a) and b), Scheme 1). In our results, these reaction sequences gave *(R)*-**2** with 85% enantiomeric excess (ee). The lower optical purity with respect to the literature data (>99%) [24], was probably due to an incomplete hydrolytic step. Despite this, we submitted the enantioenriched *(R)*-**2** to the CAL-B-catalyzed transesterification with *(R)*-1,2-butandiol **3** (step c), Scheme 1). Following the reaction course by chiral phase gas chromatographic analysis, we noted that, once the complete conversion of *(R)*-**2** to the desired *(R)*-3-hydroxybuthyl *(R)*-3-hydroxybutyrate *(R,R)*-**4** was reached, the small amount of *(S)*-**2** (7.5%) present in the starting ethyl ester remained unreacted. This prompted us to investigate the possibility to directly use racemic-**2** for the enantioselective synthesis of the ketone body ester *(R,R)*-**4**.

2.2. Synthesis of (R)-3-Hydroxybutyl (R)-3-Hydroxybutyrate from Racemic 3-Hydroxybutyrate

The possibility to produce the ketone body ester *(R,R)*-**4** starting from the racemic ethyl ester **2** was verified by reacting *(R)*-1,3-butandiol (**3**) with two equivalents of racemic-**2** in the presence of CAL-B without the addition of any solvent. The reaction was gently shaken at 30 °C under reduced pressure (80 mmHg) in order to remove of the coproduced ethanol. The formation of product **4**, as well as its stereochemistry, were periodically checked by chiral phase GC analysis. After 5 h, the diol **3** was completely converted to the expected (*R,R*)-**4** leaving the ethyl ester (*S*)-**2** unreacted (Scheme 2, route a). After removing the enzyme by filtration, the crude reaction mixture was distillated under vacuum to recover (*S*)-**2** (40% yield, >91% ee) as the distillate and (*R,R*)-**4** (48% yield, >20/1 dr) as the residue.

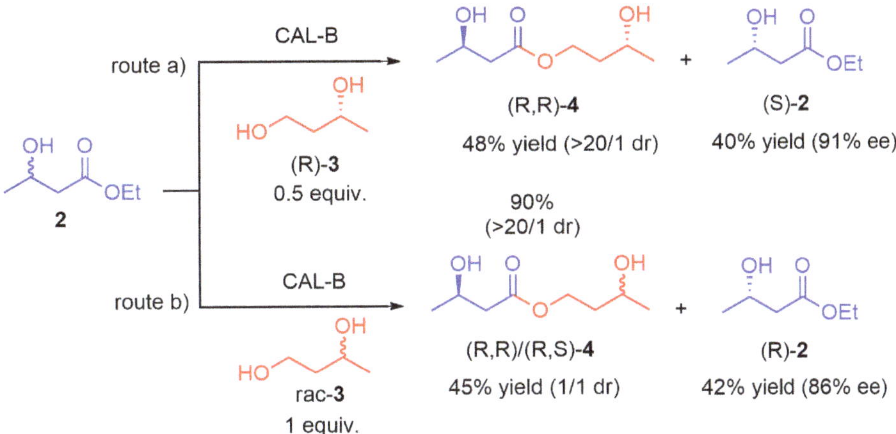

Scheme 2. CAL-B catalyzed transesterification of racemic ethyl 3-hydroxybutyrate **2** with (*R*)- or racemic 1,3-butandiol **3** (route (**a**) and (**b**), respectively). Reaction conditions: (*R*)-**2** (20 mmol), (*R*)-**3** (20 mmol, route a) or racemic-**3** (40 mmol, route b), CAL-B (0.2 g), 30 °C, 80 mm Hg, 6–8 h. Yields for routea (a) referred to the isolated products. Yields for route (b) have been deduced by gas chromatographic (GC) analysis.

The transesterification reaction between both the racemic **2** and **3**, was attempted as well (Scheme 3, route b). However, in this case, because of the distance between the reactive hydroxyl group and the chiral carbon (C3) both the enantiomers of the diol **3** reacted with comparable rates. As a result, a 1:1 mixture of (*R,R*)- and (*R,S*)-**4** was obtained (see Supplementary Materials).

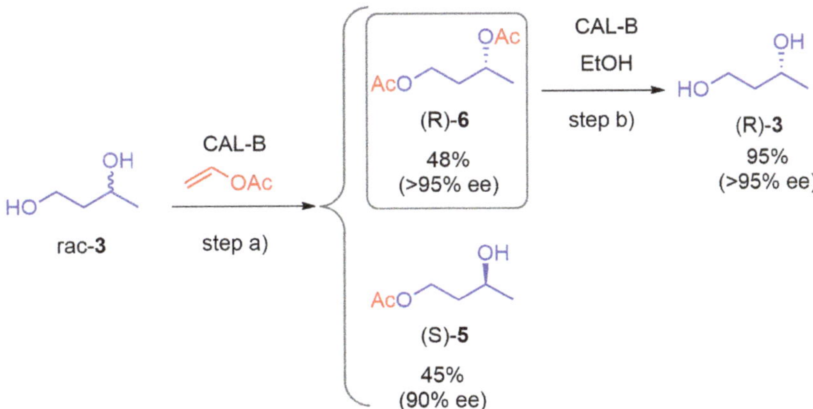

Scheme 3. The synthetic pathway for the preparation of enantiopure (*R*)-1,3-butanediol **3**. Reaction condition: step (**a**) racemic-**3** (20 mmol), vinyl acetate (30 mmol), CAL-B (0.2 g), 30 °C, 2.5 h; step (**b**) (*R*)-**6** (9.6 mmol), ethanol (28.8 mmol), CAL-B (0.2 g), cyclohexane (10 mL), 30 °C, 2h. Yields referred to isolated products.

2.3. CAL-B Catalyzed Kinetic Resolution of the 1,3-Butandiol

Once we ascertained the possibility to use ester **2** as racemate, as well as the needing of enantiomerically pure *(R)*-**3**, we engaged the study on the kinetic resolution of the racemic diol **3**.

The structural resemblance of diol **3** and ester **2**, suggested us to attempt the kinetic resolution of the former, through an enzymatic approach developed for the later [21]. On the other hand, a precedent study reported the lipase mediated kinetic resolution of racemic-3 with Chirazyme™[25]. The reaction, once again catalyzed by CAL-B, was performed in a solvent-free system with 1.5 equivalents of vinyl acetate as the acylating agent. The time course of the reaction monitored by chiral phase CG analysis (Figure 1) showed the not stereoselective esterification of the primary alcoholic group leading to the complete conversion of the racemic diol **3** to *(R)*- and *(S)*-3-hydroxybutyl acetate **5** within the first half-hour. After this, the concentration of *(S)*-**5** remained almost unvaried, while *(R)*-**5** was quickly converted to *(R)*-1,3-butandiol diacetate **6**. The reaction performed on preparative scale (1 g of racemic-3) gave after 2.5 h the complete conversion of the racemic diol **3** to an almost equimolar mixture of *(S)*-**5** and *(R)*-**6** (Scheme 3). After removing the vinyl acetate by evaporation, the crude reaction mixture was chromatographed on silica gel to separate *(S)*-**5** (45% yield, 90% ee) from *(R)*-**6** (48% yield, >95% ee).

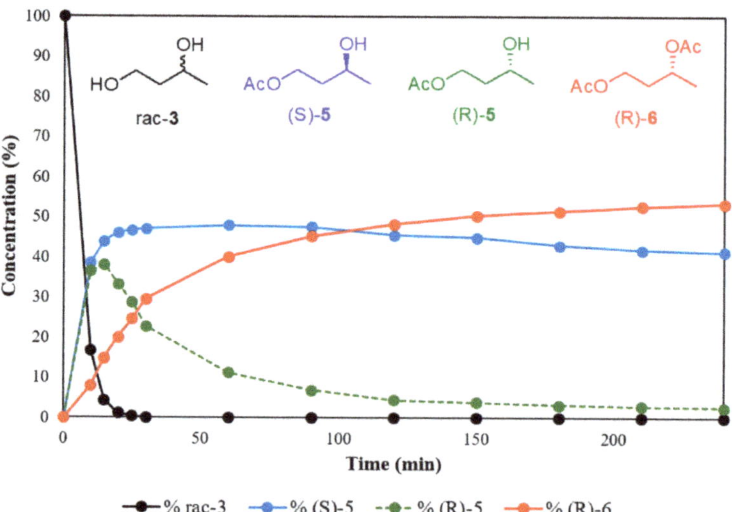

Figure 1. Time course of the CAL-B catalyzed kinetic resolution of racemic diol **3**.

The diacetate was then subjected to ethanholysis in the presence of CAL-B. The reaction was conducted in cyclohexane as the solvent since the use of pure ethanol was reported as detrimental for the stability of the enzyme [21]. After evaporation of cyclohexane, excess of ethanol and ethyl acetate coproduct, the *(R)*-**3** was obtained in 95% yield (>95 % ee) and used without further purification for the synthesis of *(R,R)*-**4**.

2.4. Inversion of Configuration of (S)-3-Hydroxybutyl Acetate 5

The overall yield of the synthetic pathway, including the preparation of the enantiopure diol *(R)*-**3**, was 38% (calculated on the starting racemic-3). Therefore, in order to increase the overall yield as well as the economy of the process, the configuration inversion of the coproducts ethyl *(S)*-hydroxybutyrate **2** and *(S)*-3-hydroxybutyl acetate **5** was then taken into account. The inversion of *(S)*- to *(R)*-**2** by mesylation of the hydroxyl group followed by S$_N$2 with cesium acetate has been recently published [26]. However, a following work reported a low selectivity of this procedure because of the formation of ethyl

3-methylacrylate as elimination by-product (Scheme 4) [24]. For this reason, we focused alternative inversion strategy, based on the tosylation of the hydroxyl group followed by S_N2 inversion with triethylammonium acetate (Scheme 4) [27]. This approach provided the expected O-acetylated *(R)*-**2** in 70% yield (88% ee), a result in line with that reported in the original study. The acetylated product was finally converted to *(R)*-**2** by ethanolysis in the presence of CAL-B as described [21].

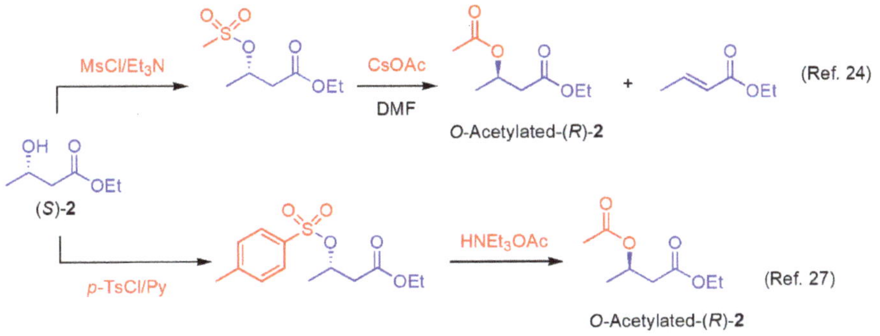

Scheme 4. Synthetic pathways for the configuration inversion of *(S)*-**2** through mesylation followed by S_N2 with cesium acetate [24] or by means of tosylation followed by S_N2 with triethylammonium acetate [27].

Once verified the efficiency of the method as well as its compatibility with the ester group, the *(S)*-3-hydroxybutyl acetate **5** was submitted to the same procedure. The reaction of *(S)*-**5** with *p*-toluenesulfonyl chloride in pyridine gave the expected tosyl derivative **9** in 96% yield (Scheme 5). The following S_N2 was performed by adding compound **9** to a solution of triethylamine and acetic acid in toluene and warming the resulting mixture for 4h at 80 °C. After aqueous workup and solvent evaporation, the residue was chromatographed to give *(R)*-**6** in 80% yield (88% ee).

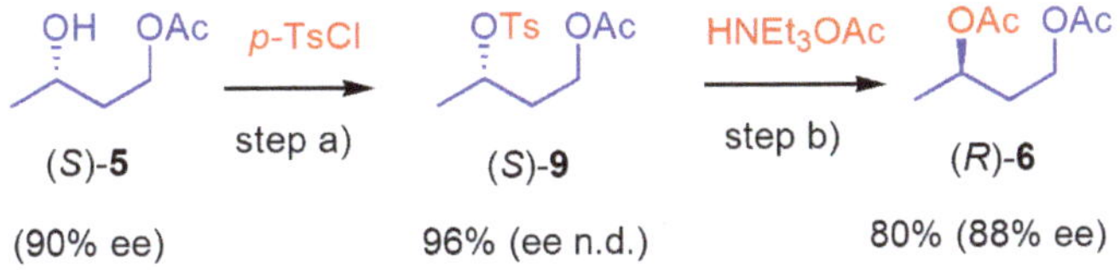

Scheme 5. Inversion of the configuration of the *(S)*-3-hydroxybutyl acetate **5**. Reaction condition: step (**a**) *(S)*-**5** (7.5 mmol), *p*-dimethylaminopyridine (0.38 mmol), pyridine (5 mL), *p*-toluenesulfonyl chloride (9.6 mmol, added at 0° C), 25 °C, 5 h; step (**b**) *(S)*-**9** (7.2 mmol), triethylamine (2.16 mmol), AcOH (13.0 mmol), toluene (5 mL), 80 °C, 4 h. The yields referred to crude product **9** and isolated product **6**. The ee of compound **9** has not determined (n.d.).

3. Materials and Methods

3.1. General Information

All commercially available reagents were used as received without further purification, unless otherwise stated. The CAL-B Novozym®435 was purchased from Novozymes (>Copenaghen, Denmark). Reactions were monitored by TLC on silica gel 60 F254 with detection by charring with phosphomolybdic acid. Flash column chromatography was performed on silica gel 60 (230–400 mesh). ^1H and ^{13}C nuclear magnetic resonance (NMR) spectra were recorded on 300 and 400 MHz Varianspectrometers (Palo Alto, CA, USA) at room temperature using $CDCl_3$ as solvent. Chemical shifts (δ) are reported in ppm relative to residual solvent signals. Optical rotations were measured at 20 ± 2 °C in the stated solvent; $[\alpha]^{20}_D$ values are given in 10^{-1} deg cm^2 g^{-1}. High-resolution mass spectra

(HRMS) were recorded in positive ion mode with an Agilent 6520 high performance liquid chromatography (HPLC)-Chip coupled with a quadrupole/time of fly-mass spectrometer (Q/TOF-MS) nanospray system unit (Santa Clara, CA, USA) to produce spectra. GC analyses were performed using a Thermo Focus gas chromatograph (Waltham, MA, USA) equipped with a flame ionization detector and a Megadex 5 column (25 m × 0.25 mm), with the temperature programs as specified.

3.2. Gas Chromatographic Analysis

Samples (5 mg) were diluted with ethyl acetate and injected (1 μL). The products were detected using the following temperature program: 70 °C for 15 min, 10° C/min up to 200 °C. R_T for ester **2**: 18.5 min; R_T for diol **3**: 20.7 min; R_T for ketone body ester *(S,R)*-**4**: 22.0 min; R_T for ketone body ester *(R,R)*-**4**: 22.0 min. For a better separation of the enantiomers, ester **2** and diol **3** were converted to the corresponding *O*-acetyl derivatives before the injection. The sample (5 mg) was diluted with acetic anhydride (20 μL) and triethylamine (5 μL) and kept at room temperature for two hours. The mixture was diluted with ethyl acetate (1 mL) and injected (1 μL) using the following temperature program: from 60 °C 2 °C/min up to 200 ° C. R_T for the acetyl derivative of *(S)*-**2**: 18.7 min; R_T for the acetyl derivative of *(2)*-**2**: 21.3 min. R_T for diacetate *(S)*-**6**: 23.1 min; R_T for diacetate *(R)*-**6**: 25.1 min.

3.3. Synthesis of (R)-3-Hydroxybutyl (R)-3-Hydroxybutyrate 4 From Racemic 3-Hydroxybutyrate

A mixture of racemic ethyl ester **2** (1 g, 7.6 mmol), *(R)*-1,3 butandiol **3** (0.34 g, 3.9 mmol) and CAL-B (70 mg) was gently shaken under reduced pressure (80 mmHg) at 30 °C for 6 h. The reaction mixture was filtered to remove the enzyme and evaporated under reduced pressure (80 mm Hg) to separate unreacted ethyl ester *(S)*-**2** as the distillate (0.4 g, 3.0 mmol), 40% yield (91% ee), from the *(R)*-3-hydroxybutyl *(R)*-3-hydroxybutyrate *(R,R)*-**4** (0.64 g, 3.6 mmol), 48% yield (>90% dr). ^{1}H NMR (300 MHz, CDCl$_3$) δ 4.34–4.25 (m, 1H, CHOH), 4.22–4.10 (m, 2H, CH$_2$OCO), 3.93–3.79 (m, 1H, CHOH), 2.45 (dd, *J* = 16.1, 3.9 Hz, 1H, CH$_2$CO$_2$), 2.39 (dd, *J* = 16.1, 8.4 Hz, 1H, CH$_2$CO$_2$), 1.82–1.63 (m, 2H, CH$_2$), 1.19 (d, *J* = 2.7 Hz, 3H, CH$_3$), 1.18 (d, *J* = 2.6 Hz, 3H, CH$_3$). ^{13}C NMR (100 MHz, CDCl$_3$) δ 172.9, 65.1, 54.6, 62.1, 43.1, 37.6, 23.5, 22.6. HRMS (ESI) m/z calcd for C$_8$H$_{17}$O$_4{}^+$: 177,1127 [M + H]$^+$; found: 177,1137.

3.4. Kinetic Resolution of Racemic-1,3-Butandiol 3

A mixture of racemic 1,3-butandiol **3** (1.8 g, 20 mmol), vinyl acetate (2.6 g, 30 mmol) and CAL-B (0.2 g) was gently shaken at 30 °C following the reaction course by chiral phase GC analysis. The reaction was stopped when the diol **3** was completely converted (about 2.5 h). The mixture was diluted with methylene chloride (10 mL) and filtered to remove the enzyme. After evaporation of the solvent the residue was chromatographed on silica gel with cyclohexane-ethyl acetate-methanol (15:4:1) as the eluent. *(S)*-3-hydroxybutyl acetate **5** (1.19 g, 9.0 mmol), 45% yield, (90% ee); [α]$_D{}^{20}$ = + 19.1 (c 2.0, CHCl$_3$), lit + 17.5 (c 1.4) [25]. ^{1}H NMR (300 MHz, CDCl$_3$) δ 4.38–4.27 (m, 1H, CHOAc), 4.16 – 4.07 (m, 1H, CHOAc), 3.95–3.88 (m, 1H, CHOH), 2.05 (s, 3H, Ac), 1.85–1.62 (m, 2H, CH$_2$), 1.22 (d, *J* = 6.2 Hz, 3H, CH$_3$). ^{13}C NMR (100 MHz, CDCl$_3$) δ 171.4, 64.6, 61.7, 37.8, 23.4, 20.9. HRMS (ESI) m/z calcd for C$_6$H$_{13}$O$_3{}^+$: 133,0865 [M + H]$^+$; found: 133,0858. *(R)*-1,3-butandiol diacetate **6** (1.67 g, 9.6 mmol), 48 % yield, (>95% ee); [α]$_D{}^{20}$ = −25.7 (c 2.0, CHCl$_3$), lit + 23.5 (c 1.4) [25]. ^{1}H NMR (300 MHz, CDCl$_3$) δ 4.96–4.83 (m, 1H, CHOAc), 4.05–3.95 (m, 2H, CH$_2$OAc), 1.93 (s, 3H, Ac), 1.92 (s, 3H, Ac), 1.86–1.67 (m, 2H, CH$_2$), 1.14 (d, *J* = 6.2 Hz, 3H, CH$_3$). ^{13}C NMR (100 MHz, CDCl$_3$) δ 170.8, 170.3, 67.7, 60.6, 34.6, 21.1, 20.7, 19.9. HRMS (ESI) m/z calcd for C$_8$H$_{15}$O$_4{}^+$: 175,0970 [M + H]$^+$; found: 175,0981.

The *(R)*-1,3-butandiol diacetate **6** (1.67 g, 9.6 mmol) was dissolved in cyclohexane (10 mL). Ethanol (1.32 g, 28.8 mmol) and CAL-B (0.2 g) were added and the mixture was gently shaken at 30 °C following the reaction course by TLC. When the diacetate **6** was fully converted to the diol **3** the reaction was filtered and evaporated to afford *(R)*-1,3-butanediol **3** (0.82 g, 9.1 mmol), 95% yield, (>95% ee).

3.5. Inversion of Configuration of (S)-3-Hydroxybutyl Acetate 5

A solution of *(S)*-3-hydroxybutyl acetate **5** (1 g, 7.5 mmol) and *p*-dimethylaminopyridine (47 mg, 0.38 mmol) in pyridine (5 mL) was cooled to 0 °C and *p*-toluenesulfonyl chloride (1.8 g, 9.6 mmol) was added in portions over 30 min. The mixture was kept at room temperature for 5 h and then diluted with water (16 mL). The white solid precipitated was filtered, washed with cold water (2 × 10 mL) and dried under vacuum at 40 °C to give compound **9** (2.06 g, 7.2 mmol), 96% yield; ^{1}H NMR (300 MHz, CDCl$_3$) δ 7.79 (d, J = 8.3 Hz, 2H, Ar), 7.32 (d, J = 8.4 Hz, 2H, Ar), 4.80–4.67 (m, 1H, CHOTs), 4.07–3.97 (m, 1H, CHOAc), 3.95–3.85 (m, 1H, CHOAc), 2.44 (s, 3H, Ts), 1.96 (s, 3H, Ac), 1.99–1.80 (m, 2H, CH$_2$), 1.34 (d, J = 6.2 Hz, 3H, CH$_3$). ^{13}C NMR (100 MHz, CDCl$_3$) δ 171.0, 145.0, 134.5, 130.1, 128.1, 77.1, 60.4, 35.8, 21.9, 21.4, 21.1. The crude compound **9** (2.06 g, 7.2 mmol) was added to a solution of triethylamine (0.22 g, 2.16 mmol) and acetic acid (0.78 g, 13 mmol) in toluene (5 mL) previously stirred at room temperature for half an hour. The mixture was heated to 80 °C, and stirred at this temperature for 4 h. After cooling to room temperature, the reaction mixture was diluted with toluene (40 mL) and was washed successively with aqueous 2 M HCl solution (20 mL) and 10% (w/v) aqueous K$_2$CO$_3$ solution (30 mL). The organic layer was separated, dried over anhydrous Na$_2$SO$_4$ and evaporated to afford the 1,3-butanediol diacetate **6** (1.0 g, 5.76 mmol), 80% yield, (88% ee).

4. Conclusions

This enzymatic methodology allows for easy access to the nutraceutical and pharmaceutical relevant *(R)*-3-hydroxybutyl *(R)*-3-hydroxybutyrate starting from cheap, racemic reagents. The ethyl 3-hydroxybutyrate was used directly in racemic form while the needed *(R)*-1,3-butandiol was obtained by enzymatic kinetic resolution of the corresponding racemate. Thanks to the configuration inversion of both the distomers *(S)*-3-hydroxybutyrate and *(S)*-1,3-butandiol, the overall yield of the process has been increased over the classical 50% normally achieved by kinetic resolution-based methodologies.

Supplementary Materials: The following are available online at https://www.mdpi.com/2073-4344/11/1/140/s1, Figure S1: ^{1}H- and ^{13}C-NMR spectra of compound **4**; Figure S2: ^{1}H- and ^{13}C-NMR spectra of compound **5**; Figure S3: ^{1}H- and ^{13}C-NMR spectra of compound **6**; Figure S4: ^{1}H- and ^{13}C-NMR spectra of compound **9**; Figure S5: Chiral phase GC of *(R,R)*-**4** and *(R,R)*/*(R,S)*-**4** mixture; Figure S6: Chiral phase GC of acetylated *(R)*-**2** from *(S)*-**2** inversion; Figure S7: Chiral phase GC of acetylated *(R)*-**3** from kinetic resolution rac-**3**; Figure S8: Chiral phase GC of *(R)*-**6** from inversion of *(S)*-**5**.

Author Contributions: Conceptualization, P.P.G. and A.F.; methodology, F.Z. and H.F.; investigation, V.V. and C.T.; resources, M.N.; funding acquisition, M.N. All authors have read and agreed to the published version of the manuscript.

Funding: This research was funded by University of Ferrara, grant FAR 2020.

Institutional Review Board Statement: Not applicable.

Informed Consent Statement: Not applicable.

Data Availability Statement: Data are contained within the article.

Acknowledgments: We gratefully thank Paolo Formaglio for NMR experiments and Francesco Presini for GC analyses.

Conflicts of Interest: The authors declare no conflict of interest.

References

1. Pawlosky, R.J.; Kashiwaya, Y.; King, T.M.; Veech, R.L. A dietary ketone ester normalizes abnormal behavior in a mouse model of Alzheimer's disease. *Int. J. Mol. Sci.* **2020**, *21*, 1044. [CrossRef] [PubMed]
2. Reichard, G.A., Jr.; Owen, O.E.; Haff, A.C.; Paul, P.; Bortz, W.M. Ketone-body production and oxidation in fasting obese humans. *J. Clin. Investig.* **1974**, *53*, 508–515. [CrossRef] [PubMed]
3. Burgess, S.C.; Iizuka, K.; Jeoung, N.H.; Harris, R.A.; Kashiwaya, Y.; Veech, R.L.; Kitazume, T.; Uyeda, K. Carbohydrate-response element-binding protein deletion alters substrate utilization producing an energy-deficient liver. *J. Biol. Chem.* **2008**, *283*, 1670–1678. [CrossRef] [PubMed]
4. Dąbek, A.; Wojtala, M.; Pirola, L.; Balcerczyk, A. Modulation of cellular biochemistry, epigenetics and metabolomics by ketone bodies. Implications of the ketogenic diet in the physiology of the organism and pathological states. *Nutrients* **2020**, *12*, 788. [CrossRef] [PubMed]
5. Cox, P.J.; Clarke, K. Acute nutritional ketosis: Implications for exercise performance and metabolism. *Extrem. Physiol. Med.* **2014**, *3*, 17–26. [CrossRef] [PubMed]
6. Shaw, D.M.; Merien, F.; Braakhuis, A.; Plews, D.; Laursen, P.; Dulson, D.K. The effect of 1,3-butanediol on cycling time-trial performance. *Int. J. Sport Nutr. Exerc. Metab.* **2019**, *29*, 466–473. [CrossRef]
7. Vanitallie, T.B.; Nonas, C.; Di Rocco, A.; Boyar, K.; Hyams, K.; Heymsfield, S.B. Treatment of Parkinson disease with diet-induced hyperketonemia: A feasibility study. *Neurology* **2005**, *64*, 728–730. [CrossRef]
8. Hashim, S.A.; Vanitallie, T.B. Ketone body therapy: From the ketogenic diet to the oral administration of ketone ester. *J. Lipid Res.* **2014**, *55*, 1818–1826. [CrossRef]
9. Veech, R.L. Ketone ester effects on metabolism and transcription. *J. Lipid Res.* **2014**, *55*, 2004–2006. [CrossRef]
10. Freeman, J.M.; Kossoff, E.H.; Hartman, A.L. The ketogenic diet: One decade later. *Pediatrics* **2007**, *119*, 535–543. [CrossRef]
11. O'Neill, B.; Raggi, P. The ketogenic diet: Pros and cons. *Atherosclerosis* **2019**, *292*, 119–126. [CrossRef] [PubMed]
12. Cole, M.A.; Murray, A.J.; Cochlin, L.E.; Heather, L.C.; McAleese, S.; Knight, N.S.; Sutton, E.; Jamil, A.A.; Parassol, N.; Clarke, K. A high fat diet increases mitochondrial fatty acid oxidation and uncoupling to decrease efficiency in rat heart. *Basic Res. Cardiol.* **2011**, *106*, 447–457. [CrossRef] [PubMed]
13. Murray, A.J.; Knight, N.S.; Cole, M.A.; Cochlin, L.E.; Carter, E.; Tchabanenko, K.; Pichulik, T.; Gulston, M.K.; Atherton, H.J.; Schroeder, M.A.; et al. Novel ketone diet enhances physical andcognitive performance. *FASEB J.* **2016**, *30*, 4021–4032. [CrossRef] [PubMed]
14. Kashiwaya, Y.; Pawlosky, R.; Markis, W.; King, T.M.; Bergman, C.; Srivastava, S.; Murray, A.; Clarke, K.; Veech, R.L. A ketone ester diet increases brain malonyl-coa an uncoupling proteins 4 and 5 while decreasing food intake in the normal wistar rat. *J. Biol. Chem.* **2010**, *285*, 25950–25956. [CrossRef]
15. Clarke, K.; Tchabanenko, K.; Pawlosky, R.; Carter, E.; Knight, N.S.; Murray, A.J.; Cochlin, L.E.; King, T.M.; Wong, A.W.; Roberts, A.; et al. Oral 28-day and developmental toxicity studies of *(R)*-3-hydroxybutyl *(R)*-3-hydroxybutyrate. *Regul. Toxicol. Pharmacol.* **2012**, *63*, 196–208. [CrossRef]
16. Clarke, K.; Tchabanenko, K.; Pawlosky, R.; Carter, E.; King, T.M.; Musa-Veloso, K.; Ho, M.; Roberts, A.; Robertson, J.; Vanitallie, T.B.; et al. Kinetics, safety and tolerability of *(R)*-3-hydroxybutyl *(R)*-3-hydroxybutyrate in healthy adult subjects. *Regul. Toxicol. Pharmacol.* **2012**, *63*, 401–408. [CrossRef]
17. Veech, R.L.; Harris, R.L.; Mehlman, M.A. Brain metabolite concentrations and redox states in rats fed diets containing 1,3-butanediol and ethanol. *Toxicol. Appl. Pharmacol.* **1974**, *29*, 196–203. [CrossRef]
18. Robertson, J.; Clarke, K.; Veech, R.L. Process for the Preparation of (3R)-Hydroxybutyl (3R)-Hydroxybutyrate by Enzymatic Enantioselective Reduction Employing lactobacillus brevis Alcohol Dehydrogenase. U.S. Patent 2012/0064611 A1, 15 March 2012.
19. Adelstein, B.; Osterhout, R.; Burk, M.J. Microorganisms and Methods for Producing (3R)-Hydroxybutyl (3R)-Hydroxybutyrate. International Patent WO 2014/190251, 27 November 2014.
20. Clarke, K.; Veech, R.L. Hydroxybutyrate Ester and Medical Use Thereof. International Patent WO 2010/021766 A1, 25 February 2010.
21. Fishman, A.; Eroshov, M.; Sheffer Dee-Noor, S.; van Mil, J.; Cogan, U.; Effenberger, R. A two-Step enzymatic resolution process for large-scale production of *(S)*- and *(R)*-Ethyl-3-Hydroxybutyrate. *Biotechnol. Bioeng.* **2001**, *74*, 256–263. [CrossRef]
22. Clarke, K.; Veech, R.L.; King, T.M. Process for Producing (R)-3-Hydroxybutyl (R)-3-Hydroxybutyrate. International Patent WO 2014/140308 A1, 18 September 2014.
23. Ulrich, S.M. Synthesis of 3-Hydroxybutyryl 3-Hydroxybutyrate and Related Compounds. International Patent WO 2019/147503 A1, 1 August 2019.
24. Budin, N.; Higgins, E.; DiBernardo, A.; Raab, C.; Li, C.; Ulrich, S. Efficient synthesis of the ketone body ester *(R)*-3-hydroxybutyryl-*(R)*-3-hydroxybutyrate and its *(S,S)* enantiomer. *Bioorg. Chem.* **2018**, *80*, 560–564. [CrossRef]
25. Izquierdo, I.; Plaza, M.T.; Rodríguez, M.; Tamayo, J.A.; Martos, A. Lipase mediated resolution of 1,3-butanediol derivatives: Chiral building blocks for pheromone enantiosynthesis. Part 3. *Tetrahedron Asymmetry* **2001**, *12*, 293–300. [CrossRef]
26. Turcu, M.C.; Kiljunenb, E.; Kanerva, L.T. Transformation of racemic ethyl 3-hydroxybutanoate into the *(R)*-enantiomer exploiting lipase catalysis and inversion of configuration. *Tetrahedron Asymmetry* **2007**, *18*, 1682–1687. [CrossRef]
27. Shi, X.-X.; Shen, C.-L.; Yao, J.-Z.; Nie, L.-D.; Quan, N. Inversion of secondary chiral alcohols in toluene with the tunable complex of R_3N–R'COOH. *Tetrahedron Asymmetry* **2010**, *21*, 277–284. [CrossRef]

Article

Sustainable Synthesis of Omega-3 Fatty Acid Ethyl Esters from Monkfish Liver Oil

Johanna Aguilera-Oviedo [1,2], Edinson Yara-Varón [1,2], Mercè Torres [2,3], Ramon Canela-Garayoa [1,2,*] and Mercè Balcells [1,2]

1   Department of Chemistry, University of Lleida, Avda. Alcalde Rovira Roure 191, 25198 Lleida, Spain; johanna.aguilera@udl.cat (J.A.-O.); edinson.yara@udl.cat (E.Y.-V.); merce.balcells@udl.cat (M.B.)
2   Centre for Biotechnological and Agrofood Developments (Centre DBA), University of Lleida, Avda. Alcalde Rovira Roure 191, 25198 Lleida, Spain; merce.torres@udl.cat
3   Department of Food Technology, University of Lleida, Avda. Alcalde Rovira Roure 191, 25198 Lleida, Spain
*   Correspondence: ramon.canela@udl.cat; Tel.: +34-973-702-841

Abstract: The search for economic and sustainable sources of polyunsaturated fatty acids (PUFAs) within the framework of the circular economy is encouraged by their proven beneficial effects on health. The extraction of monkfish liver oil (MLO) for the synthesis of omega-3 ethyl esters was performed to evaluate two blending systems and four green solvents in this work. Moreover, the potential solubility of the MLO in green solvents was studied using the predictive simulation software COnductor-like Screening MOdel for Realistic Solvents (COSMO-RS). The production of ethyl esters was performed by one or two-step reactions. Novozym 435, two resting cells (*Aspergillus flavus* and *Rhizopus oryzae*) obtained in our laboratory and a mix of them were used as biocatalysts in a solvent-free system. The yields for Novozym 435, *R. oryzae* and *A. flavus* in the one-step esterification were 63, 61 and 46%, respectively. The hydrolysis step in the two-step reaction led to 83, 88 and 93% of free fatty acids (FFA) for Novozym 435, *R. oryzae* and *A. flavus*, respectively. However, Novozym 435 showed the highest yield in the esterification step (85%), followed by *R. oryzae* (65%) and *A. flavus* (41%). Moreover, selectivity of polyunsaturated fatty acids of *R. oryzae* lipase was evidenced as it slightly esterified docosahexaenoic acid (DHA) in all the esterification reactions tested.

Keywords: omega-3 ethyl esters; monkfish liver oil; COSMO-RS; fungal resting cells; selectivity

Citation: Aguilera-Oviedo, J.; Yara-Varón, E.; Torres, M.; Canela-Garayoa, R.; Balcells, M. Sustainable Synthesis of Omega-3 Fatty Acid Ethyl Esters from Monkfish Liver Oil. *Catalysts* **2021**, *11*, 100. https://doi.org/10.3390/catal11010100

Received: 28 November 2020
Accepted: 7 January 2021
Published: 13 January 2021

Publisher's Note: MDPI stays neutral with regard to jurisdictional claims in published maps and institutional affiliations.

1. Introduction

Numerous scientific studies have demonstrated the health benefits of PUFAs, in particular those known as omega-3 [1]. The main omega-3 fatty acids are α-linolenic acid (ALA), eicosapentaenoic acid (EPA) and docosahexaenoic acid (DHA). The main health effects of omega-3 fatty acids isolated and combined as ethyl esters have been described for cardiovascular diseases (CVD) [2,3]. Beneficial effects have also been described in diseases such as diabetes [4], immune system problems [5] and cancer [6,7]. Likewise, positive interactions have even been found in diseases such as autism spectrum disorder (ASD) [8] and Alzheimer's [9]. Consequently, omega-3 fatty acid and its ethyl esters have become important as nutraceutical ingredients in functional food products and in pharmaceuticals [10,11], among other uses.

The fatty acid content of marine fish, whether oily or white, is high in omega-3 polyunsaturated fatty acids. These acids are synthesized by microalgae, and reach the fish through the food chain [12]. Fish oil, apart from being used in the food and pharmaceutical industry, is also used in agriculture and mainly in aquaculture as a feed additive [13]. They have also been used as a pesticide carrier, in paints and in leather manufacturing [14]. In addition, the fatty acid esters of simple alcohols are used in a wide variety of applications such as biodiesel fuel or biodiesel additives, lubricants, metalworking coolants, drilling and printing fluids, inks and solvents in alkyd resins [15].

13

According to the APROMAR report on aquaculture and fisheries in Spain 2019 [16], the EU consumed 13 million tons of aquatic products in 2018. There are no specific data on the consumption of monkfish in Spain; however, according to the European fisheries market report for 2019 [17], more than 50.746 tonnes of monkfish were landed in 2017. The waste and coproducts that can be obtained from the fishing industry are variable. The data presented by Erasmus et al. [18] showed that the greatest amount of monkfish waste was produced on fishing vessels (more than 60%). This species of fish has a large head that ends up being discarded with the gonads and livers [18,19]. Therefore, these byproducts or residues can be used as starting materials to prepare new commercial products within the circular economy concept [20]. Fish viscera accounts for 12 to 18% [19] of a fish and are considered animal byproducts (ABPs) not intended for human consumption. Although monkfish liver has an important culinary value in some restaurants as a gourmet dish, it is usually considered as an ABP and is discarded with the rest of the viscera in the vast majority of cases [20–22]. The monkfish, a white fish, contains ca. 30% oil, with a fatty acid profile showing the presence of DHA, EPA, gadoleic acid and oleic acid, among other fatty acids characteristic of fish oils [23].

Currently, there are various methods to determine the lipid content of biological samples. Among them, the Folch method (FM) is commonly used to determine the lipid content in fatty samples in the laboratory [24]. This method takes advantage of the solvent system consisting of chloroform/methanol in a 2:1 ratio to extract the sample. *n*-Hexane is another useful solvent for the extraction of natural products. Its low boiling point, low polarity and chemical stability make it one of the most used solvents to extract nonpolar compounds in the food industry. Nevertheless, these solvents are volatile organic compounds (VOCs) mainly sourced from nonrenewable resources. They are flammable, volatile and toxic, and are responsible for environmental pollution and the greenhouse effect [25,26]. Furthermore, these solvents are now strictly regulated by the European Registration, Evaluation, Authorization and Restriction of Chemicals (REACH). Therefore, industries have been forced to use more environmentally friendly alternatives as green solvents.

Waste fish oils are traditionally extracted by hydraulic pressing or by the use of heat or solvents [27]. The main disadvantage of these methods, apart from those already mentioned, is that the quality of the product can be affected. High temperatures used degrade the thermolabile compounds of the oils, and the use of solvents can have a negative impact on the final product [27]. Consequently, solvents must be chosen with care as they are strictly regulated for food activity by various regulations such as the Codex Alimentarius [28]. In recent years, green extraction methods have been recognized as a promising alternative to traditional organic solvents. Green extraction methods include supercritical fluid extraction (SCF-CO$_2$), microwave-assisted extraction (MAE), ultrasound-assisted extraction (UAE) and enzymatic hydrolysis. They can improve yields, product quality and omega-3 fatty acid content [27]. Therefore, the search for more environmentally friendly solvent extraction processes is a priority [29–31].

Tanzi et al. [29] extracted oil from microalgae (rich in mono and polyunsaturated fatty acids) using terpenes as green solvents (*d*-limonene, α-pinene and *p*-cymene) in order to replace *n*-hexane. The extracted mass of crude oil was higher using terpenes than *n*-hexane. In addition, de Jesus et al. [32] used 2-methyltetrahydrofuran (2-MeTHF) and cyclopentyl methyl ether (CPME) to extract lipids from microalgae *Chlorella pyrenoidosa*, yielding up to 89 mg/g of microalgae biomass. CPME, 2-MeTHF, dimethyl carbonate (DMC) and limonene (LMN) are considered green solvents and have been used in sustainable and environmentally friendly extraction processes [33]. Moreover, the extraction capacity of these solvents can be studied with computer tools, which saves time and resources in experimentation and maximizes the chances of success. COnductor-like Screening MOdel for Realistic Solvents (COSMO-RS) is a software that is used worldwide to predict the most suitable solvents for the extraction of natural products [31,33,34].

Sustainable oil extraction processes are the first step in the exploitation of fish coproducts. While the application and use of fish oils rich in omega-3 is widespread, it is usually more common to use them in the form of esters for their stability [2,10]. Free fatty acids, mainly PUFAs, are considered more susceptible to oxidation than ethyl esters [35]. As high degree unsaturated acids, EPA and DHA oxidize very easily, causing undesirable flavors, which diminish the nutritional quality and safety of food containing them [36]. Ester synthesis can be performed in one-step (transesterification) or two-step (hydrolysis followed by esterification) reactions. These reactions can be catalyzed by enzymes, which allow the development of efficient and fast processes. Lipases are widely used in industry because they have the capacity to catalyze different reactions. Moreover, several scientific articles have demonstrated the selectivity of some lipases for different fatty acids in both hydrolysis and esterification reactions. Moreno-Pérez et al. [37] studied the selectivity of two lipases (*Thermomyces lanuginosus* (TLL) and Lecitase Ultra, a phospholipase with lipolytic activity) immobilized in different supports (hydrophobic Sepabeads C18 and a Duolite anion exchanger) in the synthesis of ethyl esters of omega-3 fatty acids by the ethanolysis of sardine oil in solvent-free systems. They achieved an increase in the activity of TLL and Lecitase. Moreover, the Sepabeads support showed high selectivity for EPA ethyl ester (EPA-EE) synthesis. Castejón et al. [38] studied the enzymatic production of enriched structured triacylglycerols of EPA and DHA (STAG) from *Camelina sativa* oil by two-stage selective hydrolysis-esterification. A noteworthy selectivity of the different lipases tested towards EPA-EE compared to DHA-EE was found. Zangh et al. [35] also reported selectivity between DHA and EPA in the production of DHA-rich triacylglycerides (TAGs) using the commercial enzyme Novozym 435. Ranjan-Moharana et al. [39] described the use of phospholipase A1 for omega-3 enrichment in anchovy oil.

COSMO-RS software was used in this work to predict the extractive potential of monkfish liver oil (MLO) of four alternative green solvents (2-MeTHF, CPME, DMC and LMN). Afterward, the experimental extraction of MLO was carried out to develop a comprehensive, sustainable and environmentally friendly process. The Folch method (a mixture of chloroform and methanol) was used as a reference. Moreover, two systems of agitation were tested (roller mixer and ULTRA-TURRAX®). The recovered MLO was used to prepare the fatty acid ethyl esters (FAEEs) using three biocatalysts, a commercial enzyme (Novozym 435) and two resting cells (*Rhizopus oryzae* and *Aspergillus flavus*) for 24, 48 and 72 h. In addition, the selectivity of these biocatalysts for the different fatty acids present in the MLO was studied. In this sense, we developed a process that potentiates the use of fish coproducts (participating in the circular economy) to synthesize products with industrial potential, such as the ethyl esters of PUFAs, which are significant for various applications, such as cosmetics, food pharmacy and others.

2. Results and Discussion

2.1. COSMO-RS Prediction

A COSMO-RS simulation was conducted to determine the relative solubility of the four main TAGs of MLO in the targeted solvents. ^{1}H-NMR oil analysis was used to determine that TAGs were the main lipids present in the fish oil and to determine the main fatty acids present in the oil. GC-FID analysis confirmed that these TAGs were mainly composed of long carbon chains such as palmitic acid (C16:0), oleic acid (C18:1n9), EPA (C20:5n3) and DHA (C22:6n3). Therefore, we decided to use the four main fatty acids in the oil to define four TAG structures: TAG-1(R1 (C16:0); R2 (C22:6n3); R3 (C16:0)); TAG-2 (R1 (C18:1n9); R2 (C20:5n3); R3 (C16:0)); TAG-3 (R1 (C18:1n9); R2 (C22:6n3); R3 (C18:1n9)); TAG-4 (R1 (C22:6n3); R2 (C22:6n3); R3 (C22:6n3)) as models for COSMO-RS analysis (Figure 1). These main components were modeled with ChemSketch software and used for the predictive study. COSMO-RS integrates a quantum chemistry approach that allows the calculation of several properties, such as the relative solubility of a compound in several solvents. This means that the analysis of the σ profile and the σ potential of the components of the mixture (TAGs and solvents) provides important information about the molecules that can be used to predict possible interactions in the fluid phase.

Compounds   Solvents

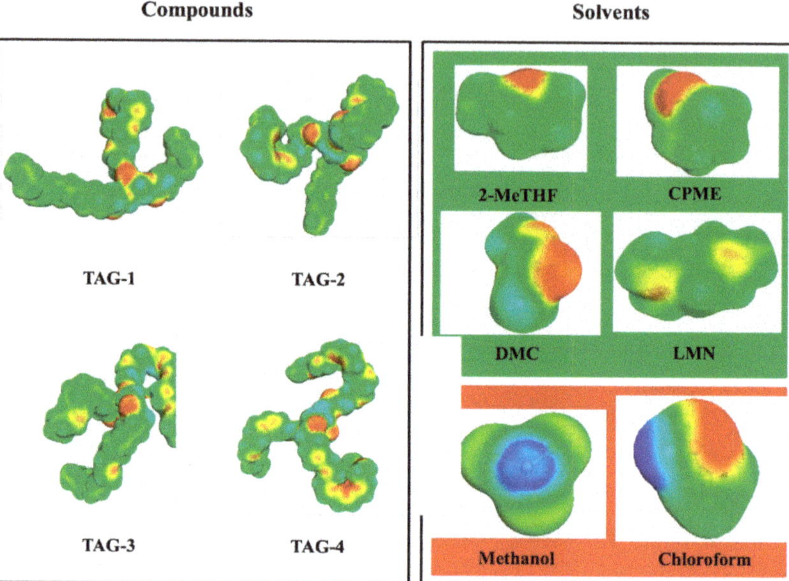

Figure 1. Model of α-surfaces by COSMO-RS of compounds and solvents used in the theoretical study. Compounds (TGAs): TAG-1 (R1 (C16:0); R2 (C22:6n3); R3 (C16:0)); TAG-2 (R1 (C18:1n9); R2 (C20:5n3); R3 (C16:0)); TAG-3 (R1 (C18:1n9); R2 (C22:6n3); R3 (C18:1n9)); TAG-4 (R1 (C22:6n3); R2 (C22:6n3); R3 (C22:6n3)). Green solvents (green color): 2-methyltetrahydrofuran (2-MeTHF), cyclopentyl methyl ether (CPME), dimethyl carbonate (DMC) and limonene (LMN). Reference solvent (red color): Folch reagent (FR), chloroform/methanol (2:1, *v/v*).

Table 1 shows the solubility of the model TAGs from MLO in the solvents used in this study. The solubility is expressed in log10(x_solub) (best solubility is set to 0, and all solvents are given relative to the best solvent) and percentage of probability of solubility for a better understanding of the results. The solvent used as a reference was the Folch reagent (FR), a mixture of chloroform and methanol (2:1, *v/v*), considered to be the most reliable method for full recovery of total lipids [40]. Three of the green solvents tested, 2-MeTHF, CPME and LMN, showed a higher probability of solubility (60–100%) than the reference solvent for TAG-1 to TAG-3 and TAG-4 was similar. Even though DMC presented a low probability of solubility (0–20%) for three of the four model TAGs, this green solvent showed better solubility percentage than the FR. Finally, taking into consideration the theoretical results obtained by the COSMO-RS computational predictive method, we decided to perform the experimental study using the four green solvents with the potential to replace the FR solvent in the extraction of lipids (2-MeTHF, CPME, DMC and LMN).

2.2. Monkfish Liver Oil (MLO) Extraction

The solid-liquid extraction of oil from fresh monkfish liver using five different solvents was performed by maceration using a roller mixer and an ULTRA-TURRAX® system. Folch reagent [24] was used until exhaustion (until no color was observed in the solvent) to determine the maximum content of oil in the liver. A percentage of 39.0% *w/w* was the maximum content of oil in the fresh material, with 49.8% moisture content. This was the first determination of the oil content in monkfish liver as far as we know. This value was higher than the percentage reported for tuna liver, with an oil yield of 17.5% [41], or from salmon byproducts (head, frame and viscera), with an oil content ranging from 13.09 to 19.2% [42]. Ciriminna et al. [43] reported 1.5% of oil content from anchovy heads.

Table 1. COSMO-RS relative solubility (log10(x_solub)) and probability of solubility of triacylglycerides (TAGs) from monkfish liver oil using four different green solvents and Folch reagent (FR) as a reference solvent.

Solvent	TAG 1		TAG 2		TAG 3		TAG 4	
	Log10 (x_solub)	Probability (%)	Log10 (x_solub)	Probability (%)	Log10 (x_solub)	Probability (%)	Log10 (x_solub)	Probability (%)
2-MeTHF	0.0000	100.00	0.0000	100.00	0.0000	100.00	0.0000	100.00
CPME	0.0000	100.00	0.0000	100.00	0.0000	100.00	0.0000	100.00
DMC	−0.9721	10.66	−0.8726	13.41	−0.6799	20.90	0.0000	100.00
LMN	0.0000	100.00	0.0000	100.00	0.0000	100.00	0.0000	100.00
FR	−1.7547	1.76	−1.5751	2.66	−1.4406	3.63	0.0000	100.00

Green color: high probability of solubility (60–100%). Yellow color: medium probability of solubility (20–60%). Red color: low probability of solubility (0–20%). Compounds (triacylglycerides): TAG-1 (R1 (C16:0); R2 (C22:6n3); R3 (C16:0)); TAG-2 (R1 (C18:1n9); R2 (C20:5n3); R3 (C16:0)); TAG-3 (R1 (C18:1n9); R2 (C22:6n3); R3 (C18:1n9)); TAG-4 (R1 (C22:6n3); R2 (C22:6n3), R3 (C22:6n3)). Green solvents: 2-methyltetrahydrofuran (2-MeTHF), cyclopentyl methyl ether (CPME), dimethyl carbonate (DMC) and limonene (LMN). Reference solvent: Folch reagent (FR), chloroform/methanol (2:1, *v/v*).

2.2.1. Extraction in the Two Systems: Roller Mixer (RM) and ULTRA-TURRAX® (UT)

The MLO extraction was carried out following Scheme 1. Two types of agitation (RM and UT) with different extraction times were used. Different solvents were also used (FR or green solvents).

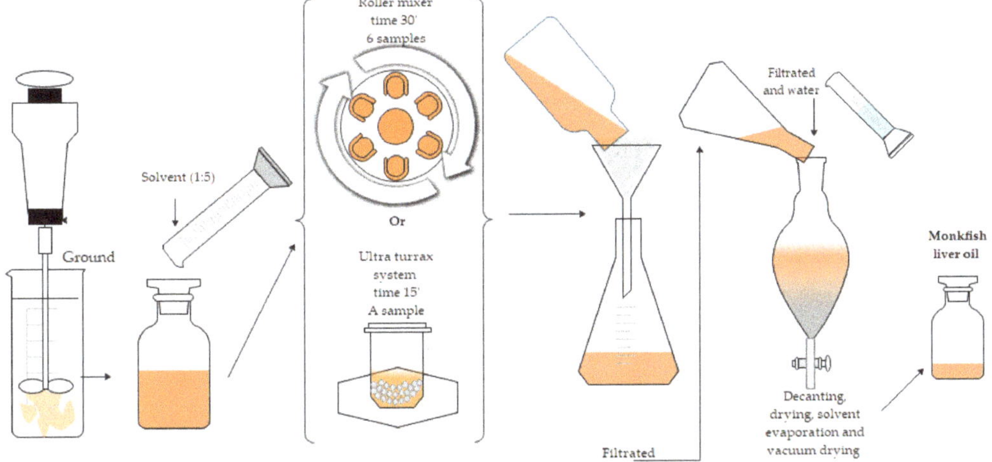

Scheme 1. General procedure to the extraction of monkfish liver oil using two systems; Roller Mixer (RM) and ULTRA-TURRAX® (UT).

Table 2 summarizes the extraction yields obtained with the different solvents and the two agitation systems. The green solvents used correspond to an ester (DMC), two ethers (2-MeTHF and CPME) and a terpene (LMN). The extraction capacity of these green solvents was compared with the extraction capacity of a conventional extraction method (Folch reagent (FR)), which used a mixture of two traditional solvents (chloroform-methanol) in a 1:2 *v/v* ratio. Extraction yields were higher when the RM was used. All green solvents showed extraction yields between 96 and 100% of the maximum oil content in monkfish liver using RM. These extraction yields were higher than those by the FR solvent (89%). Authors such as Fang et al. [41] reported the importance of agitation in their tuna oil extraction experiments. No differences were observed on the NMR spectra of the MLO samples extracted with different solvents and stirring methods (RM or UT) (Figures S1 and S2).

Table 2. Monkfish liver oil extraction with four green solvents and a conventional solvent (reference solvent) using two blending systems (mean ± standard deviations, $n = 2$).

Solvent	Roller Mixer		ULTRA-TURRAX® System	
	Oil Yield (OY) (g Per 100 g FM)	OY Compared to Maximum Oil Content (%)	Oil Yield (OY) (g Per 100 g FM)	OY Compared to Maximum Oil Content (%)
Reference (FR)	39.0	100	39.0	100
2-MeTHF	39.0 ± 0.9	100 ± 3.0	33.9 ± 1.5	87.0 ± 2.2
CPME	39.0 ± 2.4	100 ± 0.7	39.0 ± 0.3	100 ± 2.2
DMC	38.6 ± 1.9	99.0 ± 0.4	29.3 ± 0.2	75.0 ± 5.0
LMN	37.4 ± 1.7	96.0 ± 6.7	32.0 ± 2.6	82.0 ± 4.3
FR	34.5 ± 1.5	89.0 ± 1.5	29.1 ± 1.5	75.0 ± 1.4

FM, fresh material; Reference (FR), maximum oil content in monkfish liver; solvents (2-methyltetrahydrofuran (2-MeTHF), cyclopentyl methyl ether (CPME), dimethyl carbonate (DMC), limonene (LMN) and Folch reagent (FR)).

This system is operated at 4000 rpm. CPME and 2-MeTHF extract the highest quantity of fish oil in the two extraction systems evaluated, which coincide with the COSMO-RS prediction. In the theoretical study, DMC shows a lower probability of oil solubility; however, in the experimental extraction, it shows better results than expected, e.g., 99% in RM. LMN presents worse extraction yields than those predicted by COSMO-RS. Nevertheless, this lower percentage of extraction could be a consequence of the harsh conditions used to recover the oil from the solution (90 °C/0.3 mbar), which could help the evaporation of part of the more volatile compounds present in the fish oil.

Considering the results obtained using the RM system, it can be said that the evaluated green solvents could be interesting alternatives to replace conventional solvents such as hexane, chloroform and methanol to develop more environmentally friendly extractive processes for oil samples [26,30]. It is particularly important to consider the ideal solvent and the best agitation method to choose the best extraction method of the MLO. An ideal alternative solvent must fulfill the following requirements: (a) it is not considered a VOC; (b) it has low toxicity for humans; (c) it has a limited impact on the environment (is eco-friendly); (d) it is obtained from renewable resources; (e) it has a high dissolving power; (f) it is easy to recover; (g) it does not change the setup process significantly [33].

In this sense, the information published in several solvent guides [25,44–47] was considered in this work. Various characteristics and properties of the solvents were evaluated according to the criteria of risk, life cycle, cost, production sources and MLO extraction performance (see Table S1) [44–47].

Considering the above information, the most appropriate solvent for MLO extraction is be 2-MeTHF. This solvent is derived from renewable resources such as corncobs and bagasse [25] and yields up to 100% with the RM extraction method. It is one of the cheapest green solvents (only beaten by LMN), and its score with respect to possible risks is acceptable (ranging from 4 to 6, except for an environmental air risk of 8).

Another interesting solvent is CPME, which provides excellent extraction yields and has a similar risk score to 2-MeTHF. However, this solvent is a byproduct of the synthesis of artificial caucho from crude oil. Moreover, it is the most expensive of all solvents studied. DMC is a solvent that presents high extraction yields with RM agitation and low risks to health and the environment (score between 1 and 5). However, its low extraction yields with UT, price (the second-highest) and the fact that it is prepared by chemical synthesis from methane gas eliminates it from consideration as an ideal solvent. LMN is acceptable in terms of risk assessment as this solvent has a similar score to CPME. However, according to its properties and what has been corroborated in experimentation, it is difficult to evaporate and presents the lowest extraction yields. Therefore, it is not considered a suitable solvent for MLO extraction in this work [25,44–47].

2.2.2. Analysis of the Extracted Oil

The extracted MLO was analyzed by ^{1}H-NMR and GC-FID (Table 3). The ^{1}H-NMR spectrum of MLO showed signals between 0.99 and 1.1 ppm, corresponding to omega-3 fatty acids (Figures S1 and S2). In addition, other signals corresponding to PUFA, MUFA and saturated fatty acids (SFAs) were also observed. These signals coincided with the signals reported by Catrin et al. and Bratu et al. [48,49], who determined the presence of omega-3 fatty acids in fish oil. Specific signals of DHA at 2.4 ppm corresponding to hydrogen linked to both α-carbon (=C–C–CH$_2$–COOR) and allyl-carbon (=C–CH$_2$–C–COOR) were also observed (Figure S3) [50].

Table 3. Fatty acid profile (%) of monkfish liver oil (n = 3).

Common Name.	Common Symbol	FA (%*w/w*)
Saturate Fatty Acid (SFAs)		
Hendecanoic	C10:0	0.2 ± 0.3
Lauric	C12:0	0.3 ± 0.4
Myristic	C14:0	1.9 ± 3.3
Palmitic	C16:0	15.6 ± 0.8
Margaric	C17:0	1.1 ± 0.1
Stearic	C18:0	4.2 ± 0.1
Arachidic	C20:0	1.3 ± 0.2
Behemic	C22:0	3.5 ± 0.5
Tetracosenoic	C24:0	1.7 ± 0.2
Monosaturated Fatty Acids (MUFAs)		
Myristoleic	C14:1n5	0.7 ± 0.4
cis-10-Pentadecenoic	C15:1n5	0.5 ± 0.3
Palmitelaidic	C16:1n7t	0.4 ± 0.0
Palmitoleic	C16:1n7c	7.1 ± 0.2
cis-10-Heptadecenoic	C17:1n7	0.4 ± 0.1
Elaidic	C18:1n9t	1.1 ± 0.1
Oleic	C18:1n9c	21.1 ± 0.5
Vaccenic	C18:1n7	5.4 ± 0.3
Gadoleic	C20:1n9	3.9 ± 0.8
Erucic	C22:1n9	0.7 ± 0.1
Nervonic	C24:1n9	0.8 ± 0.0
	Σ MUFAs	**43.7**
Polyunsaturated Fatty Acids (PUFAs)		
all cis-9,12-Hexadecatrienoic	C16:2n4	0.5 ± 0.1
all cis-6,9,12-Hexadecatrienoic	C16:3n4	0.2 ± 0.1
Linoelaidic	C18:2n6t	0.3 ± 0.1
Linoleic	C18:2n6c	1.0 ± 0.1
α-Linoleic	C18:3 n3	0.2 ± 0.1
γ-Linolenic	C18:3n6	0.4 ± 0.0
Stearidonic	C18:4n3	0.6 ± 0.1
cis-11,14-Eicosadienoic	C20:2n6	0.5 ± 0.1
cis-11,14,17-Eicosatrienoic	C20:3n3	0.4 ± 0.1
all cis-8,11,14-Eicosatrienoic	C20:3n6	0.1 ± 0.1
Arachidonic	C20:4n6	1.1 ± 0.1
Juniperonic	C20:4n3	0.6 ± 0.1
Eicosapentaenoic (EPA)	C20:5n3	4.4 ± 0.3
cis-13,16-Docosadienoic	C22:2n6	0.2 ± 0.1
Adrenic	C22:4n6	0.3 ± 0.0
Clupadonic	C22:5n3	0.5 ± 0.0
Docosahexaenoic (DHA)	C22:6n3	15.2 ± 0.2
	Σ PUFAs	**26.5**

The fatty acid profile of the extracted oils was determined by GC-FID (Table S1 and Figure S4). MLO contained 29.8% SFAs, 43.7% monounsaturated fatty acids (MUFAs) and 26.5% PUFAs. Oleic acid was the main fatty acid with 21.1% of the total, and has been found to have similar beneficial effects on human health as omega-3 fatty acids [51]. Furthermore, the presence of 5.4% of vaccenic acid, an omega-7 isomer of the oleic acid, was determined worthy of consideration due to its association with a low risk of cardiovascular disease [52,53]. Other MUFAs present were gadoleic acid (C20:1n9), characteristic in fish oils, and erucic acid (C22:1n9), which has also been found in fish liver [41,54]. DHA and EPA were the main PUFAs in monkfish oil, with 15.2 and 4.4% of the total, respectively. The oil content of 39% *w/w* of fresh monkfish liver and its composition confirms the potential that this byproduct can be used as a raw material to obtain products with high added value.

2.3. Enzymatic Preparation of Fatty Acid Ethyl Esters (FAEEs)

Two procedures were performed for the production of FAEEs using MLO and ethanol: (a) a one-step procedure based on a transesterification reaction (Figure 2), and (b) a two-step procedure based on sequential hydrolysis and esterification reactions (Figure 3). A commercial enzyme (Novozym 435), two resting cells (*R. oryzae* and *A. flavus*) and two mixtures of these resting cells (1:1 and 7:3 *R. oryzae-A. flavus*) were used as biocatalysts in a solvent-free medium.

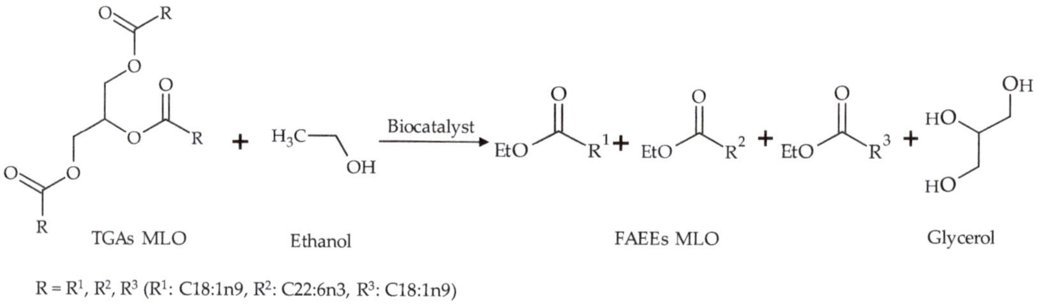

R = R^1, R^2, R^3 (R^1: C18:1n9, R^2: C22:6n3, R^3: C18:1n9)

Figure 2. Schematic reaction diagram of the synthesis of ethyl esters in the one-step transesterification reaction.

A. HYDROLYSIS

B. ESTERIFICATION

R = R^1, R^2, R^3 (R^1: C18:1n9; R^2: C22:6n3, R^3: C18:1n9)

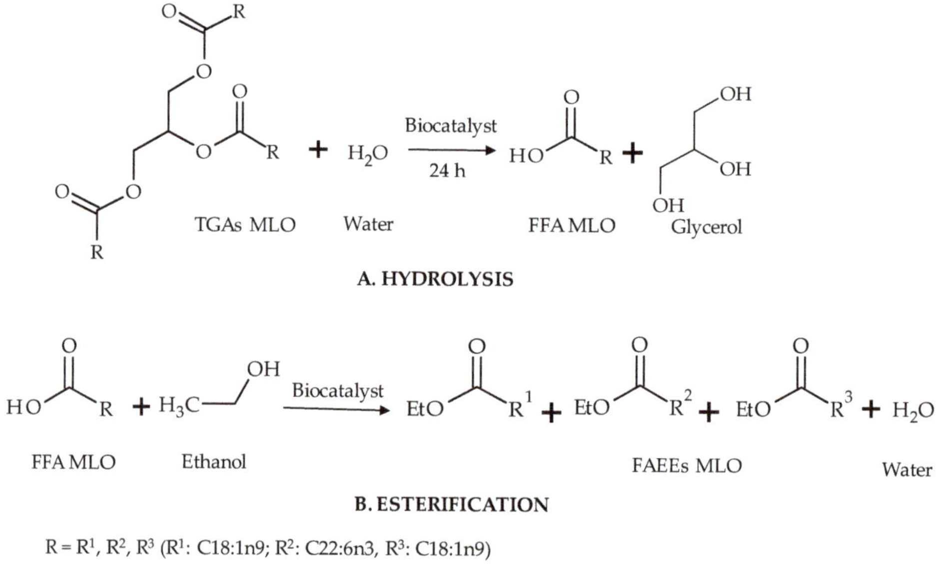

Figure 3. Schematic reaction diagram of the synthesis of ethyl esters in the two-step reaction (**A.** hydrolysis followed by **B.** esterification reaction).

2.3.1. One-Step Synthesis: Transesterification Reactions

The results of the one-step (transesterification) and two-step reactions (hydrolysis and esterification) are shown in Table 4. The commercial enzyme showed the highest yield in the transesterification reaction. Yields were 44%, 61% and 63% for 24, 48 and 72 h, respectively. Transesterification yield with *R. oryzae* was the highest (53%) for 24 h, although the yield for 72 h was slightly lower (61%) than the yield achieved with Novozym (63%). *A. flavus* showed the lowest yields (46% for 72 h). Therefore, the resting cells from *A. flavus* should not be considered an alternative to the commercial enzyme for this transesterification reaction. Moreover, two mixtures of the fungal resting cells (1:1 and 7:3 *R. oryzae-A. flavus*) were used for the transesterification reaction. Only the 7:3 mixture led to a 57% yield after 72 h of reaction, a lower yield than that achieved using only *R. oryzae*. Considering these low yields, the fatty acid ethyl esters profiles of these reactions were not determined.

Table 4. Yields (%) of the different reactions performed to prepare monkfish liver oil ethyl esters. (mean ± standard deviations, $n = 3$).

Biocatalyst	One-Step				Two-Step		
	Transesterification (%)			Hydrolysis (%)	Esterification (%)		
	24 h	48 h	72 h	24 h	24 h	48 h	72 h
Novozym 435	44.0 ± 2.8	61.0 ± 1.5	63.0 ± 0.4	83.1 ± 3.3	54.0 ± 0.5	70.0 ± 3.1	85.0 ± 1.4
R. oryzae	53.0 ± 4.1	54.0 ± 5.0	61.0 ± 2.3	88.1 ± 1.7	42.0 ± 2.7	55.0 ± 1.7	65.0 ± 4.0
A. flavus	32.0 ± 2.7	38.0 ± 3.4	46.0 ± 0.6	93.2 ± 0.0	37.0 ± 5.1	39.0 ± 3.1	41.0 ± 2.0
R. oryzae-A. flavus (1:1)	34.0 ± 4.5	38.0 ± 3.6	45.0 ± 1.7	87.9 ± 5.6	32.0 ± 3.2	37.0 ± 4.5	37.0 ± 4.5
R. oryzae-A. flavus (7:3)	38.0 ± 4.2	45.0 ± 1.7	57.0 ± 4.2	95.7 ± 0.3	41.0 ± 0.6	34.0 ± 4.7	42.0 ± 1.4

The main FAEEs obtained in the one-step reaction studied using the three biocatalysts are shown in Figure 4. The commercial enzyme and the fungal resting cells led to different fatty acid ethyl esters contents. The results corresponding to the percentage of each fatty acid is always referred to as the initial content of the corresponding fatty acid in the MLO. Thus, taking into account the total content of each fatty acid in the MLO (Table 3), and specifically the PUFAs (DHA, 15.2% and EPA, 4.4%), we can conclude that the commercial enzyme esterified 90% of DHA and 100% of EPA. This result shows that this enzyme does not discriminate between the different fatty acids to synthesize the FAEEs. Similar results have previously been described for this commercial enzyme [38,39], which indicates that it is a suitable biocatalyst for the synthesis of these omega-3 EEs. In contrast, the resting cells showed the lowest yields for the esterification of PUFAs, mainly DHA. *A. flavus* showed the highest yield of DHA-EE (38%) of the two resting cells studied; *R. oryzae* only esterified 22% of DHA present in the fish oil after a 72 h reaction; no differences were observed for the esterification of EPA. This finding suggested that lipases from *R. oryzae* discriminated between the different PUFAs present in the MLO. The selectivity of enzymes has been reported by different authors who suggested that some lipases could be selective for certain types of fatty acid depending on their chain length, the solvent used in its extraction and purification, the method of immobilization used for the enzyme and the reaction conditions (temperature and time) [54–57]. The selectivity showed by the resting cells from *R. oryzae* could facilitate the separation of the DHA from the mixture of FAEEs.

2.3.2. Two-Step Synthesis: Hydrolysis and Esterification Reactions

Hydrolysis, the first step of this study, was carried out using the three previous biocatalysts and two mixtures of the two resting cells (1:1 and 7:3) for 24 h (Figure 3A). The resting cells of *R. oryzae* mixed with *A. flavus* in a 7:3 ratio showed the highest percentage of hydrolysis (>95% free fatty acid (FFA)). The second biocatalyst presenting a high hydrolysis percentage was the resting cells of *A. flavus* (>93% FFA) (Figure 5). All resting cells studied showed a higher percentage of hydrolysis than the commercial lipase (83%), indicating that the resting cells could be a cheap alternative to immobilized commercial biocatalysts for these reactions. These hydrolysis percentages can be considered high when compared to those reported by Aranthya et al. [58] for the hydrolysis of fish oil with the enzymes of *Cryptococcus* sp., which yielded 25 and 66.5% of FFA for 24 and 72 h, respectively. Furthermore, it is worth noting that MLO contains more DHA than hydrolyzed cod, sardine, salmon and shark liver oils [58]. A scaling up of the hydrolytic process was carried out using the best reaction conditions achieved in the previous experiment. Starting from 25 g of fish oil, a mixture of *R. oryzae* and *A. flavus* (7:3) allowed the preparation of hydrolyzed monkfish liver oil (HMLO) with 97.8% FFA.

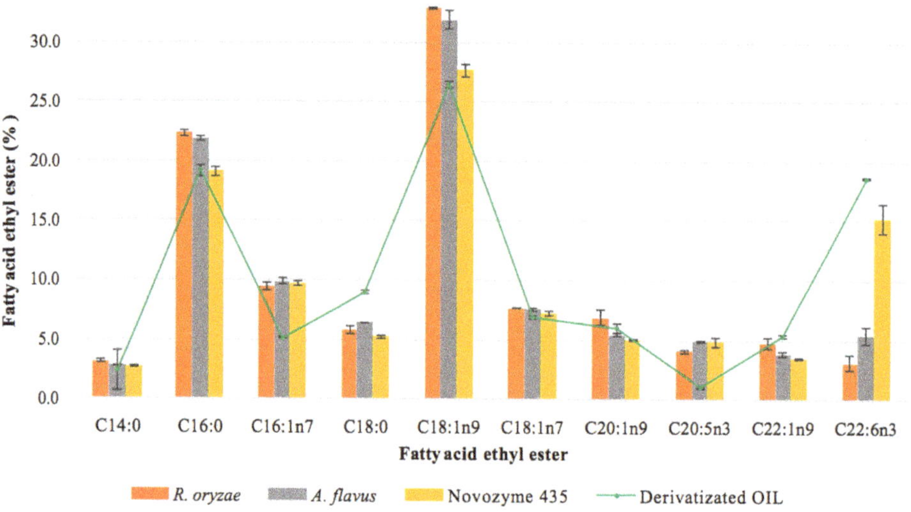

Figure 4. Effect of the biocatalyst on the profile of the main fatty acid ethyl esters obtained in reaction 72 h. OIL: total content of the corresponding ethyl ester in the monkfish liver oil.

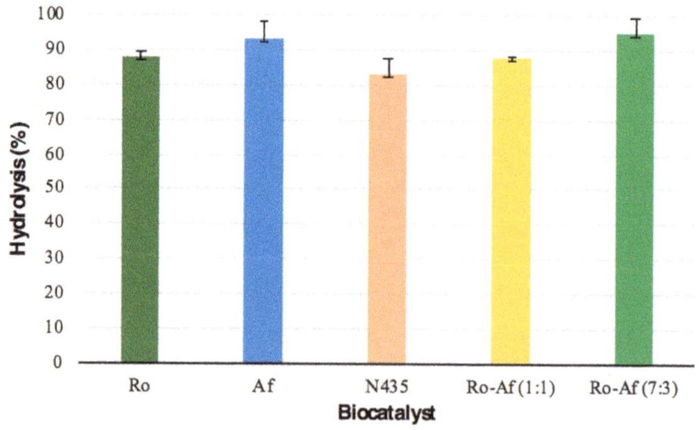

Figure 5. Hydrolysis percentage of monkfish liver oil obtained with the different enzymes (mean ± standard deviations, *n* = 3, reaction 24 h). Ro, *R. oryzae*; Af, *A. flavus*; N435, Novozym 435.

The esterification reaction (Figure 3B) was performed using the hydrolyzed monkfish liver oil (HMLO) obtained in the previous experiment. In these experiments, the same positive correlation was found between reaction times and esterification performance as in the transesterification. The commercial enzyme showed the highest yield in the esterification reaction. Yields were 54, 70 and 85% for 24, 48 and 72 h, respectively. Esterification yield with *R. oryzae* was 65% for 72 h. *A. flavus* again showed the lowest yields (41% for 72 h). These results confirm that *A. flavus* resting cells are not very active in the esterification of the fatty acids from monkfish oil. Regarding the mixtures of the fungal resting cells, the yields were lower than the ones achieved with *R. oryzae* as in the transesterification reaction (see Table 4).

In relation to the FAEEs profiles (Figure 6), the results were similar to those shown in the one-step reactions. There was an enrichment of MUFAs such as oleic acid, gadoleic acid and vaccenic acid, which have been found beneficial to health [51]. Vaccenic acid, a source of omega-7, has been positively correlated with the presence of DHA and EPA [52,53]. The increase of palmitic acid and stearic acid observed in the assays could be due to the fatty acids present in the sunflower oil used in the production of the resting cells as lipase inductors. The microorganisms themselves can also synthesize different fatty acids, *Aspergillus* sp. produces long-chain fatty acids (C16:0, C16:1n7, C17:0, C18:0, C18:1n9, C18:2, C18:3 and C20:0). This fatty acid profile has been used to discriminate fungal species belonging to the *Aspergillus* genus [59]. In addition, the presence of some fatty acids, such as gadoleic acid (C20:1n9), may depend on the type of substrate used in the culture medium [60]. Nevertheless, the increase of these fatty acids in the final crude of the reaction cannot be higher than 4% considering the percentage of biocatalysts used (10%) and the percentage of fatty acids present in the resting cells (40%).

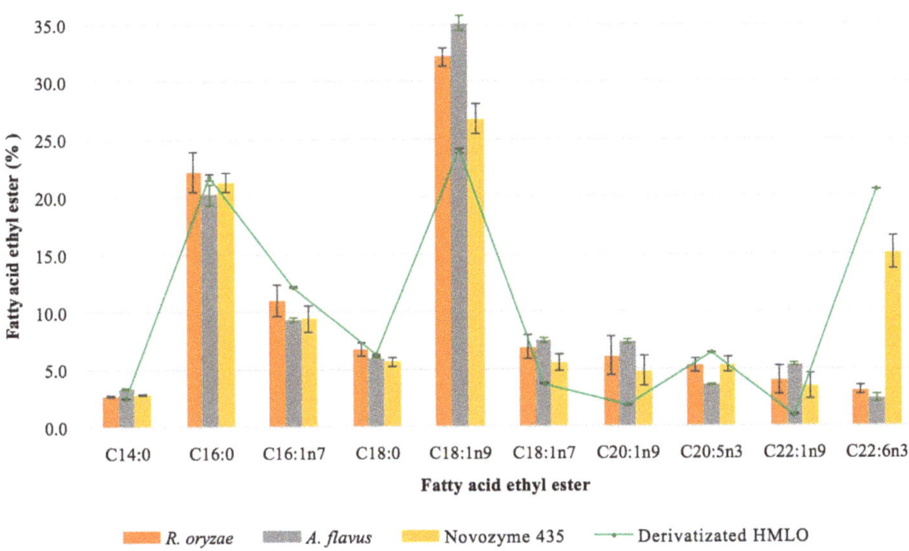

Figure 6. Effect of the biocatalyst on the percentage of the main ethyl esters obtained in a 72 h two-step reaction. Hydrolyzed monkfish liver oil (HMLO) corresponds to the total content of the corresponding ethyl ester in the hydrolyzed monkfish liver oil.

EPA behaved similarly to one-step reactions, in which the commercial enzyme esterified up to 100% of EPA contained in HMLO, *R. oryzae* esterified 70% and *A. flavus* esterified 60%. For DHA, Novozym 435 esterified up to 90% of the DHA contained in HMLO, while the fungal resting cells did not exceed 20%. *A. flavus* lipase esterified up to 19% of the DHA content, while *R. oryzae* lipase esterified up to 16%. These low yields seem to confirm the hypothesis of DHA selectivity shown by *R. oryzae*. Due to the selectivity of *R. oryzae* lipase, DHA should be found as free fatty acid within the esterified material. The possibility of isolating these free fatty acids is also interesting as it has been shown that PUFAs ingested in the form of free fatty acids may be more bioavailable than FAEEs and therefore more assimilable in human metabolism [61].

To confirm the above-mentioned hypothesis and that DHA was present as an acid in the esterified fraction, these samples were subjected to total esterification through chemical catalysis (a derivatization process) using H_2SO_4. Figure 7 compares the DHA-EE obtained in the one and two-step reactions at different times (24, 48 and 72 h) using *R. oryzae* or applying chemical catalysis.

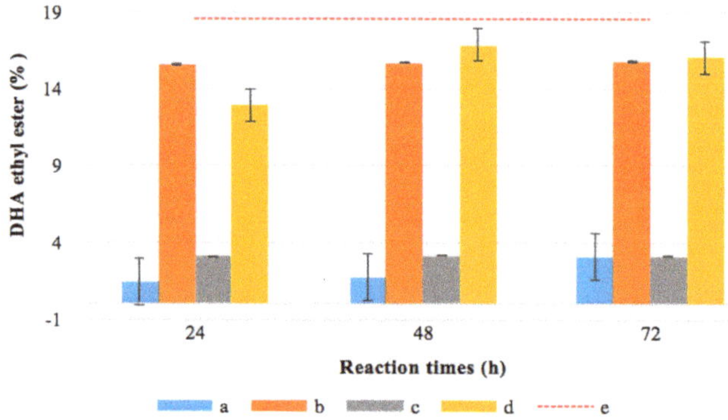

Figure 7. Percentage of docosahexaenoic acid (DHA) before and after total chemical esterification for 24, 48 and 72 h using *R. oryzae* lipase and chemical catalysis. (**a**) Blue color: ester obtained before chemical catalysis from one-step esterification with *R. oryzae*. (**b**) Orange color: ester obtained after applying chemical catalysis to the crude resulting from one-step transesterification. (**c**) Grey color: ester obtained before catalysis from two-step esterification. (**d**) Yellow color: ester obtained after applying chemical catalysis to the crude resulting from the two-step esterification. (**e**) Red dotted line: total DHA content in the MLO.

As expected, chemical catalysis achieved more than 90% esterification of the DHA contained in both monkfish liver oil and the hydrolyzed monkfish liver oil. In the three reaction periods studied, it was evidenced that the lipases from *R. oryzae* were able to effectively discriminate between the DHA and the rest of the fatty acids present in MLO and HMLO. In the HMLO, the DHA indicated it was present as free acid since a 97.8% hydrolysis was achieved.

The selectivity demonstrated by the resting cells of this strain of *R. oryzae* has not previously been reported to our knowledge. However, Ashjaria et al. [62] did study the selectivity of lipases isolated from *R. oryzae* and immobilized by different methods in the hydrolysis of fish oil. All immobilized biocatalysts discriminated between EPA and DHA in favor of EPA.

3. Materials and Methods

3.1. Reagents and Solvents

Chloroform (purity 99%), 2-methyltetrahydrofuran, deuterated chloroform (99.9 atom % D), anhydrous sodium sulfate (Na_2SO_4), oleic acid (purity 90%), potassium hydrogen phosphate (K_2HPO_4) and magnesium sulfate ($MgSO_4$) were purchased from Sigma-Aldrich (Sigma-Aldrich Química SA, Madrid, Spain and St. Louis, MO, USA). Cyclopentyl methyl ether (purity 99%) was from ZEON Corporation (Tokyo, Japan). Dimethyl carbonate (purity 99%), hexane (HPLC grade) and limonene (96%) were purchased from Acros Organics (Fair Lawn, NJ, USA). Ethanol absolute was purchased from Scharlau (Scharlab, Barcelona, Spain). Methanol was purchased from Fisher Scientific (Madrid, Spain). The culture media agar potato dextrosa (PDA) and yeast extract (EY) were provided by Scharlau Microbiology (Scharlab, Barcelona, Spain). Lipase-B from *Candida antarctica* (Novozym 435) was a gift sample from Novozymes A/S (Bagsvaerd, Denmark). The *Rhizopus oryzae* (CECT20476) and *Aspergillus flavus* (CECT20475.2.1) strains were housed in the Spanish Type Culture Collection (CECT) (Burjassot, Valencia, Spain). Monkfish liver was supplied by Congelados y Especialidades Barrufet SL (Barcelona, Spain), and sunflower oil was bought at the market.

3.2. Computational Method

This study was performed by a theoretical procedure using a computational predictive method (COSMO-RS) by considering the technical properties of the solvents and via experimentation. The comparison was made considering the amount of monkfish liver oil extracted and the technical parameters of the solvents used.

COSMO-RS Procedure

The Conductor-like Screening Model for Real Solvents (COSMO-RS) developed by Klamt and coworkers [63] is as known as a powerful method for molecular description and solvent screening based on the result of quantum chemical calculations for an understanding of the dissolving mechanism. COSMO-RS combines quantum chemical considerations (COSMO) and statistical thermodynamics (RS) to determine and predict thermodynamic properties without experimental data. In the first step, the molecule is embedded into a virtual conductor. In such an environment, the molecule induces a polarization charge density on its surface (σ-surface). Calculation can then be performed for each molecule of interest (Figure 1). The second step uses the statistical thermodynamic calculation. Blue is used to represent strongly positive polar regions, and red represents very negative polar surfaces. Green and yellow correspond to lower polarities. The thermodynamics of the molecular interactions that are based on the obtained σ-profile are then used to calculate the chemical potential of the surface segment (σ-potential) [26,31,35]. According to the composition of the monkfish liver oil determined by ^{1}H-RMN and GC-FID, we decided to use four different models of TAG as models in the COSMO-RS analysis using the four main fatty acids from the oil.

In this work, the prediction of the relative solubility of the main triacylglycerides from monkfish liver oil in four green solvents 2-MeTHF, CPME, DMC and LMN, as well as FR, that is, a mixture of chloroform and methyl alcohol (2:1, v/v) was made by implementing this COSMO-RS model in COSMOtherm software (BIOVIA COSMOthermX19; Dassault Systèmes, France). The chemical structures of the solvents and solutes were mutually transformed into their simplified molecular-input line entry syntax (SMILES) notations, which were subsequently used to calculate the solubility parameters of solvents and compounds. The relative solubility was calculated using the following equation [63,64].

$$log_{10}(x_j) = log_{10}\left[\exp\left(\left(\mu_j^{pure} - \mu_j^{Solvent} - \Delta G_{j,\,fusion}\right)/RT\right)\right] \tag{1}$$

μ_j^{pure}: chemical potential of pure compound j (J/mol); $\mu_j^{Solvent}$: chemical potential of j at infinite dilution (J/mol); $\Delta G_{j,\,fusion}$: free energy of fusion of j (J/mol); x_j: solubility of j (g/g solvent); R: gas constant; T: temperature (K).

Relative solubility is always calculated in infinite dilution. The logarithm of the best solubility is set to 0 and all other solvents are given relative to the best solvent. A solvent with a log10 (x_j) value of −1.00 yields a solubility that is decreased by a factor of 10 compared to the best solvent. Additionally, the logarithm is transformed into the probability of solubility and expressed in a percentage.

3.3. Monkfish Liver Oil Extraction

3.3.1. Extraction in a Roller Mixer

The monkfish liver oil was extracted by solid-liquid extraction with FR (chloroform-methanol 2:1 *v/v*) as a reference solvent and using the four green solvents (2-MeTHF, DMC, CPME and LMN). An amount of 20 g of monkfish liver previously defrosted and ground was extracted with 100 mL of each solvent at a ratio of 1:5 *w/v*. The mixture was stirred at 60 rpm for 30 min in a roller mixer. The sample was filtered through paper, and the solid residue was washed two times with fresh solvent. Solvents were then joined, a solution of 1% *v/v* of NaCl was added to the extract obtained with FR, and the mixture was shaken vigorously and left to stand for 2 h. The organic phase was recovered and dried with anhydrous sodium sulfate. It was filtered and the solid was washed with chloroform. Final solutions were evaporated under vacuum in a rotary evaporator. Samples were dried under vacuum for 2 h and were finally stored at $-20\,^{\circ}\mathrm{C}$ until analyzed (see Scheme 1). All experiments were carried out in triplicates.

To perform a better comparison of the amount of monkfish liver oil obtained by every solvent, we determined the maximum content of oil in the monkfish liver by extracting one sample five times with the FR (until no color was observed in the solvent). Solvents were joined and processed as indicated above.

3.3.2. Extraction in an ULTRA-TURRAX® System

An amount of 2 g of ground monkfish liver, 2 g of ceramics balls and 10 mL of the solvents (green solvents or FR) were added to the ULTRA-TURRAX system. The mixture was mechanically stirred at 4000 rpm for 15 min at room temperature. The mixture was then filtered and centrifuged at 5000 rpm for 5 min to separate the supernatant. The organic phase was recovered and dried with anhydrous sodium sulfate. The mixture was filtered and then the solvent was evaporated in a rotary evaporator and the oil was dried under vacuum. The samples were stored at $-20\,^{\circ}\mathrm{C}$ until analysis. All experiments were carried out in triplicate (see Scheme 1).

3.3.3. Preparation of Resting Cells

R. oryzae and *A. flavus* cells were grown in a synthetic liquid medium containing 2 g of asparagine, 1 g of K_2HPO_4, 0.5 g of $MgSO_4$, 5 mg of thiamine hydrochloride, 1.45 mg of Fe $(NO_3)_3 \cdot 9H_2O$, 0.88 mg of $ZnSO_4 \cdot 7H_2O$ and 0.235 mg of $MnSO_4 \cdot H_2O$ per litter of distilled water. The initial pH of the medium was adjusted to pH 6.0. Next, 250 mL aliquots of the medium were sterilized at 121 $^{\circ}$C for 15 min, and 1% (*v/v*) of refined sunflower oil was added aseptically. The medium was inoculated with 2.5 mL of a *R. oryzae* and *A. flavus* spore suspension (1–4×10^6 spores/mL) and then incubated at 28 $^{\circ}$C for 5 d using an orbital shaker at 200 rpm. Mycelium was harvested from the culture medium using a Büchner funnel and washed with distilled water followed by acetone. It was then dried under vacuum for 18 h and ground to a powder. The *R. oryzae* (CECT20476) and *A. flavus* (CECT20475.2.1) strains were housed in the Spanish Type Culture Collection (CECT). The enzymatic units (U) were determined beforehand on the basis of the enzymatic esterification rate of ethyl oleate from oleic acid and ethanol. The specific activity of the resting cells of *R. oryzae* was 1.14 *U* and *A. flavus* was 0.57 *U*.

3.4. Enzymatic Preparation of Ethyl Esters

3.4.1. One-Step Synthesis: Transesterification Reaction

A 1:3.2 mixture of monkfish liver oil (0.453 g; 0.5 mmol) and ethanol (0.0736 g; 1.6 mmol) was added to a reaction vial (5 mL) fitted with a PTFE-lined cap that contained 0.045 g of each biocatalyst (10% *w/w* based on the weight of monkfish liver oil), a commercial enzyme (Novozym 435) or resting cells (*R. oryzae* and *A. flavus*), or their mixtures (1:1 and 7:3, *R. oryzae*-*A. flavus*), respectively (see Figure 2). The mixture was stirred (220 rpm) continuously at atmospheric pressure and 28 $^{\circ}$C. Reaction progress was evaluated for 24, 48 and 74 h. Samples were collected, filtered and the solvent was evaporated. An aliquot of

20 mg of the crude material of the reaction was dissolved in deuterated chloroform and the resulting solution was analyzed by NMR [65–69] and GC-FID. Experiments were carried out in triplicate.

3.4.2. Two-Step Synthesis: Hydrolysis and Esterification Reactions

In the first step, biocatalytic hydrolysis reactions (see Figure 3) were performed using the commercial enzyme, the two resting cells and their mixtures (1:1 and 7:3, *R. oryzae*/*A. flavus*). A mixture of monkfish liver oil (0.453 g; 0.5 mmol), biocatalyst (0.045 g; 10% *w/w* based on the weight of monkfish liver oil) and water (0.453 g; 25 mmol) was stirred (220 rpm) at 28 °C for 24 h. The sample was filtrated, centrifuged and the supernatant was collected. The resulting hydrolysate material was analyzed by NMR and GC-FID, with this process determining which of the tests offered the best results in terms of percentage of free fatty acids. With the better biocatalyst, the hydrolysis was scaled up to obtain a minimum of 30 mL of hydrolysate, with which the esterification studies were carried out. The hydrolysate material was analyzed by NMR and GC-FID to determine the hydrolysis degree and the fatty acids profile. In the second step, the hydrolyzed oil was used to perform the esterification reactions using the same biocatalysts indicated above. A mixture of hydrolyzed monkfish liver oil (0.300 g; 1 mmol), biocatalyst (0.03 g, 10% *w/w* based on the weight of hydrolyzed monkfish liver oil) and ethanol (0.147 g; 3.2 mmol) were stirred (220 rpm) at 28 °C. The esterification reactions were conducted for 24, 48 or 72 h. The reaction products were analyzed by NMR [65–69] and GC-FID. All the experiments were carried out in triplicate.

3.5. Analytical Methods

Preliminary analysis of monkfish liver oil, hydrolyzed oil and ethyl esters were carried out by ^{1}H NMR. Spectra were recorded with a MERCURYplus NMR Spectrometer Systems VARIAN 400 MHz magnet using deuterated chloroform (99.9 atom % D) as solvent. The fatty acid profiles of the initial oil (as methyl and ethyl esters) and the ethyl esters after synthesis reactions were analyzed using an Agilent 6890 series gas chromatograph (Barcelona, Spain) coupled to a flame ionization detector (FID). The chromatographic column was a 30 m × 0.25 mm fused silica capillary coated with a 0.25 µm film thickness (50%-cyanopropyl)-methylpolysiloxane (DB-23; Agilent J&W, Madrid, Spain). The temperature program used was 180 °C for 1 min, followed by an increase of 20 °C per minute until the final temperature of 270 °C was reached, which was then maintained for 20 min. A splitless mode of 20 mL/min was applied for 9 s. Hydrogen was used as carrier gas at a constant pressure. The injection volume was 1 µL. The injection system was maintained at 270 °C and the FID at 280 °C.

4. Conclusions

During this study, different methods for MLO extraction were studied, using a traditional solvent (FR) and green solvents (2-MeTHF, CPME, DMC and LMN). In addition, two agitation systems (Roller Mixer and Ultra Turrax®) were studied. The roller mixer was the best agitation system, obtaining higher extraction percentages with all the solvents studied. All the green solvents tested showed extraction yields greater than or equal to the FR single extraction with both agitation systems. Analysis of the tested solvents (taking into account the properties of the solvents, costs, toxicity and safety use, risks to human health, environmental risks, percentage of extraction and resources) showed that 2-MeTHF was the best option for MLO extraction under the studied parameters. The lipidic profile of the MLO illustrated that it could be considered as a suitable source for obtaining PUFAs and DHA.

The results obtained in the esterification trials showed that the percentages of MLO esters in one-step transesterification were 63, 61 and 46% using Novozym 435, *R. oryzae* lipase and *A. flavus* lipase, respectively. In the two-step reactions, yields were 85, 65 and 41% using the commercial enzyme, *R. oryzae* and *A. flavus*, respectively. Consequently, the latter was definitely not a good biocatalyst for these reactions. Moreover, *R. oryzae* resting cells (CECT20476) showed the lowest yields of DHA-EE in one or two-processed steps, suggesting selectivity towards this fatty acid. The resting cells of the filamentous fungi used in this study, unlike commercial enzymes, did not undergo any kind of extraction, purification and immobilization process. Therefore, they were very cheap biocatalysts compared to commercial ones. Furthermore, several authors reported that enzymatic activity and selectivity of lipases could be affected by the immobilization process. Similarly, it was clear that commercial lipase esterified the omega-3 fatty acids optimally, with yields above 90% for DHA and 100% for EPA contained in the MLO. The resting cells of the *R. oryzae* and *A. flavus* tested showed a good percentage of hydrolysis (88% and 93%, respectively), higher than those of the commercial enzyme (61%). These results open a new path to studying the enrichment of PUFA using these resting cells.

Supplementary Materials: The following are available online at https://www.mdpi.com/2073-4344/11/1/100/s1, Figure S1: 1H-NMR spectrum of monkfish liver oil (MLO) extracted with Roller mixer (RM) and different solvents; Figure S2: 1H-NMR spectrum of hydrolysed monkfish liver oil (HMLO) extracted with ULTRA-TURRAX® (UT) and different solvents; Figure S3: 1H-NMR spectrum of MLO extracted by Folch method (FM) and with RM agitation; Figure S4: Chromatogram (GC-FID on DB-23, 30 m) chemically esterified monkfish liver oil; Table S1: Green solvent properties, extraction yields (MLO), substance information and cost analysis.

Author Contributions: The individual contributions of each author is the following: R.C.-G. and E.Y.-V. conceptualization; J.A.-O. performed the experiments; R.C.-G. and M.T. methodology; J.A.-O. and E.Y.-V. validation, formal analysis and investigation; M.B. resources; J.A.-O. writing—original draft preparation; E.Y.-V. and R.C.-G. writing—review and editing, E.Y.-V. and R.C.-G. supervision; M.B. funding acquisition. All authors have read and agreed to the published version of the manuscript.

Funding: This research was funded by Interreg POCTEFA program (EFA253/16 BIOPLAST) and the PhD student aid program "Jade Plus" financed by Banco Santander of which Johanna Aguilera Oviedo is a beneficiary.

Institutional Review Board Statement: Not applicable.

Informed Consent Statement: Not applicable.

Data Availability Statement: Data is contained within the article or supplementary material.

Acknowledgments: This work was supported in part by the Spanish government through the Interreg POCTEFA program (EFA253/16 BIOPLAST) and the PhD student aid program "Jade Plus" financed by Banco Santander of which Johanna Aguilera Oviedo is a beneficiary. The authors would like to thank the Vice-rector of Research of the University of Lleida for the scholarship granted to Johanna Aguilera Oviedo. The authors would also like to thank Congelados y Especialidades Barrufet SL for supplying the monkfish livers.

Conflicts of Interest: The authors declare no conflict of interest.

References

1. Punia, S.; Sandhub, K.S.; Siroha, A.K.; Dhull, S.B. Omega 3-metabolism, absorption, bioavailability and health benefits—A Review. *Pharma Nutr.* **2019**, *10*, 100162. [CrossRef]
2. Shahidi, F.; Ambigaipalan, P.S. Omega-3 Polyunsaturated Fatty Acids and Their Health Benefits. *Annu. Rev. Food Sci. Technol.* **2018**, *9*, 345–381. [CrossRef]
3. Von Schacky, C. A review of omega-3 ethyl esters for cardiovascular prevention and treatment of increased blood triglyceride levels. *Vasc. Health Risk Manag.* **2006**, *2*, 251–262. [CrossRef]
4. Golpour, P.; Nourbakhsh, M.; Mazaherioun, M.; Janani, L.; Yaghmaei, P. Improvement of NRF2 gene expression and antioxidant status in patients with type 2 diabetes mellitus after supplementation with omega-3 polyunsaturated fatty acids: A double-blind randomised placebo-controlled clinical trial. *Diabetes Res. Clin. Pract.* **2020**, *162*, 108120. [CrossRef]
5. Calder, P.C. Immunomodulation by omega-3 fatty acids. *Prostaglandins Leukot. Essent. Fat. Acids* **2007**, *77*, 327–335. [CrossRef]
6. Volpato, M.; Hull, M.A. Omega-3 polyunsaturated fatty acids as adjuvant therapy of colorectal cancer. *Cancer Metastasis Rev.* **2018**, *37*, 545–555. [CrossRef]
7. Nindrea, R.D.; Aryandono, T.; Lazuardi, L.; Dwiprahasto, I. Protective Effect of Omega-3 Fatty Acids in Fish Consumption Against Breast Cancer in Asian Patients: A Meta-Analysis. *Asian Pac. J. Cancer Prev.* **2019**, *20*, 327–332. [CrossRef]

8. Mazahery, H.; Conlon, C.A.; Beck, K.L.; Mugridge, O.; Kruger, M.C.; Stonehouse, W.; Camargo, C.A., Jr.; Meyer, B.J.; Tsang, B.; Von Hurst, P.R. Inflammation (IL-1β) Modifies the Effect of Vitamin D and Omega-3 Long Chain Polyunsaturated Fatty Acids on Core Symptoms of Autism Spectrum Disorder—An Exploratory Pilot Study. *Nutrients* **2020**, *12*, 661. [CrossRef]
9. Swanson, D.; Block, B.; Mousa, S.A. Omega-3 Fatty Acids EPA and DHA: Health Benefits Throughout Life. *Adv. Nutr.* **2012**, *3*, 1–7. [CrossRef]
10. Kwatra, B.A. Review on Potential Properties and Therapeutic Applications of DHA and EPA. *Int. J. Pharm. Pharm. Res.* **2019**, *16*, 140–176.
11. Sahena, F.; Zaidul, I.S.M.; Jinap, S.; Saari, N.; Jahurul, H.A.; Abbas, K.A.; Norulaini, N.A. PUFAs in Fish Extraction, Fractionation, Importance in Health. *Compr. Rev. Food Sci. Food Saf.* **2009**, *8*, 59–74. [CrossRef]
12. Sargent, J.R.; Bell, M.V.; Bell, J.G.; Henderson, R.J.; Tocher, D.R. Origins and functions of n-3 polyunsaturated fatty acids in marine organisms. In *Phospholipids: Characterization, Metabolism and Novel Biological Applications*; Cevc, G., Paltauf, F., Eds.; AOCS Press: Champaign, IL, USA, 1995; pp. 248–258.
13. Espinosa, S.; Diaza, M.S.; Brignolea, E.A. Food additives obtained by supercritical extraction from natural sources. *J. Supercrit. Fluids* **2008**, *45*, 213–219. [CrossRef]
14. Pike, I.H.; Jackson, A. Fish oil: Production and use now and in the future. *Lipid Technol.* **2010**, *22*, 59–61. [CrossRef]
15. Angulo, B.; Fraile, J.M.; Gil, L.; Herrerías, C.I. Comparison of chemical and enzymatic methods for the transesterification of waste fish oil Fatty Ethyl Esters with different Alcohols. *ACS Omega* **2020**, *5*, 1479–1487. [CrossRef]
16. Informe Realizado por la Asociación Empresarial de Acuicultura de España (APROMAR). 2019, pp. 1–91. Available online: http://apromar.es/sites/default/files/2019/InformeAcui/APROMAR%20Informe%20ACUICULTURA%202019%20v-1-2.pdf (accessed on 15 August 2020).
17. Informe EL MERCADO PESQUERO DE LA UE 2019. 2019, p. 1. Available online: https://www.eumofa.eu/documents/20178/314856/ES_El+mercado+pesquero+de+la+UE_2019.pdf/ (accessed on 29 August 2020).
18. Erasmus, V.N.; Kadhila, T.; Gabriel, N.N.; Thyberg, K.L.; Ilungu, S.; Machado, T. Assessment and quantification of Namibian seafood waste production. *Ocean Coast. Manag.* **2021**, *199*, 105402. [CrossRef]
19. Iñarra, B.; Bald, C.; San Martín, D.; Orive, M.; Cebrián, M.; Zufía, J. Guía para la Valorización de Subproductos de la Acuicultura. AZTI, Derio. 2018, pp. 1–44. Available online: https://www.azti.es/wp-content/uploads/2018/12/AZTI_guia_VALACUI101218online.pdf (accessed on 15 August 2020).
20. Fisher, R.A.; DuPaul, B. *A Fisherman's Guide: Getting the Most out of Monkfish. Marine. Resource Advisory No. 37*; Virginia Institute of Marine Science, College of William and Mary: Gloucester Point, VA, USA, 1990; pp. 1–7. [CrossRef]
21. *Manipulación de la Pesca del día en Embarcaciones de Red de Enmalle*; Servicio Central de Publicaciones Gobierno Vasco, Departamento de Medio Ambiente, Planificación Territorial y Pesca; ISBN 978-84-457-3186-433-34. Colección ITSASO 37; 2011; pp. 33–34. Available online: http://www.euskadi.net/ejgvbiblioteka (accessed on 15 July 2020).
22. Guía de Buenas Prácticas de Higiene para Buques Palangreros y Buques Factoría Congeladores. Organización de Productores de Buques Congeladores de Merlúcidos, Cefalópodos y Especies Varias. OPPC-3. 2017, pp. 78–80. Available online: http://www.arvi.org/publicaciones/GUIA_DE_BUENAS_PRACTICAS_DE_HIGIENE.PDF (accessed on 15 July 2020).
23. Pacetti, D.; Alberti, F.; Boselli, E.; Frega, N.G. Characterisation of furan fatty acids in Adriatic fish. *Food Chem.* **2010**, *122*, 209–215. [CrossRef]
24. Folch, A.J.; Lees, M.; Sloane-Stanlet, G.H. Simple method for the isolation and purification of total lipides from animal tissues. *J. Biol. Chem.* **1957**, *226*, 497–509. Available online: http://www.jbc.org/ (accessed on 16 May 2019). [CrossRef]
25. Cascant, M.M.; Breil, B.; Garrigues, S.; De La Guardia, M.; Fabiano-Tixier, A.S.; Chemat, F. A green analytical chemistry approach for lipid extraction: Computation methods in the selection of green solvents as alternative to hexane. *Anal. Bioanal. Chem.* **2017**, *409*, 3527–3539. [CrossRef]
26. Yara-Varón, E.; Fabiano-Tixier, A.S.; Balcells, M.; Canela- Garayoa, R.; Billy, A.; Chemat, F. Is it possible to substitute hexane with green solvents for extraction of carotenoids? A theoretical versus experimental solubility study. *RSC Adv.* **2016**, *6*, 27750–27759. [CrossRef]
27. Ivanovs, K.; Blumberga, D. Extraction of fish oil using green extraction methods: A short review. *Energy Procedia* **2017**, *128*, 477–483. [CrossRef]
28. Codex Alimentarius. NORMA PARA ACEITES DE PESCADO: CXS 329-2017. Adoptada en. 2017. Available online: www.codexalimentarios.org (accessed on 20 August 2020).
29. Tanzi, D.C.; Vian, M.A.; Ginies, C.; Elmaataoui, M.; Chemat, F. Terpenes as Green Solvents for Extraction of Oil from Microalgae. *Molecules* **2012**, *17*, 8196–8205. [CrossRef] [PubMed]
30. Chemat, F.; Vian, M.A.; Ravi, H.K.; Khadhraoui, B.; Hilali, S.; Perino, S.; Fabiano-Tixier, A.S. Review of Alternative Solvents for Green Extraction of Food and Natural Products: Panorama, Principles, Applications and Prospects. *Molecules* **2019**, *24*, 3007. [CrossRef]
31. Yara-Varón, E.; Selka, A.; Fabiano-Tixier, A.S.; Balcells, M.; Canela-Garayoa, R.; Bily, A.; Touaibiac, M.; Chemat, F. Solvent from forestry biomass. Pinane a stable terpene derived from pine tree byproducts to substitute n-hexane for the extraction of bioactive compounds. *Green Chem.* **2016**, *18*, 6596–6608. [CrossRef]
32. De Jesus, S.S.; Ferreira, G.F.; Moreira, L.S.; Wolf, M.R.M.; Filho, R.M. Comparison of several methods for effective lipid extraction from wet microalgae using green solvents. *Renew. Energy* **2019**, *143*, 130–141. [CrossRef]

33. Yara-Varón, E.; Li, Y.; Balcells, M.; Canela-Garayoa, R.; Fabiano-Tixier, A.S.; Chemat, F. Review. Vegetable Oils as Alternative Solvents for Green Oleo-Extraction, Purification and Formulation of Food and Natural Products. *Molecules* **2017**, *22*, 1474. [CrossRef]

34. Sicaire, A.G.; Vian, M.; Fine, F.; Joffre, F.; Carré, P.; Tostain, S.; Chemat, F. Alternative Bio-Based Solvents for Extraction of Fat and Oils: Solubility Prediction, Global Yield, Extraction Kinetics, Chemical Composition and Cost of Manufacturing. *Int. J. Mol. Sci.* **2015**, *16*, 8430–8453. [CrossRef]

35. Zhang, Z.; Liu, F.; Ma, X.; Huang, H.; Wang, Y. Two-Stage Enzymatic Preparation of Eicosapentaenoic Acid (EPA) And Docosahexaenoic Acid (DHA) Enriched Fish Oil Triacylglycerols. *J. Agric. Food Chem.* **2018**, *66*, 218–227. [CrossRef]

36. Miyashita, K.; Uemura, M.; Hosokawa, M. Effective Prevention of Oxidative Deterioration of Fish Oil: Focus on Flavor Deterioration. *Annu. Rev. Food Sci. Technol.* **2018**, *9*, 209–226. [CrossRef]

37. Moreno-Perez, S.; Machado, D.F.; Pires, J.; Luna, P.; Señorans, F.J.; Guisan, J.M.; Fernandez-Lorente, G. Critical Role of Different Immobilized Biocatalysts of a Given Lipase in the Selective Ethanolysis of Sardine Oil. *J. Agric. Food Chem.* **2017**, *65*, 117–122. [CrossRef]

38. Castejón, N.; Señoráns, F.J. Strategies for Enzymatic Synthesis of Omega-3 Structured Triacylglycerols from Camelina sativa Oil Enriched in EPA and DHA. *Eur. J. Lipid Sci. Technol.* **2019**, *121*, 1800412. [CrossRef]

39. Ranjan-Moharana, T.; Byreddy, A.R.; Puri, M.; Barrow, C.; Rao, N.M. Selective Enrichment of Omega-3 Fatty Acids in Oils by Phospholipase A1. *PLoS ONE* **2016**, *11*, e0151370. [CrossRef] [PubMed]

40. Axelsson, M.; Gentili, F. A single-step method for rapid extraction of total lipids from green microalgae. *PLoS ONE* **2014**, *9*, e89643. [CrossRef] [PubMed]

41. Fang, Y.; Gu, S.; Liu, S.; Zhang, J.; Ding, Y.; Liu, J. Extraction of oil from high-moisture tuna liver by subcritical dimethyl ether: Feasibility and optimization by the response surface method. *RSC Adv.* **2018**, *8*, 2723–2733. [CrossRef]

42. Routray, W.; Dave, D.; Ramakrishnan, V.V.; Murphy, W. Production of High Quality Fish Oil by Enzymatic Protein Hydrolysis from Cultured Atlantic Salmon By-Products: Investigation on Effect of Various Extraction Parameters Using Central Composite Rotatable Design. *Waste Biomass Valor* **2018**, *9*, 2003–2014. [CrossRef]

43. Ciriminna, R.; Scurria, A.; Fabiano-Tixier, A.S.; Lino, C.; Avellone, G.; Chemat, F.; Pagliaro, M. Omega-3 Extraction from Anchovy Fillet Leftovers with Limonene: Chemical, Economic, and Technical Aspects. *ACS Omega* **2019**, *4*, 15359–15363. [CrossRef] [PubMed]

44. Alder, C.M.; Hayler, J.D.; Henderson, R.K.; Redman, A.M.; Shukla, L.; Shuster, L.E.; Sneddon, H.F. Updating and further expanding GSK's solvent sustainability guide. *Green Chem.* **2016**, *18*, 3879. [CrossRef]

45. Henderson, R.K.; Jimenez-Gonzalez, C.; Constable, D.J.C.; Alston, S.R.; Inglis, G.G.A.; Fisher, G.; Sherwood, J.; Binks, S.P.; Curzons, A.D. Expanding GSK's solvent selection guide-embedding sustainability into solvent selection starting at medicinal chemistry. *Green Chem.* **2011**, *13*, 854–862. [CrossRef]

46. Prat, D.; Wells, A.; Hayler, J.; Sneddon, H.; McElroy, C.R.; Abou-Shehadad, S.; Dunne, P.J. CHEM21 selection guide of classical-and less classical-solvent. *Green Chem.* **2016**, *18*, 288. [CrossRef]

47. De Jesus, S.; Filho, R.M. Recent advances in lipid extraction using green solvents. *Renew. Sustain. Energy Rev.* **2020**, *133*, 110289. [CrossRef]

48. Catrin, E.T.; Brecker, L.; Wagner, K.H. ^{1}H NMR spectroscopy as tool to follow changes in the fatty acids of fish oils. *Eur. J. Lipid Sci. Technol.* **2008**, *110*, 141–148. [CrossRef]

49. Bratu, A.; Mihalache, M.; Hanganu, A.M.; Chira, N.A.; Todaşcă, M.C.; Roşca, S. Quantitative determination of fatty acids from fish oils using GC-MS method and ^{1}H NMR spectroscopy. *U.P.B. Sci. Bull.* **2013**, *75*, 24–31.

50. Nestor, G.; Bankefors, J.; Schlechtriem, G.; Brannas, E.; Pickova, J.; Sandstrom, C. High-Resolution ^{1}H Magic Angle Spinning NMR Spectroscopy of Intact Arctic Char (Salvelinus Alpinus) Muscle. Quantitative Analysis of n-3 Fatty Acids, EPA and DHA. *J. Agric. Food Chem.* **2010**, *58*, 10799–10803. [CrossRef] [PubMed]

51. Gavahian, M.; Khaneghah, A.M.; Lorenzo, J.M.; Munekata, P.E.S.; Garcia-Mantrana, I.; Collado, M.C.; Meléndez-Martínez, A.J.; Barba, F.J. Health benefits of olive oil and its components: Impacts on gut microbiota antioxidant activities, and prevention of noncommunicable diseases. *Trends Food Sci. Technol.* **2019**, *88*, 220–227. [CrossRef]

52. Djoussé, L.; Matsumoto, C.; Hanson, N.Q.; Weir, N.L.; Tsai, M.Y.; Gaziano, J.M. Plasma cis-vaccenic acid and risk of heart failure with antecedent coronary heart disease in male physicians. *Clin. Nutr.* **2014**, *33*, 478–482. [CrossRef]

53. Field, C.J.; Blewett, H.H.; Proctor, S.; Vine, D. Human health benefits of vaccenic acid. *Appl. Physiol. Nutr. Metab.* **2009**, *34*, 979–991. [CrossRef] [PubMed]

54. Sissener, N.H.; Ørnsrud, R.; Sanden, M.; Frøyland, L.; Remø, S.; Lundebye, A.K. Erucic Acid (22:1n-9) in Fish Feed, Farmed, and Wild Fish and Seafood Products. *Nutrients* **2018**, *10*, 1443. [CrossRef]

55. Shimada, Y.; Maruyama, K.; Sugihara, A.; Moriyama, S.; Tominaga, Y. Purification of Docosahexaenoic Acid from Tuna Oil by a Two-Step Enzymatic Method: Hydrolysis and Selective Esterification. *JAOCS* **1997**, *74*, 1441. [CrossRef]

56. Gudmundur, G.; Haraldsson, G.G.; Kristinsson, B.; Sigurdardottir, R.; Gudmundson, G.G.; Breivikb, H. The Preparation of Concentrates of Eicosapentaenoic Acid and Docosahexaenoic Acid by Lipase-Catalyzed Transesterification of Fish Oil with Ethanol. *JAOCS* **1997**, *74*, 1419–1424.

57. Xu, Y.; Wang, D.; Mu, X.Q.; Zhao, K.A.; Zhang, K.C. Biosynthesis of ethyl esters of short-chain fatty acids using whole-cell lipase from *Rhizopus chinesis* CCTCC M201021 in non-aqueous phase. *J. Mol. Catal. B Enzym.* **2002**, *18*, 29–37. [CrossRef]

58. Aarthya, M.; Saravananb, P.; Ayyaduraia, N.; Gowthamana, M.K.; Kaminia, N.R. A two-step process for production of omega 3-polyunsaturated fatty acid concentrates from sardine oil using *Cryptococcus* sp. MTCC 5455lipase. *J. Mol. Catal. B Enzym.* **2016**, *125*, 25–33. [CrossRef]
59. Fraga, M.E.; Santana-N, D.M.; Gatti, J.M.; Direito, G.M.; Cavaglieri, L.R.; Rosa-R, C.A. Characterization of Aspergillus species based on fatty acid profiles. *Mem. Inst. Oswaldo Cruz Rio Janeiro* **2008**, *103*, 540–544. [CrossRef] [PubMed]
60. Asci, F.; Aydin, B.; Akkus, G.U.; Unal, A.; Erdogmus, S.F.; Korcan, S.E.; Jahan, I. Fatty acid methyl ester analysis of Aspergillus fumigatus isolated from fruit pulps for biodiesel production using GC-MS spectrometry. *Bioengineered* **2020**, *11*, 408–415. [CrossRef] [PubMed]
61. Schuchardt, J.P.; Neubronner, J.; Kressel, G.; Merkel, M.; Schacky, C.V.; Hahn, A. Moderate doses of EPA and DHA from re-esterified triacylglycerols but not from ethyl-esters lower fasting serum triacylglycerols in stat in-treated dyslipidemic subjects: Results from a six-month randomized controlled trial. *Prostaglandins Leukot Essent Fat. Acids* **2011**, *85*, 381–386. [CrossRef] [PubMed]
62. Ashjaria, M.; Mohammadic, M.; Badria, R. Selective concentration of eicosapentaenoic acid and docosahexaenoic acid from fish oil with immobilized/stabilized preparations of *Rhizopus oryzae* lipase. *J. Mol. Catal. B Enzym.* **2015**, *122*, 147–155. [CrossRef]
63. Klamt, A. Prediction of the mutual solubilities of hydrocarbons and water with COSMO-RS. *Fluid Phase Equilib.* **2003**, *206*, 223–235. [CrossRef]
64. *BIOVIA COSMOtherm 2020 User Guide*; Dassault Systèmes: Vélizy-Villacoublay, France, 2019.
65. Canela-Garayoa, R.; Yara-Varón, E.; Balcells, M.; Torres, M.; Eras, J. Nuclear Magnetic Resonance Spectroscopy: An Alternative Fast Tool for Quantitative Analysis of the Solvent-free Ethanolysis of Coconut Oil Using Fungal Resting Cells. *New Biotechnol.* **2014**, *31*, S89. [CrossRef]
66. Prakash, R.; Aulakh, S.S. Transesterification of used edible and non-edible oils to alkyl esters by Aspergillus *sp.* as a whole cell catalyst. *J. Basic Microbiol.* **2011**, *51*, 607–613. [CrossRef]
67. De Jesus, M.D.P.M.; De Melo, L.N.; Da Silva, J.P.V.; Crispim, A.C.; Figueiredo, I.M.; Bortoluzzi, J.H.; Meneghetti, S.M.P. Evaluation of proton nuclear magnetic resonance spectroscopy for determining the yield of fatty acid ethyl esters obtained by transesterification. *Energy Fuels* **2015**, *29*, 7343–7349. [CrossRef]
68. Di Pietro, M.E.; Mannu, A.; Mele, A. NMR Determination of Free Fatty Acids in Vegetable Oils. *Processes* **2020**, *8*, 410. [CrossRef]
69. Knothe, G.; Kenar, J.A. Determination of the fatty acid profile by ^{1}H-NMR spectroscopy. *Eur. J. Lipid Sci. Technol.* **2004**, *106*, 88–96. [CrossRef]

Article

Conformational Landscapes of Halohydrin Dehalogenases and Their Accessible Active Site Tunnels

Miquel Estévez-Gay [1], Javier Iglesias-Fernández [1,*,†] and Sílvia Osuna [1,2,*]

[1] CompBioLab Group, Institut de Química Computacional i Catàlisi (IQCC) and Departament de Química, Universitat de Girona, c/Maria Aurèlia Capmany 69, 17003 Girona, Catalonia, Spain; miquel.estevez@udg.edu

[2] ICREA, Passeig Lluís Companys 23, 08010 Barcelona, Catalonia, Spain

[*] Correspondence: jiglesiasfrn@gmail.com (J.I.-F.); silvia.osuna@udg.edu (S.O.)

[†] Present address: Nostrum Biodiscovery, Carrer de Baldiri Reixac, 10–12, 08028 Barcelona, Catalonia, Spain.

Received: 12 November 2020; Accepted: 28 November 2020; Published: 1 December 2020

Abstract: Halohydrin dehalogenases (HHDH) are industrially relevant biocatalysts exhibiting a promiscuous epoxide-ring opening reactivity in the presence of small nucleophiles, thus giving access to novel carbon–carbon, carbon–oxygen, carbon–nitrogen, and carbon–sulfur bonds. Recently, the repertoire of HHDH has been expanded, providing access to some novel HHDH subclasses exhibiting a broader epoxide substrate scope. In this work, we develop a computational approach based on the application of linear and non-linear dimensionality reduction techniques to long time-scale Molecular Dynamics (MD) simulations to study the HHDH conformational landscapes. We couple the analysis of the conformational landscapes to CAVER calculations to assess their impact on the active site tunnels and potential ability towards bulky epoxide ring opening reaction. Our study indicates that the analyzed HHDHs subclasses share a common breathing motion of the halide binding pocket, but present large deviations in the loops adjacent to the active site pocket and N-terminal regions. Such conformational differences affect the available tunnels for epoxide binding to the active site. The superior activity of the HHDH G subclass towards bulkier substrates is explained by the additional structural elements delimiting the active site region, its rich conformational heterogeneity, and the substantially wider and frequently observed active site tunnels. This study therefore provides key information for HHDH promiscuity and engineering.

Keywords: Halohydrin dehalogenases; conformational dynamics; active site tunnels; molecular dynamics simulations

---

1. Introduction

Enzymes are highly efficient in accelerating the chemical reactions under biologically controlled conditions, and can provide synthetically useful building blocks with high selectivity and specificity. The ability of enzymes of accelerating additional side reactions, i.e., they present catalytic promiscuity, is thought to play a key role in the evolution of enzymes towards new functions [1,2]. The appearance of novel enzyme functionalities through evolution has been attributed to the fine-tuning of the conformational ensemble present in solution, whose relative stabilities can be tuned by mutations [3–7]. Many of these pre-existing conformations can play a key role in recognizing and binding the substrate and/or releasing the product, in conferring the enzyme the catalytic promiscuity, and in some cases in regulating the operating allosteric communication. These additional conformations of the enzyme can present deviations in the available tunnels for accessing the active site, thus playing a role in the enzyme catalytic activity. Indeed, the engineering of some flexible loops gating substrate access to the active site and contributing to product release were shown to be key for boosting the catalytic activity of some enzymes [8,9]. The enzyme conformational landscape therefore plays a crucial role in its function, promiscuity, regulation, and evolution.

Halohydrin dehalogenases (HHDHs) perform a cofactor independent dehalogenation reaction for degrading halogenated compounds. They are highly valuable biocatalysts as they exhibit promiscuous epoxide ring-opening catalytic activity in the presence of small nucleophiles, thus giving access to novel carbon–carbon, carbon–oxygen, carbon–nitrogen, and carbon–sulfur bonds [10,11]. Some biocatalytic examples of HHDH-catalyzed reactions include their application for obtaining statin side chains precursors more efficiently, enantiopure epihalohydrins, oxazolidinones, tertiary and beta-substituted alcohols [12–18]. As shown by the solved X-ray structures, the active site of HHDH is composed by a binding site for the epoxide and a spacious halide binding pocket that can accommodate linear monovalent anions as nucleophiles (see Figure 1) [19,20]. HHDHs feature a conserved catalytic triad composed by Ser-Tyr-Arg, which catalyzes epoxide formation and subsequent halide release [19,20]. The promiscuous epoxide ring-opening reaction usually occurs at the less-hindered carbon via S_N2 mechanism [20], and the range of epoxides accepted by HHDHs is usually limited to terminal epoxides [21], although some recent examples of HHDH accepting sterically more demanding epoxides have been reported [22] (see Scheme 1). In a recent paper by the Schallmey lab, novel HHDHs were identified following a database mining approach, with six phylogenetic subtypes of HHDH ranging from A through G characterized [23]. Particularly useful is HheG, as it represents the first example of HHDH able to accept with synthetically useful activities bulky cyclic epoxides as substrates [22,23]. Interestingly, HheG has also been recently found to exhibit high activity towards sterically demanding di-substituted epoxides, whereas the A-F subtypes present activity only towards methyl-disubstituted epoxide substrates [24] (see Scheme 1). Unfortunately, structure (i.e., conformational dynamics)–activity relationships are not available for this family of enzymes.

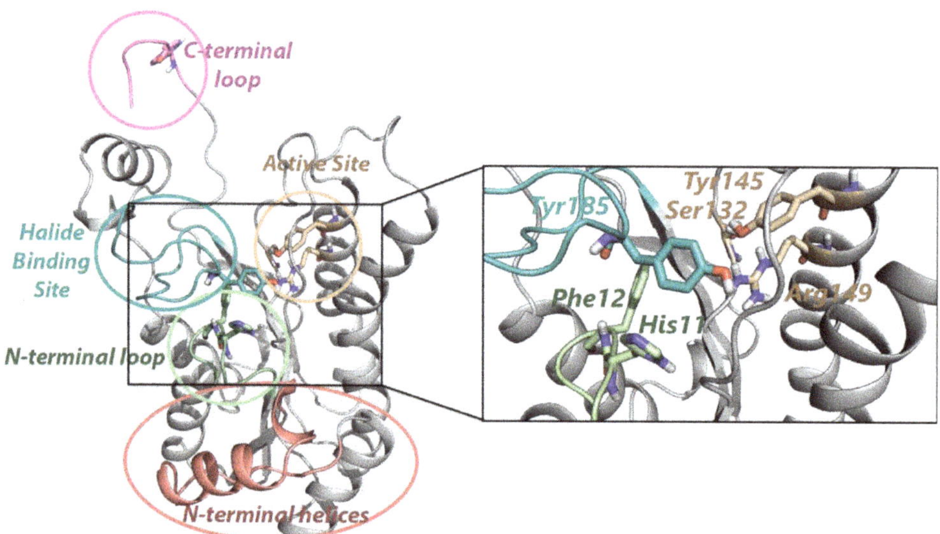

Figure 1. Halohydrin dehalogenase (HHDH) distinct structural elements and zoom of the active site and halide binding pockets based on the HheC structure. Active site residues are highlighted in wheat color, halide-binding site in teal, N-terminal loop in light green, C-terminal loop in purple and N-terminal 6–7 helices in salmon. In the active site zoom, potential residues blocking the accessible active site tunnels are depicted using the same color scheme.

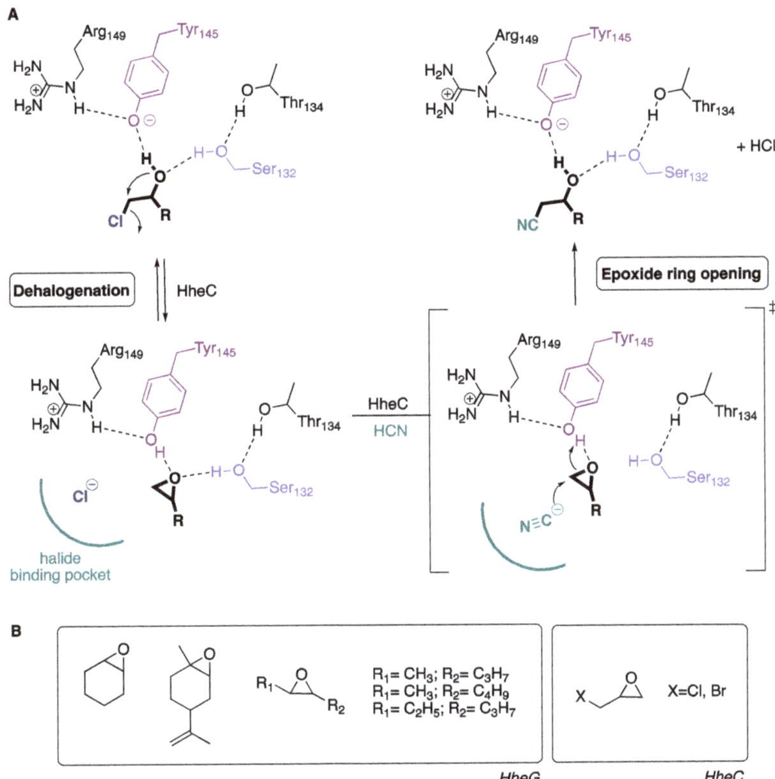

Scheme 1. (**A**). Reaction scheme of the two-step HHDH catalyzed enzymatic reaction: (1, left) dehalogenation for epoxide formation, followed by the promiscuous (2, right) enantioselective epoxide-ring opening reaction by a nucleophile. Numbering of the residues is based on HheC. (**B**). Representative bulky epoxide substrates accepted by the HheG variant: cyclohexene oxide, limonene epoxide [22], and racemic di-substituted trans-epoxides [24]. Examples of epoxide substrates accepted by HheC are also displayed: epichlorohydrin and epibromohydrin [11].

Computational techniques and, in particular, molecular dynamics (MD) simulations are particularly useful in elucidating the ensemble of thermally accessible enzyme conformations by integrating Newton's laws of motion [25]. This enables the reconstruction of the enzyme conformation landscape and assess how this is shifted by ligand binding, sequence differences between protein family members, and/or the introduction of mutations in the enzyme active site or at distal positions [7,26]. Recovery of time-dependent dynamical descriptors, such as volume cavities, solvent-accessible surface area, or changes in internal tunnels/channels is also possible by post-processing the highly dimensional MD datasets [4,27]. Particularly useful is the application of dimensionality reduction techniques for automatically identifying key enzymatic states from MD simulations and account for as much information as possible. These methods can be broadly classified into linear and non-linear dimensionality reduction techniques and have been successfully used to identify key states in MD simulations [28–33]. Combinations of linear and non-linear methods have also been proposed to take advantage of both approximations, with the time-lagged t-Distributed Stochastic Neighbor Embedding (t-SNE) [33] as a clear example [34]. In this direction, we have previously developed a computational protocol based on the combination of the linear time-Independent Component Analysis (tICA) [35] and t-SNE [33] for elucidating the conformational ensemble of *Candida rugosa* lipase and its accessible tunnels for substrate binding to the active site [27].

In this work, intrigued by the observed differences in the catalytic activity and substrate scope of the different subclasses of HHDHs, we characterized the intrinsic conformational landscapes of HHDHs and correlate them with changes in available tunnels for substrate binding and product release in A2, B, C, D2, and G of HHDH subclasses. Extensive MD simulations, followed by dimensionality reduction techniques and tunnel analysis with the CAVER package, provide a clear rationalization of substrate preferences of the studied HHDH subclasses. Our protocol based on projecting MD data into a linear and non-linear reduced space ensures an extensive characterization of the HHDH dynamical ensemble and elucidates how it is modified in the different subclasses. This approach not only provides clear insights of how the available tunnels for substrate binding and product release are altered, but also derives interesting data for HHDH evolution and engineering. This is the first comprehensive study that evaluates the conformational dynamics and associated changes in tunnel accessibility to the active site in different phylogenetic subclasses of HHDH.

2. Results

2.1. Conformational Landscapes of Halohydrin Dehalogenases (HHDHs)

Our study starts with the evaluation of the conformational landscapes of the different Halohydrin dehalogenase (HHDH) variants from the subclasses A2, B, C, D2, and G, followed by a rationalization of how dynamics affect their promiscuity towards epoxide ring opening [22–24]. All analyzed HHDH subclasses are tetrameric, they share the catalytic triad (Ser, Tyr and Arg) and present the halide binding residues located in the loop that connects the 6th β-strand and 9th α-helix (see Figure 1). They also exhibit some structural differences: two α-helices (2nd and 3rd, residues 32–55 according to HheC numbering) are found near the N-terminal part of the protein, with the exception of B and D2; and in G the 2nd α-helix is disordered. HheG also presents the distinctive feature of presenting an additional α-helix in the halide binding site loop, which might potentially broaden the active site entrance tunnel. In the particular case of HheC, the C-terminal part of the opposite monomer is positioned close to the active site and halide binding pockets, which interacts with some halide binding site sidechain residues. Such large structural differences among the subclasses studied might lead to substantial deviations in the HHDH conformational landscapes.

We evaluated the conformational landscapes of A–D, G HHDHs by performing Molecular Dynamics (MD) simulations with an accumulated simulation time of 1.25 microseconds for each system, in explicit water solvent using AMBER (see Methods) [36]. The obtained high dimensional MD dataset was then further analyzed by combining linear and non-linear dimensionality reduction techniques (see Figure 2). We first applied the linear method time-Independent Component Analysis (tICA) [35], which focuses on those motions that most rarely occur, i.e., the slowest kinetically relevant processes. Similarly to the linear Principal Component Analysis (PCA) method [28], tICA allows for a direct identification of relevant descriptors with limited complexity on the dynamics represented, thus requiring more data dimensions to represent the same data variance. To capture more data variance with a reduced number of descriptors, we combined tICA with the non-linear t-Distributed Stochastic Neighbor Embedding (t-SNE) [33] method. This method is used to represent high dimensional data into a 2D or 3D visually appealing low dimensional space by approximating the probability distribution of points in the high dimensional data into the reduced space. In this way, similar data points are placed together with high probability in the reduced space whereas dissimilar data is located distant. By following this tICA-t-SNE methodology, the most relevant conformational states sampled along the MD simulations for each HHDH subclass were revealed (see Figure 2 and Figure S1).

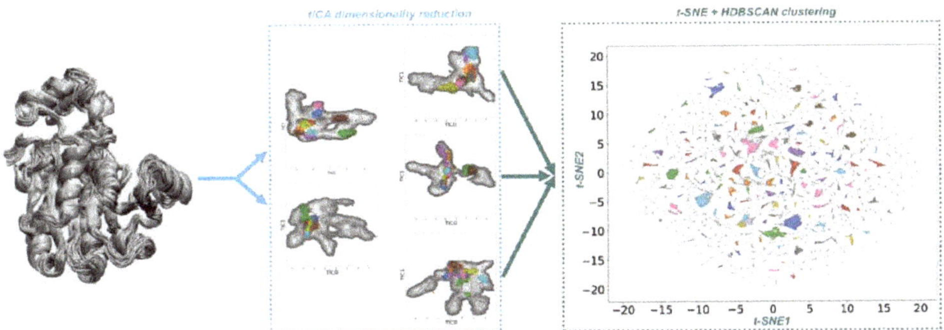

Figure 2. Computational protocol used to reconstruct the conformational landscapes of the different HHDH subclasses. It is based on a two-step process consisting of: first applying to the MD dataset the linear time-Independent Component Analysis (tICA) [35], followed by the application of the non-linear t-distributed Stochastic Neighbor Embedding (t-SNE) [33] method. In this fashion, the high dimensional MD dataset is reduced into a 2D space, that is subsequently clustered using HDBScan [37].

The evaluation of the enzyme conformational dynamics by means of the mentioned tICA-t-SNE methodology indicated that all HHDH subclasses have a moderate to high degree of flexibility, exploring conformations outside the main energy minima (see Figure 3). This is especially true for HheA2 and HheG, which display the most flexible behavior among variants. It is worth mentioning that for all analyzed HHDH subclasses, the projection of the t-SNE most populated clusters onto the corresponding tICA space agrees with well-defined energy minima, confirming that the t-SNE dimensionality reduction method faithfully represents protein dynamics (Figure 3). For all HHDH subclasses explored, the slowest conformational change corresponds to a 'breathing' motion of the protein, involving a coordinated conformational change of both catalytic and halide binding sites. The impact of this breathing motion into substrate binding or product release will be evaluated below.

Analysis of the structural differences observed for HheA2 clusters indicates that conformational changes mainly occur in the halide binding site (Residues 170–210), and the loop located close to the catalytic Tyr146 (Residues 80–95) (Figure 3). The slowest conformational changes (as represented by tIC1, tIC2 and tIC3) correspond to different conformations of the α-helix contained in the halide binding region, and loop motions within the active site (tIC2). The conformational changes observed in the case of the HheD2 variant are similar to those observed for HheA2, however, the catalytic (130–150) and halide binding (170–190) residues display a lower degree of flexibility compared to HheA2.

As mentioned earlier, a distinctive feature of HheC and HheG is the presence of a flexible region close to the N-terminal part of the protein, comprised by an α-helix (Residues 32–36) in HheC and a disordered loop (Residues 30–50) in HheG. The HheC most populated conformations mainly involve coordinated motions of the N-terminal flexible α-helix and the halide binding pocket region. The HheG disordered character of the loop region comprised by residues 30 to 50, which is close to the halide binding residue Tyr13, is the slowest conformational change (as described by tIC0 and tIC1 dimensions). The large structural variability of this protein region is likely involved in substrate accessibility and the presence/absence of lateral access channels, which most likely confer the enzyme the ability to accept bulkier epoxide substrates (see below). As explained before for the D2 and A2 subclasses, HheG also presents a 'breathing' or coordinated motion of the catalytic and halide binding domains.

Contrary to the previous HHDH variants, our MD simulations indicate that HheB displays a quite different conformational behavior, with a tight packing of the protein structure and only minor rearrangements of the α-helix located above the halide binding site (residues 170–190). Most populated conformations only display minor rearrangements on the halide binding region. Interestingly, most visited t-SNE clusters for HheB fall into a narrow region of the tICA space, thus explaining the observed conformational rigidity.

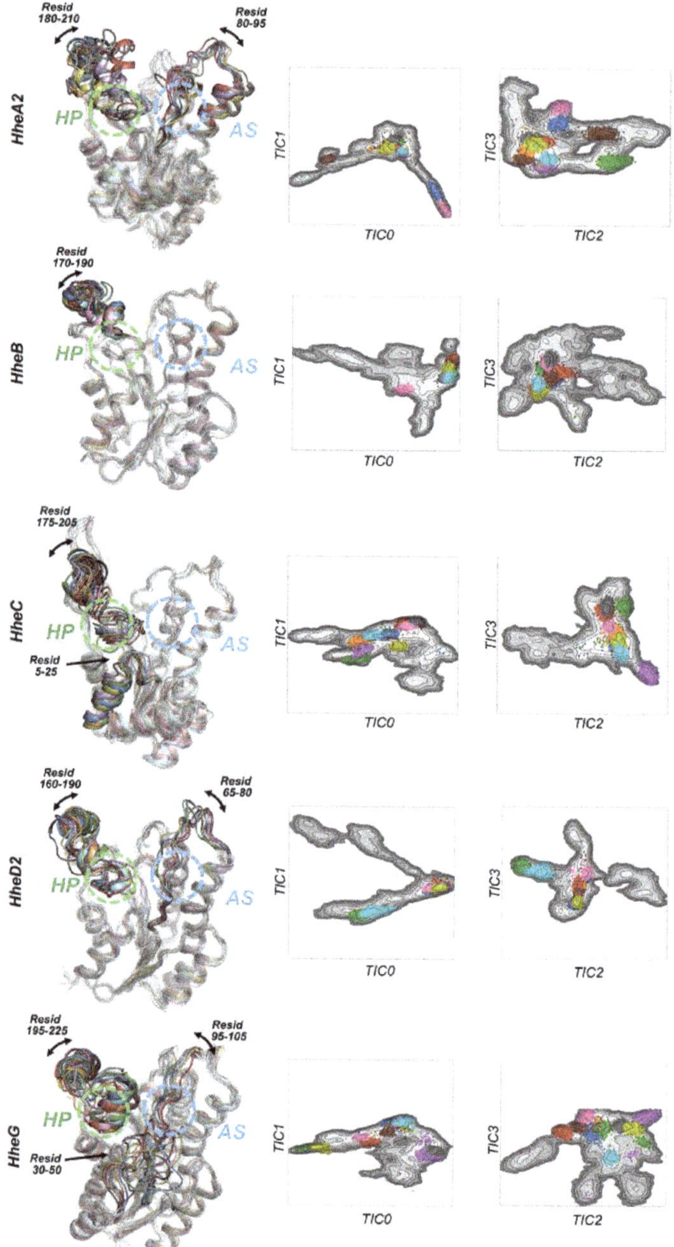

Figure 3. Representation of the 10 most populated MD conformations as described by the t-SNE technique for the different HHDH subclasses analyzed: HheA2, HheB, HheC, HheD2, and HheG. The 10 different conformations (each one colored differently) are projected on the tICA conformational landscapes. The most flexible parts of the enzymes are marked and numbered accordingly. The location of the active site (AS) and halide binding pocket (HP) are marked with a green and blue discontinuous circle, respectively.

2.2. Active Site Accessibility Tunnels of Halohydrin Dehalogenases (HHDHs)

The analysis of the conformational landscapes of the A–D, G HHDHs has revealed major differences among subclasses in the flexibility of the halide binding site region and loops located at the vicinity of the active site pocket. Such conformational changes may impact the available tunnels for substrate accessibility to the active site, thus regulating the enzyme ability to accept bulky epoxide substrates for the industrially relevant promiscuous reactivity. Tunnel analysis with the CAVER software [38] was performed for all t-SNE clusters of each HHDH studied system (see Figures 2 and 3). The average bottleneck radius of the computed tunnel (BR, i.e., narrower region of the tunnel) for each cluster was calculated (see Tables 1 and 2).

Table 1. Mean tunnel bottleneck radius (BR, in Å) for each HHDH system computed on a representative structure of each cluster center (see Methods).

HHDH	Tunnel T1	Tunnel T2	Tunnel T3
HheA2	1.8 ± 0.4	1.6 ± 0.6	n.d. [1]
HheB	1.9 ± 0.6	1.8 ± 0.8	n.d. [1]
HheC	2.0 ± 0.3	1.3 ± 0.2	1.0 ± 0.02
HheD2	1.8 ± 0.5	1.7 ± 0.4	n.d. [1]
HheG	2.2 ± 0.4	1.9 ± 0.5	1.8 ± 0.5 Å

[1] Not detected.

Table 2. Computed tunnel frequency for each HHDH subclass (see Methods).

HHDH	Tunnel T1	Tunnel T2	Tunnel T3
HheA2	92.4%	12.3%	n.d. [1]
HheB	97.6%	25.7%	n.d. [1]
HheC	96.9%	77.5%	36.2%
HheD2	88.0%	71.1%	n.d. [1]
HheG	97.6%	91.8%	65.8%

[1] Not detected.

One major tunnel (named T1) was identified in all analyzed HHDH variants (see Figure 4). In some cases, two additional tunnels (T2–T3) were also found, although with different probabilities. T1 defines the direct vertical path from the active site to the bulk solvent, and is shaped by the active and halide binding sites. This tunnel was found in 92.4%, 97.6%, 96.9%, 88.0%, and 97.6% of the clustered MD frames for HHDH subclasses A2, B, C, D2, and G, respectively. Contrary to what we initially expected, the C-terminal part from the neighbor chain in HheC does not affect T1 formation, thus not blocking substrate accessibility to the active site (see Figure S2). The analysis of the average bottleneck radius (BR) of T1 in the different HHDH subclasses indicates that in A2, B, C, and D it is ca. 1.90 Å, whereas it is slightly larger in the case of HheG (ca. 2.2 Å). This larger BR for T1 observed in G together with its high frequency (98%) explains the higher catalytic activity of HheG with substantially bigger epoxide substrates [22]. The BR of T1 is located close to the active site in all HHDHs, and thus contrary to what one might have initially expected, T1 is not directly affected by the 'breathing' motions of the α-helix contained in the halide binding region and the loop close to the catalytic residues described above.

T2 and T3 are lateral tunnels delimitated by the position of the sidechains of some blocking residues (H11, F12, I84, Y185, F186 and the N-terminal loop mentioned above, numbering according to HheC). Some deviations in the T2 exit to the bulk solvent can be found depending on the position of the N-terminal part and the H11, and F12 sidechain conformations. If the N-terminal part is not blocking the direct exit of T2, the shape and exit of T2 is regulated by the rotation of H11 and F12 sidechains. Alternatively, the longer tunnel T2′ located under the halide binding site loop (residues 180–183 in HheC) can be formed. Tunnel T2 was found in 12.3%, 25.7%, 77.5%, 71.1% and 91.8% of the MD frames for HHDH subclasses A2, B, C, D2, and G, respectively. T2 is therefore hardly found in subclasses A2 and B, more visited in C and D2, and highly frequent in HheG. The analysis of the mean BR reveals that in A2, T2 is hardly present and when formed has a narrow mean BR value of ca. 1.6 Å. In B2, the average BR is expanded to ca. 1.8 Å although with a large variability and low frequency (26%). In contrast, T2 is highly frequent in C (78%) but it is likely too narrow to allow the access of bulky substrates (mean BR of 1.3 Å). The high frequency of T2 in HheD2 and HheG (72% and 92%, see above), is combined with larger average BR values of ca. 1.7 Å in the case of D2, and ca. 1.9 Å in G. These findings of a high frequency of T2 combined with wider BR are in line with the higher activity of HheD2 and HheG with larger di-substituted epoxide substrates [24].

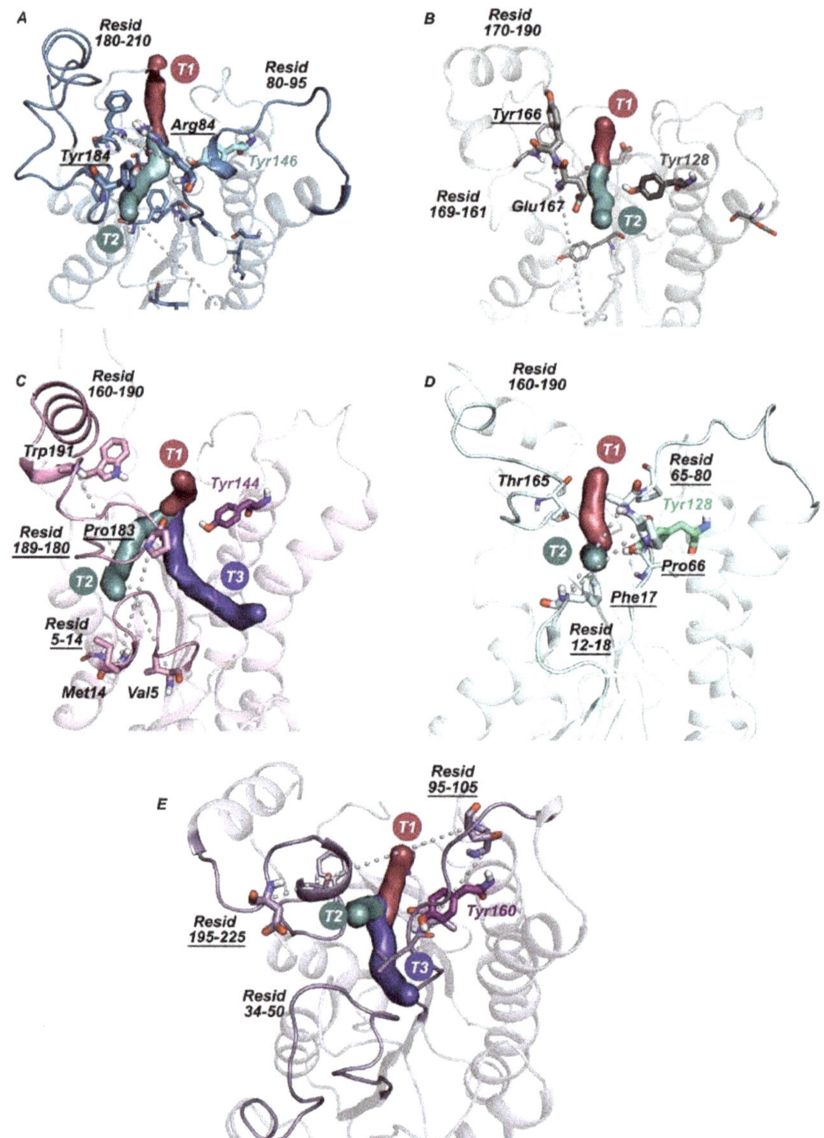

Figure 4. Representation of the three major tunnels that exist in (**A**) HheA2, (**B**) HheB, (**C**) HheC, (**D**) HheD2, and (**E**) HheG: T1 shown in raspberry, T2 in teal, and T3 in dark blue. The key elements that determine T2 formation in the different subclasses are highlighted.

We hypothesized that these large differences in the prevalence of T2 in the analyzed HHDH classes might be related to its associated conformational changes and/or structural variations. To that end, we relied on random forest classifiers [39] to elucidate the key heavy atom distances that modulate T2 formation (see Figure 4, Figures S3–S8). In A2, T2 formation is directly affected by the side-chain conformation of Tyr184 located at the halide binding loop (180–210) and Arg84 contained in the loop close to the catalytic Tyr146 (Residues 80–95). As shown in Figure 3, these regions correspond to the most flexible parts of HheA2, and thus T2 formation is directly related to the enzyme conformational dynamics. In B, T2 formation depends on the side-chain conformation of Tyr166 located at the halide binding loop (see Figure 4B). In fact, T2 is only observed when Tyr166 is displaced out of the active site pocket (see Figure 4B). HheB is the most rigid HHDH analyzed (see Figure 3), and the halide binding region containing Tyr166 does not exhibit a high flexibility, thus explaining the low frequency of T2 in B.

In HheC, T2 is substantially more frequently observed (78%), however, as shown in Figure 4C, T2 follows a slightly different path. In C, T2 formation is dependent on the conformation of the halide binding loop containing Pro183 (189–180) and the positioning of the N-terminal loop 5–14. These regions also correspond to the most flexible parts of the enzyme (see Figure 3), and therefore the formation of T2 in C is also related to its conformational dynamics. The high frequency of T2 in HheD2 is explained by its different location with respect to the previously mentioned cases (see Figure 4D). T2 is rather short and depends on the side-chain conformation of Phe17 at the N-terminal loop 12–18, and the adjacent loop close to the catalytic Tyr128 (65–80). The latter loop exhibits a quite high flexibility (see Figure 3), however, the N-terminal loop is rather rigid. As observed in Figure 4D, T2 in HheG is located between the halide binding loop 195–225, and the loop located close to the catalytic Tyr160 (95–105). However, given the wider active site pocket of HheG a longer distance between the latter loops is observed (this distance is 7.6 ± 2.3 Å in HheG, whereas 5.0 ± 2.4, 6.4 ± 3.2, 4.6 ± 1.4, 6.6 ± 2.3 for HheA2, HheB, HheC, HheD2, respectively), which makes T2 less dependent on their adopted conformation thus explaining the high prevalence (92%) of this tunnel in G. Thanks to the additional α-helix in the halide binding loop in HheG, T2 is not hampered by the N-terminal loop as observed in the other HHDH subclasses.

Tunnel T3 is only observed in those HHDH subclasses with a rather flexible N-terminal region (HheC and HheG, see Figures 3 and 4). The presence of T3 is therefore related to the positioning of the N-terminal loop and its associated conformational dynamics. T3 was found only in 36.19% of the HheC MD clusters, whereas it was found in 65.8% of the HheG MD dataset in line with the higher flexibility of the N-terminal loop observed in the G subclass. The analysis of the average BR indicates that although in HheC T3 is observed in 37% of the analyzed structures it is too narrow (ca. 1.0 Å) to allow the access of the epoxide substrate to the active site. In contrast, T3 is highly frequent in HheG and presents a substantially larger average BR of ca. 1.8 Å. This observation is again in line with the higher ability of HheG to accommodate bulkier epoxide substrates for the industrially relevant promiscuous reactivity [22].

3. Discussion

The repertoire of Halohydrin dehalogenases (HHDH) has been recently expanded, which provides access to some novel HHDH subclasses. These novel enzymes present substantial structural similitudes, although large differences especially in loops and alpha-helices located at the vicinity of the active site of the enzyme are also present. The analysis of the conformational landscapes of HHDH by means of linear and non-linear dimensionality reduction techniques has revealed that a common feature of all analyzed HHDH is a high flexibility of the alpha-helix and loop containing the halide binding pocket, i.e., 'breathing motion'. Apart from this common motion, there are some conformational differences among the analyzed subclasses: HheA2 and HheD2 exhibit a high flexibility of the loop located close to the catalytic Tyr; HheC has instead a high mobility of the N-terminal loop, and HheG is the most conformationally rich HHDH displaying a large mobility of the N-terminal loop, the loop located adjacent to the catalytic Tyr and the already mentioned halide binding pocket.

The characterization of the accessible tunnels at the ensemble of conformations explored by means of CAVER has evidenced some relevant deviations in the available active site tunnels, some of them clearly influenced by the conformational dynamics of the HHDH subclass. All analyzed enzymes present a major tunnel (named T1) that vertically connects the active site pocket to the bulk solvent through the cavity formed between the loop located close to the catalytic Tyr and the halide binding pocket. In contrast to what we originally expected, the formation of T1 is independent from the above-mentioned breathing motion of the halide binding pocket and has a high occurrence in all HHDH (which ranges from ca. 88–98%). The bottleneck radius (BR) of T1 is located close to the active site residues, thus not being substantially affected by the halide binding pocket conformation. The computed BR for T1 is ca. 1.9 Å for all HHDH, except in HheG that is broadened to ca. 2.2 Å. Interestingly, dramatic differences are observed in the case of the secondary tunnel T2. The length, the exact positioning, BR values, and the frequency of T2 is subclass-dependent. In A2 and B, T2 is hardly present and is mostly dependent on the conformation of a Tyr residue (185 in A2 and 167 in B) contained in the halide binding pocket. In HheC, T2 has a high frequency and is located between the halide binding pocket (179–190) and the N-terminal loops (5–14), which present a rather high flexibility. T2 is substantially shorter in HheD2 and is situated between the rather rigid N-terminal and the substantially more flexible loop situated close to the catalytic Tyr. Thanks to the additional α-helix in the halide binding loop in HheG, T2 formation is less affected by the conformation of the halide binding region, the N-terminal and the loop situated close to the catalytic Tyr. The BR of T2 ranges from 1.3 Å for HheC to 1.9 Å for HheG. Finally, a third secondary tunnel (T3) is also present in the case of HheC and HheG that present a substantially more flexible N-terminal region. T3 is, however, less observed in HheC and exhibits a substantially narrower BR value (1.0 Å for C and 1.8Å for G).

4. Materials and Methods

Systems Set-Up. MD simulations were carried out on the structures of A2, B, C, and G HHDH subclasses with protein database (PDB) codes 1ZMO, 4ZD6, 1ZMT, and 5O30, respectively, as initial structures. The X-ray structure for D2 is not released yet (made available by the Schallmey lab). Protonation states of enzyme residues were assigned based on pKa values provided by the H++ server (http://biophysics.cs.vt.edu/H++) [40]. The enzymes were then solvated in a pre-equilibrated cubic box with a 10 Å buffer of transferable intermolecular potential with 3 points (TIP3P) [41] water molecules, resulting in the addition of approximately 27,000 solvent molecules per protein variant. The systems were neutralized by the addition of approximately 32 explicit counter ions (Na+). All simulations were done using the Amber 99SB force field (ff99SB-ildn) [42].

MD Simulations. The graphics processing unit (GPU) version of pmemd in Amber16 was used for the MD simulations. A two-stage geometry optimization was performed, first minimizing the positions of solvent molecules and ions, by imposing harmonic positional restraints of 500 kcal mol^{-1} Å^{-2} on solute molecules, followed by an unrestrained minimization. Afterwards, a gradual heating of the systems was performed by increasing the temperature 50 K along six 20 ps sequential MD simulations (0–300 K) under constant volume and periodic boundary conditions. Harmonic restraints of 10 kcal/mol were applied to the solute, and the Langevin equilibration scheme was used to control and equalize the temperature. The time step was kept at 1 fs during the heating stages, allowing potential inhomogeneities to self-adjust. Each system was then equilibrated without restraints for 2 ns with a 2 fs time step at a constant pressure of 1 atm and temperature of 300 K. After equilibration in the isothermal-isobaric ensemble (NPT), 5 replicas of 250 ns were run for each system (i.e., 1.25 µs per HHDH subclass) in the canonical ensemble (NVT) yielding a total MD simulation time for all systems of 6.25 µs.

MD dimensionality reduction. MD simulation trajectories were post-process with the pyemma2 software package [43]. C-alpha coordinates of the aligned protein subclasses at each nanosecond of MD simulation were used as initial features, resulting in 182,250,000, 168,000,000, 189,000,000, 168,000,000, 192,750,000 extracted values (features x frames x replicas) for the A2, B, C, D2, and G HHDH subclasses, making the statistical analysis unfeasible. Subsequently, the time-lagged Independent Component Analysis (t-ICA) [35], with a lag time τ set to obtain the minimum number of reduced dimensions, was applied to reduce the dimensionality of the initial MD features. The linear method t-ICA focuses on those motions that most rarely occur, i.e., the slowest kinetically relevant processes. Conversely to linear methods, non-linear techniques have the advantage of capturing more data variance with less descriptors, although at the cost making the biophysical interpretations of such reduced descriptors challenging. These methods include the recently proposed variational autoencoders [31,32], and the t-Distributed Stochastic Neighbor Embedding (t-SNE) [33], among others. After applying t-ICA, we further reduced the dimensionality of the data by applying the t-SNE method to the 20 most informative t-ICA dimensions. These 20 most informative t-ICA dimensions describe the 25% of the total variance. The resulting 2D t-SNE space was clustered with the HDBSCAN algorithm [37], with a minimum cluster size of 200 and other default parameters, resulting in 133, 126, 134, 124, 119 clusters for the A2, B, C, D2, and G variants, respectively. By applying the t-SNE dimensionality reduction, less than 75% of the variance was lost.

Tunnel analysis of MD simulation. CaverAnalyst [44] was used to compute substrate entry channels for the 10 most populated HDBSCAN clusters of each HHDH variant. For each t-SNE cluster, the nearest MD snapshot was extracted with the Mdtraj software [45] for the analysis of accession tunnels, thus spanning the whole dynamical space of the enzyme. The parameters used for the tunnel search were 4 Å shell depth, 2.5 Å shell radius, clustering threshold value of 3.5 and a 1 Å minimum probe radius were used as tunnel search parameters.

Decision trees and feature importance. Decision trees are supervised learning methods and, therefore, require a set of training examples for which the output or label is known. They infer relations from training instances by asking a series of questions about the input in a tree-shaped hierarchy. For categorical data, yes/no questions can be asked regarding the presence/absence of a particular input, whereas for real-valued features, such as atomic distances, threshold values are normally used. Here, we defined as input features all possible minimum distances between residues defining the shape of the corresponding tunnel and the presence/absence of the studied tunnel as a target feature. We used a Python pipeline to standardize the input data and select the best Random Forest parameters for the classification. MD data was randomly split into a training set (80%) and test set (20%). We used Python packages Numpy [46], Pandas, Scikit-Learn [47], and Matplotlib for data manipulation, machine-learning, and visualization. Pyemma2 [43] was used for feature extraction from MD simulations, PCA, and tICA dimensionality reduction.

Formula for computing the tunnel frequency (f) for each HHDH subclass (Table 2):

$$f = \frac{\sum_{i=1}^{n} \delta_i p_i}{M} \cdot 100 \quad \delta_i \Rightarrow \begin{array}{l} 0 \text{ if tunnel not present} \\ 1 \text{ if tunnel present} \end{array}$$

where M is the total number of frames, p_i is the number of frames in the cluster and n is the number of clusters of each system.

5. Conclusions

The exploration of the conformational landscape of the different HHDH subclasses coupled to the active site tunnel calculations has indicated that the superior activity of HheG towards bulky epoxide substrates is due to the presence of some additional structural elements adjacent to the active site pocket, its higher conformational heterogeneity, and the presence of highly prevalent active site tunnels that present bottleneck radius of ca. 1.9 Å. This is unique to the G subclass, as the rest of the analyzed HHDH are conformationally more restricted and present a reduced number of narrower active site tunnels. Altogether, our study has shown how the HHDH structural dissimilarities influence their conformational landscape, thus impacting their associated active site tunnels, and in turn, their catalytic promiscuity. By means of extensive MD simulations and CAVER analysis, this work has provided key information for rationalizing HHDH promiscuity and for further engineering.

Supplementary Materials: The following are available online at http://www.mdpi.com/2073-4344/10/12/1403/s1, Figure S1: tSNE and HDBSCAN representations, Figure S2: T1 representation as tetramer, Figures S3–S7: Random Forest Classifier for HheA2, HheB, HheC, HheD2, and HheG, Figure S8: Most important contacts computed using Random Forest Classifier.

Author Contributions: M.E.-G. has performed all simulations and analysis, M.E.-G., J.I.-F. and S.O. have discussed the results, and written the manuscript. All authors have read and agreed to the published version of the manuscript.

Funding: We thank the Generalitat de Catalunya for the emerging group CompBioLab (2017 SGR-1707) and Spanish MINECO for project PGC2018-102192-B-I00. J.I.F. was supported by the European Community for Marie Curie fellowship (H2020-MSCA-IF-2016-753045) and Juan de la Cierva-Incorporación fellowship (IJCI-2017-34129). S.O. is grateful to the funding from the European Research Council (ERC) under the European Union's Horizon 2020 research and innovation program (ERC-2015-StG-679001).

Conflicts of Interest: The authors declare no conflict of interest.

References

1. Tokuriki, N.; Tawfik, D.S. Protein Dynamism and Evolvability. *Science* **2009**, *324*, 203–207. [CrossRef] [PubMed]

2. Tawfik, O.K.; Dan, S. Enzyme Promiscuity: A Mechanistic and Evolutionary Perspective. *Annu. Rev. Biochem.* **2010**, *79*, 471–505. [CrossRef] [PubMed]

3. Campbell, E.C.; Correy, G.J.; Mabbitt, P.D.; Buckle, A.M.; Tokuriki, N.; Jackson, C.J. Laboratory evolution of protein conformational dynamics. *Curr. Opin. Struct. Biol.* **2018**, *50*, 49–57. [CrossRef] [PubMed]

4. Maria-Solano, M.A.; Serrano-Hervás, E.; Romero-Rivera, A.; Iglesias-Fernández, J.; Osuna, S. Role of conformational dynamics in the evolution of novel enzyme function. *Chem. Commun.* **2018**, *54*, 6622–6634. [CrossRef]

5. Petrović, D.; Risso, V.A.; Kamerlin, S.C.L.; Sanchez-Ruiz, J.M. Conformational dynamics and enzyme evolution. *J. R. Soc. Interface* **2018**, *15*. [CrossRef]

6. Jiménez-Osés, G.; Osuna, S.; Gao, X.; Sawaya, M.R.; Gilson, L.; Collier, S.J.; Huisman, G.W.; Yeates, T.O.; Tang, Y.; Houk, K.N. The role of distant mutations and allosteric regulation on LovD active site dynamics. *Nat. Chem. Biol.* **2014**, *10*, 431–436. [CrossRef]

7. Romero-Rivera, A.; Garcia-Borràs, M.; Osuna, S. Role of Conformational Dynamics in the Evolution of Retro-Aldolase Activity. *ACS Catal.* **2017**, *7*, 8524–8532. [CrossRef]

8. Nestl, B.M.; Hauer, B. Engineering of Flexible Loops in Enzymes. *ACS Catal.* **2014**, *4*, 3201–3211. [CrossRef]

9. Pavlova, M.; Klvana, M.; Prokop, Z.; Chaloupkova, R.; Banas, P.; Otyepka, M.; Wade, R.C.; Tsuda, M.; Nagata, Y.; Damborsky, J. Redesigning dehalogenase access tunnels as a strategy for degrading an anthropogenic substrate. *Nat. Chem. Biol.* **2009**, *5*, 727. [CrossRef]

10. de Vries, E.J.; Janssen, D.B. Biocatalytic conversion of epoxides. *Curr. Opin. Biotechnol.* **2003**, *14*, 414–420. [CrossRef]

11. Hasnaoui-Dijoux, G.; Majerić Elenkov, M.; Lutje Spelberg, J.H.; Hauer, B.; Janssen, D.B. Catalytic Promiscuity of Halohydrin Dehalogenase and its Application in Enantioselective Epoxide Ring Opening. *ChemBioChem* **2008**, *9*, 1048–1051. [CrossRef] [PubMed]

12. Fox, R.J.; Davis, S.C.; Mundorff, E.C.; Newman, L.M.; Gavrilovic, V.; Ma, S.K.; Chung, L.M.; Ching, C.; Tam, S.; Muley, S.; et al. Improving catalytic function by ProSAR-driven enzyme evolution. *Nat. Biotechnol.* **2007**, *25*, 338–344. [CrossRef] [PubMed]

13. Schallmey, A.; Schallmey, M. Recent advances on halohydrin dehalogenases-from enzyme identification to novel biocatalytic applications. *Appl. Microbiol. Biotechnol.* **2016**, *100*, 7827–7839. [CrossRef] [PubMed]

14. Schallmey, M.; Floor, R.J.; Hauer, B.; Breuer, M.; Jekel, P.A.; Wijma, H.J.; Dijkstra, B.W.; Janssen, D.B. Biocatalytic and Structural Properties of a Highly Engineered Halohydrin Dehalogenase. *ChemBioChem* **2013**, *14*, 870–881. [CrossRef] [PubMed]

15. Wan, N.-W.; Liu, Z.-Q.; Huang, K.; Shen, Z.-Y.; Xue, F.; Zheng, Y.-G.; Shen, Y.-C. Synthesis of ethyl (R)-4-cyano-3-hydroxybutyrate in high concentration using a novel halohydrin dehalogenase HHDH-PL from Parvibaculum lavamentivorans DS-1. *RSC Adv.* **2014**, *4*, 64027–64031. [CrossRef]

16. Assis, H.M.S.; Bull, A.T.; Hardman, D.J. Synthesis of Chiral Epihalohydrins Using Haloalcohol Dehalogenase A from Arthrobacter Erithii H10a. *Enzyme Microb. Technol.* **1998**, *22*, 545–551. [CrossRef]

17. Elenkov, M.M.; Tang, L.; Meetsma, A.; Hauer, B.; Janssen, D.B. Formation of Enantiopure 5-Substituted Oxazolidinones through Enzyme-Catalysed Kinetic Resolution of Epoxides. *Org. Lett.* **2008**, *10*, 2417–2420. [CrossRef]

18. Molinaro, C.; Guilbault, A.-A.; Kosjek, B. Resolution of 2,2-Disubstituted Epoxides via Biocatalytic Azidolysis. *Org. Lett.* **2010**, *12*, 3772–3775. [CrossRef]

19. de Jong, R.M.; Tiesinga, J.J.W.; Rozeboom, H.J.; Kalk, K.H.; Tang, L.; Janssen, D.B.; Dijkstra, B.W. Structure and mechanism of a bacterial haloalcohol dehalogenase: A new variation of the short-chain dehydrogenase/reductase fold without an NAD(P)H binding site. *EMBO J.* **2003**, *22*, 4933–4944. [CrossRef]

20. de Jong, R.M.; Tiesinga, J.J.W.; Villa, A.; Tang, L.; Janssen, D.B.; Dijkstra, B.W. Structural Basis for the Enantioselectivity of an Epoxide Ring Opening Reaction Catalyzed by Halo Alcohol Dehalogenase HheC. *J. Am. Chem. Soc.* **2005**, *127*, 13338–13343. [CrossRef]

21. Elenkov, M.M.; Hauer, B.; Janssen, D.B. Enantioselective Ring Opening of Epoxides with Cyanide Catalysed by Halohydrin Dehalogenases: A New Approach to Non-Racemic β-Hydroxy Nitriles. *Adv. Synth. Catal.* **2006**, *348*, 579–585. [CrossRef]

22. Koopmeiners, J.; Diederich, C.; Solarczek, J.; Voß, H.; Mayer, J.; Blankenfeldt, W.; Schallmey, A. HheG, a Halohydrin Dehalogenase with Activity on Cyclic Epoxides. *ACS Catal.* **2017**, *7*, 6877–6886. [CrossRef]

23. Schallmey, M.; Koopmeiners, J.; Wells, E.; Wardenga, R.; Schallmey, A. Expanding the Halohydrin Dehalogenase Enzyme Family: Identification of Novel Enzymes by Database Mining. *Appl. Environ. Microbiol.* **2014**, *80*, 7303–7315. [CrossRef] [PubMed]

24. Calderini, E.; Wessel, J.; Süss, P.; Schrepfer, P.; Wardenga, R.; Schallmey, A. Selective Ring-Opening of Di-Substituted Epoxides Catalysed by Halohydrin Dehalogenases. *ChemCatChem* **2019**, *11*, 2099–2106. [CrossRef]

25. Orozco, M. A theoretical view of protein dynamics. *Chem. Soc. Rev.* **2014**, *43*, 5051–5066. [CrossRef]

26. Osuna, S. The challenge of predicting distal active site mutations in computational enzyme design. *Wiley Interdiscip. Rev. Comput. Mol. Sci.* **2020**, e1502. [CrossRef]

27. Wang, L.; Marciello, M.; Estévez-Gay, M.; Rodriguez, P.E.D.S.; Morato, Y.L.; Iglesias-Fernández, J.; Huang, X.; Osuna, S.; Filice, M.; Sanchez, S. Enzyme Conformation Influences the Performance of Lipase-powered Nanomotors. *Angew. Chem. Int. Ed.* **2020**, *59*, 21080–21087. [CrossRef]

28. Mu, Y.; Nguyen, P.H.; Stock, G. Energy landscape of a small peptide revealed by dihedral angle principal component analysis. *Proteins* **2005**, *58*, 45–52. [CrossRef]

29. Ferguson, A.L.; Panagiotopoulos, A.Z.; Kevrekidis, I.G.; Debenedetti, P.G. Nonlinear dimensionality reduction in molecular simulation: The diffusion map approach. *Chem. Phys. Lett.* **2011**, *509*, 1–11. [CrossRef]

30. Ceriotti, M.; Tribello, G.A.; Parrinello, M. Simplifying the representation of complex free-energy landscapes using sketch-map. *Proc. Natl. Acad. Sci. USA* **2011**, *108*, 13023–13028. [CrossRef]

31. Hernández, C.X.; Wayment-Steele, H.K.; Sultan, M.M.; Husic, B.E.; Pande, V.S. Variational encoding of complex dynamics. *Phys. Rev. E* **2018**, *97*, 062412. [CrossRef] [PubMed]

32. Mardt, A.; Pasquali, L.; Wu, H.; Noé, F. VAMPnets for deep learning of molecular kinetics. *Nat. Commun.* **2018**, *9*, 5. [CrossRef] [PubMed]

33. Zhou, H.; Wang, F.; Tao, P. t-Distributed Stochastic Neighbor Embedding Method with the Least Information Loss for Macromolecular Simulations. *J. Chem. Theory Comput.* **2018**, *14*, 5499–5510. [CrossRef] [PubMed]

34. Spiwok, V.; Kříž, P. Time-Lagged t-Distributed Stochastic Neighbor Embedding (t-SNE) of Molecular Simulation Trajectories. *Front. Mol. Biosci.* **2020**, *7*. [CrossRef] [PubMed]

35. Pérez-Hernández, G.; Paul, F.; Giorgino, T.; De Fabritiis, G.; Noé, F. Identification of slow molecular order parameters for Markov model construction. *J. Chem. Phys.* **2013**, *139*, 015102. [CrossRef]

36. Case, D.A.; Darden, T.A.; Cheatham, T.E.; Simmerling, C.L.; Wang, J.; Duke, R.E.; Luo, R.; Crowley, M.; Walker, R.C.; Zhang, W.; et al. *AMBER 16, University of California, San Francisco*; United States of America: Washington, DC, USA, 2016.

37. Campello, R.J.G.B.; Moulavi, D.; Sander, J. *Density-Based Clustering Based on Hierarchical Density Estimates*; Springer: Berlin/Heidelberg, Germany, 2013; pp. 160–172.

38. Chovancova, E.; Pavelka, A.; Benes, P.; Strnad, O.; Brezovsky, J.; Kozlikova, B.; Gora, A.; Sustr, V.; Klvana, M.; Medek, P.; et al. CAVER 3.0: A tool for the analysis of transport pathways in dynamic protein structures. *PLoS Comput. Biol.* **2012**, *8*, e1002708. [CrossRef]

39. Breiman, L. Random Forests. *Mach. Learn.* **2001**, *45*, 5–32. [CrossRef]

40. Anandakrishnan, R.; Aguilar, B.; Onufriev, A.V. H++ 3.0: Automating pK prediction and the preparation of biomolecular structures for atomistic molecular modeling and simulations. *Nucleic Acids Res.* **2012**, *40*, W537–W541. [CrossRef]

41. Jorgensen, W.L.; Chandrasekhar, J.; Madura, J.D.; Impey, R.W.; Klein, M.L. Comparison of simple potential functions for simulating liquid water. *J. Chem. Phys.* **1983**, *79*, 926–935. [CrossRef]

42. Maier, J.A.; Martinez, C.; Kasavajhala, K.; Wickstrom, L.; Hauser, K.E.; Simmerling, C. ff14SB: Improving the Accuracy of Protein Side Chain and Backbone Parameters from ff99SB. *J. Chem. Theory Comput.* **2015**, *11*, 3696–3713. [CrossRef]

43. Scherer, M.K.; Trendelkamp-Schroer, B.; Paul, F.; Pérez-Hernández, G.; Hoffmann, M.; Plattner, N.; Wehmeyer, C.; Prinz, J.-H.; Noé, F. PyEMMA 2: A Software Package for Estimation, Validation, and Analysis of Markov Models. *J. Chem. Theory Comput.* **2015**, *11*, 5525–5542. [CrossRef] [PubMed]

44. Jurcik, A.; Bednar, D.; Byska, J.; Marques, S.M.; Furmanova, K.; Daniel, L.; Kokkonen, P.; Brezovsky, J.; Strnad, O.; Stourac, J.; et al. CAVER Analyst 2.0: Analysis and visualization of channels and tunnels in protein structures and molecular dynamics trajectories. *Bioinformatics* **2018**, *34*, 3586–3588. [CrossRef] [PubMed]

45. McGibbon, R.T.; Beauchamp, K.A.; Harrigan, M.P.; Klein, C.; Swails, J.M.; Hernández, C.X.; Schwantes, C.R.; Wang, L.-P.; Lane, T.J.; Pande, V.S. MDTraj: A Modern Open Library for the Analysis of Molecular Dynamics Trajectories. *Biophys. J.* **2015**, *109*, 1528–1532. [CrossRef] [PubMed]

46. Harris, C.R.; Millman, K.J.; van der Walt, S.J.; Gommers, R.; Virtanen, P.; Cournapeau, D.; Wieser, E.; Taylor, J.; Berg, S.; Smith, N.J.; et al. Array programming with NumPy. *Nature* **2020**, *585*, 357–362. [CrossRef] [PubMed]

47. Pedregosa, F.; Varoquaux, G.; Gramfort, A.; Michel, V.; Thirion, B.; Grisel, O.; Blondel, M.; Prettenhofer, P.; Weiss, R.; Dubourg, V.; et al. Scikit-learn: Machine Learning in Python. *J. Mach. Learn. Res.* **2011**, *12*, 2825–2830.

Publisher's Note: MDPI stays neutral with regard to jurisdictional claims in published maps and institutional affiliations.

 *catalysts*

Article

Biocatalysis at Extreme Temperatures: Enantioselective Synthesis of both Enantiomers of Mandelic Acid by Transesterification Catalyzed by a Thermophilic Lipase in Ionic Liquids at 120 °C

Jesús Ramos-Martín, Oussama Khiari, Andrés R. Alcántara * and Jose María Sánchez-Montero *

Department of Chemistry in Pharmaceutical Sciences, Pharmacy Faculty,
Complutense University of Madrid (UCM), Ciudad Universitaria, Plaza de Ramon y Cajal, s/n., 28040 Madrid,
Spain; rivasfarma@hotmail.com (J.R.-M.); oussama.khiari@gmail.com (O.K.)
* Correspondence: andalcan@ucm.es (A.R.A.); jmsm@ucm.es (J.M.S.-M.);
  Tel.: +34-91-394-1820 (A.R.A.); +34-91-394-1839 (J.M.S.-M.)

Received: 15 August 2020; Accepted: 11 September 2020; Published: 14 September 2020

Abstract: The use of biocatalysts in organic chemistry for catalyzing chemo-, regio- and stereoselective transformations has become an usual tool in the last years, both at lab and industrial scale. This is not only because of their exquisite precision, but also due to the inherent increase in the process sustainability. Nevertheless, most of the interesting industrial reactions involve water-insoluble substrates, so the use of (generally not green) organic solvents is generally required. Although lipases are capable of maintaining their catalytic precision working in those solvents, reactions are usually very slow and consequently not very appropriate for industrial purposes. Increasing reaction temperature would accelerate the reaction rate, but this should require the use of lipases from thermophiles, which tend to be more enantioselective at lower temperatures, as they are more rigid than those from mesophiles. Therefore, the ideal scenario would require a thermophilic lipase capable of retaining high enantioselectivity at high temperatures. In this paper, we describe the use of lipase from *Geobacillus thermocatenolatus* as catalyst in the ethanolysis of racemic 2-(butyryloxy)-2-phenylacetic to furnish both enantiomers of mandelic acid, an useful intermediate in the synthesis of many drugs and active products. The catalytic performance at high temperature in a conventional organic solvent (*iso*octane) and four imidazolium-based ionic liquids was assessed. The best results were obtained using 1-ethyl-3-methyl imidazolium tetrafluoroborate (EMIMBF$_4$) and 1-ethyl-3-methyl imidazolium hexafluorophosphate (EMIMPF$_6$) at temperatures as high as 120 °C, observing in both cases very fast and enantioselective kinetic resolutions, respectively leading exclusively to the (*S*) or to the (*R*)-enantiomer of mandelic acid, depending on the anion component of the ionic liquid.

Keywords: *Geobacillus thermocatenolatus*; lipases; ethanolysis; ionic liquids; kinetic resolution; mandelic acid

---

1. Introduction

Employing biocatalysts in organic chemistry, either alone [1,2] or combined with chemical catalysts [3,4] for developing selective transformations has become an common tool in the last years [5,6]. This is based on the extremely enzymatic precision (chemo-, regio- and stereoselectivity) acquired when applied in biotransformations not only at lab, but also at industrial scale [7–11], being used mainly in pharma industry [12–17]. Moreover, moving from chemical catalysis to biocatalysis leads to an increase in process sustainability—given that biocatalysis and green chemistry usually go hand-in-hand [2,17–19].

One of the green credentials of biocatalysis derives from the fact that biotransformations can be conducted under very mild reaction conditions, e.g., atmospheric pressure, room temperature or aqueous media. However, harsh conditions required for many industrial processes—such as high temperature and/or the use of organic (co)solvents—may impede the use of some enzymes. To address these drawbacks, using thermotolerant biocatalysts obtained from thermophilic organisms is an excellent alternative [20–23], as these thermozymes can efficiently work at very high temperatures [24,25] and are generally very resistant to organic solvent-promoted denaturation [26,27]. Among all the arsenal of enzymes available for being used in biotransformations, lipases (triacylglycerol hydrolases, EC 3.1.1.3) are one of the most frequently applied, as they are easily available, do not need cofactors and display a wide range of substrate recognition [28–31]. Furthermore, their ability for working in almost anhydrous organic solvents allows conducting reactions in the sense of synthesis instead of hydrolysis, therefore favoring the transformation of many organic compounds, which are generally water-insoluble and thus reverting the original enzymatic selectivity [32–34].

The use of lipases from thermophiles has been frequently reported [23,35,36]. Among them, the term "thermoalkaline (TA) lipases" describes some enzymes resistant not only to temperature (70–80 °C) but also to the presence of alkaline media (pH values between 8 and 10) [36]. These enzymes possess a peculiar feature in their 3D structure due to the presence of a relatively large lid domain (around 70 residues) formed by two alpha-helices (α6 and α7) [37], so that the opening of this lid domain upon exposing the active site requires a significant conformational change [38]. One of the most representative examples of TA lipases is that one from *Geobacillus* (formerly *Bacillus* [39]) *thermocatenulatus*. From this microorganism, Schmidt-Dannert et al. [40–42] described two lipases, namely BTL1 and BTL2. The latter was crystallized in its open form by Carrasco-Lopez et al. [37,43]. The stereoselectivity of BTL2 towards 29 chiral substrates was initially tested by Liu et al. [44], reporting only good results for the acylation of 1-phenylethanol and 1-phenylpropanol with vinyl acetate (as well as for the hydrolysis of the corresponding esters). Later, this enzyme—immobilized on different supports via diverse methodologies—has been extensively tested on different substrates, some of them chiral [45–64], generally with moderate results. Additionally, chemical [55,58,65–67] and genetic [56,68–71] modifications of BTL2 lipase for improving its catalytic behavior (typically, to reduce the steric hindrance around the active site) have been also reported. Finally, different papers in recent literature have employed the reported 3D structure of BTL2 for performing molecular simulations aiming to rationalize its catalytic performance and stability [38,72–74].

Although TA lipases are very resistant to organic solvents and high temperatures, the boiling point of the solvents clearly limits the maximum operational temperature. In this sense, the use of RTILs (room-temperature ionic liquids) can be very convenient, as they display very high boiling points and have been proven to be compatible with enzymatic catalysis [75–78]. RTILs (organic salts consisting of an organic cation and a polyatomic inorganic anion, liquid under 100 °C), are broadly regarded as green solvents [79], as they have extremely high enthalpies of vaporization (making them effectively nonvolatile), as well as high chemical and thermal stabilities and remarkable solvating power, so that they can be safely used at high temperature. However, the large number of steps required for their synthesis, sometimes demanding the use of non-renewable crude oil sources and some toxic intermediates, may alter their consideration as eco-friendly solvents [80]. On the other hand, the high cost of RTILs can be compensated by their easy recovery from the reaction media (by simply extracting with (green) organic solvents), which allows to reuse them several times [78].

Remarkably, as their rigidity is higher, thermophilic enzymes are more stereoselective when used at the optimum reaction temperature of the corresponding mesophilic counterparts, which usually is far below their own ideal value [23]. As temperature increases, reaction rate also increases as the thermozyme is approaching its optimal value, but at the expense of a stereoselectivity fall (because of the higher enzymatic flexibility) down to the total loss of activity and enantioselectivity once a certain maximum temperature is overpassed. Finding a thermophilic enzyme retaining its activity and stereodiscrimination capability at very high temperature would be highly desirable.

To illustrate this point, we present the results obtained in the kinetic resolution of 2-(butyryloxy)-2-phenylacetic acid via ethanolysis catalyzed by BTL2 using different ionic liquids (1-butyl-3-methyl imidazolium tetrafluoroborate (BMIMBF$_4$), 1-butyl-3-methyl imidazolium hexafluorophosphate (BMIMPF$_6$), 1-ethyl-3-methyl imidazolium tetrafluoroborate (EMIMBF$_4$) and 1-ethyl-3-methyl imidazolium hexafluorophosphate (EMIMPF$_6$) at high temperatures (90 and 120 °C), comparing the results with those obtained with a conventional organic solvent (*iso*octane).

2. Results

The kinetic resolution of racemic 2-(butyryloxy)-2-phenylacetic acid (R, S)-**1** via ethanolysis to yield pure enantiomers of mandelic acid (R) or (S)-**2** was selected as test reaction to check the performance of BTL2, as depicted in Scheme 1a.

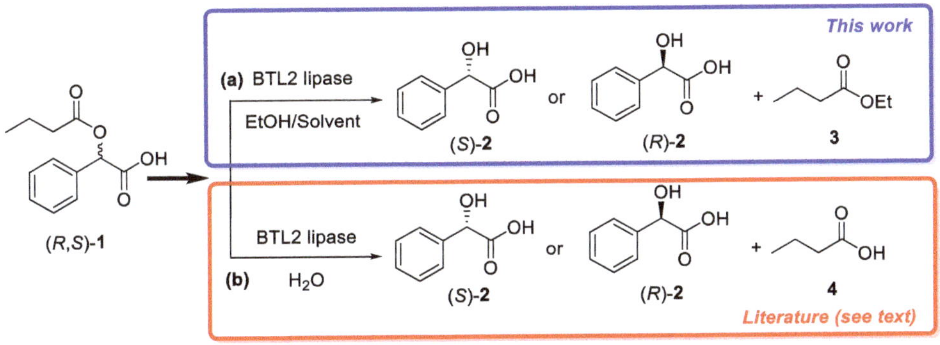

Scheme 1. Kinetic resolution of 2-(butyryloxy)-2-phenylacetic acid (R, S)-**1** via ((**a**), this work) BTL2-catalyzed ethanolysis or ((**b**), literature) BTL2-catalyzed hydrolysis.

(R)-mandelic acid and analogs are key synthons in the preparation of several drugs or biologically active compounds, being in the core of semi-synthetic antibiotics (cephalosporins as cefamandole [81] or penicillins as MA-6-APA II [82]) or anticholinergic drugs (such as oxybutynin [83] and homatropine [84]). Moreover, derivatives of (R)-mandelic acid have been also used as chiral synthons in the preparation of some drugs with different therapeutic activities: platelet/antithrombotic agents (clopidogrel [85]), vasodilator (cyclandelate [86]), antitumor (complex of *cis*-[Pt{2-(α-hydroxy)benzylbenzimidazole))2Cl2] [87], antiobesity [88,89] or CNS-stimulant dopaminergic agents ((R)-pemoline [90]). Conversely, (S)-mandelic acid is used for the production of non-steroidal anti-inflammatory drugs such as deracoxib and celecoxib [91].

In this paper, we present the first reported example of kinetic resolution of (R, S)-**1**, leading to enantiopure mandelic acid, by ethanolysis catalyzed by BTL2 (Scheme 1a) at very high reaction temperature. Selection of (R, S)-**1** as model substrate was based in many previous studies in which BTL2 (mainly immobilized) had been used for catalyzing its stereoselective hydrolysis (Figure 1b) [45,48,49,51,56,58]. In these papers, only low to moderate enantioselectivity is generally observed in the hydrolysis of (R, S)-**1**, and, remarkably, the enzymatic stereobias, leading to either (R)-**2** or (S)-**2**, depends on the type of support and the methodology used for the immobilization.

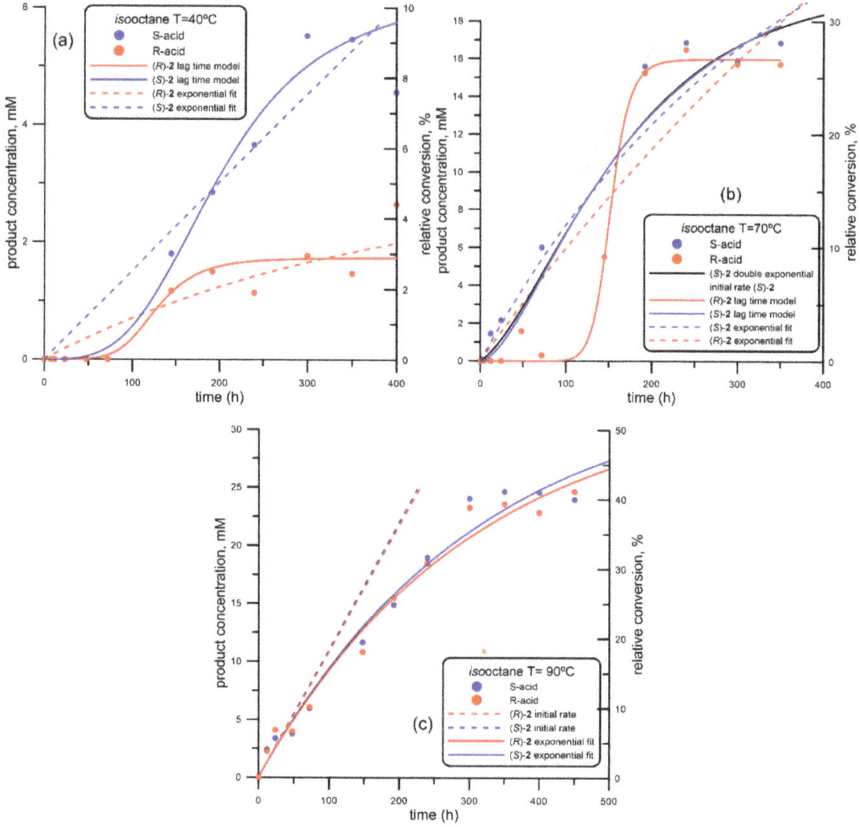

Figure 1. Progress curve of the lipase-form *Geobacillus thermocatenulatus* (BTL2)-catalyzed production of both enantiomers of mandelic acid ((*R*)-acid, in red; (*S*)-acid, in blue) via ethanolysis of racemic 2-(butyryloxy)-2-phenylacetic acid (*R, S*)-**1**, using isooctane as organic solvent at different temperatures. (**a**) 40 °C; (**b**) 70 °C; (**c**) 90 °C.

2.1. Ethanolysis of (R, S)-1 in Isooctane at Different Temperatures

Nevertheless, hydrolysis is not a good alternative for checking the real thermotolerance of a lipase, as the stability of these enzymes is much higher when they are working on organic solvents [26,33,75]. Thus, we decided to use a water-free reaction media, selecting EtOH as nucleophile instead of water and a water-insoluble organic solvent (Figure 1a). Opting for EtOH was based on our previous studies on esterification of phthalic acids with BTL2 [92], and it has been very recently substantiated by Shehata et al. [38]; indeed, these authors have published a molecular dynamics (MD) simulation of the effect of different polar and nonpolar solvents on the thermostability and lid-opening of BTL2, reporting that the open (active) conformation of BLT2 is more stable in EtOH than in MeOH and even water. Additionally, this same study revealed that the overall lipase structure became more stable in nonpolar organic solvents, while it was destabilized in polar solvents except EtOH. Thus, it seems reasonable to use EtOH as nucleophile. On the other hand, *iso*octane (2,2,4-trimethylpentane) was the classical organic solvent selected for comparative reasons, as we had previously described its effectiveness in lipase-promoted catalysis [93–98]. This solvent is considered usable according to the Pfizer solvent selection guide for medicinal chemistry [99].

Thus, following the experimental procedure described in Section 4.2, the ethanolysis of (*R, S*)-**1** was tested at three different temperatures (40, 70 and 90 °C). The progress curves are depicted in Figure 1.

As can be seen from Figure 1a, the reaction at 40 °C proceeds very slowly, reaching a global conversion of around 10% for (*S*)-**2** after 400 h and 3–4% for the (*R*)-**2** counterpart. In addition, a very strong lag-time is observed for both enantiomers, not detecting any trace of ethanolysis of (*R, S*)-**1** in the first 75 h. A similar pattern was observed in the generation of (*R*)-**2** at 70 °C (Figure 1b); in any case, no lag-time was observed for (*S*)-**2** at that temperature and also for both enantiomers at 90 °C (Figure 1c), following a typical exponential grow. Thus, all the progress curves were adjusted using the program INRATE implemented inside SIMFIT fitting package (version 7.6, Release 9), a free-of-charge Open Source software for simulation, curve fitting, statistics and plotting [100] (accessible at https://simfit.org.uk/simfit.html). Using this program, data were fitted either to lag-time kinetics or to standard single exponential growing model. From these mathematical fittings, several parameters were calculated (shown in Table 1) and used to quantify the activity and enantioselectivity of BTL2 in the kinetic resolution of (*R, S*)-**1** via ethanolysis.

Table 1. Quantitative assessment of the ethanolysis of racemic 2-(butyryloxy)-2-phenylacetic acid (*R, S*)-**1** catalyzed by BTL2 using *isooctane* as organic solvent at different temperatures.

Medium	T (°C)	V_S [1]	V_R [1]	VS/VR	t_{MAX} [4]	[C] max [5]	P [6]	[(S)-2] [7]	[(R)-2] [7]	E [7]
#1 isooctane	40	0.015 [2] 0.027 [3]	0.008 [2] 0.014 [3]	1.9 [2] 1.9 [3]				0	0	nd
#2 isooctane	70	0.083 [2]	0.063 [2] 0.332 [3]	1.32 [2]	24	2.26	0.09	1.44	0	>200
#3 isooctane	90	0.110 [2]	0.108 [2]	1.02 [2]				4.0	3.75	nd

[1] initial rate (mM/h). [2] single exponential model. [3] lag-time model. [4] higher reaction time (h) at which only one enantiomer is detected. [5] concentration (mM) of the only isomer detected at that higher reaction time. [6] productivity (mM acid/h) at the higher reaction time. [7] enantiomeric ratio, calculated at 12 h.

As commented before, reaction was extremely slow at 40 °C; increasing the temperature up to 70 °C (Figure 1b), the reaction rate increased very markedly, and the catalytic performance for both enantiomers was clearly different: while the generation (*S*)-**2** was detected from the early reaction stages and follows a single exponential model, it is not until 100 h when (*R*)-**2** was clearly detected, quickly growing after this point to reach similar conversion values than those observed for (*S*)-**2** after 200 h. When the temperature was increased up to 90 °C (Figure 1c), both enantiomers were produced by a similar pattern, at the same initial rate (Table 1, entry #3) and reaching similar conversion degrees (around 40%) after 500 h, with no enantioselectivity at all. Higher temperatures were not tested as it would mean approaching the boiling point of *isooctane* (99.6 °C)

Overall, best results are those obtained at 70 °C at short reaction times. In fact, inside the time interval from 0 to 75 h, the enantioselectivity is almost perfect, although at the expenses of a low overall conversions (around 10%, Figure 1b). Nevertheless, as results with this organic solvent were quite unsatisfactory, room-temperature ionic liquids (RTILs) were subsequently tested.

2.2. Ethanolysis of (R, S)-1 in RTILs at Different Temperatures

As commented in the Introduction, RTILs are fully compatible with enzymatic catalysis [75–78]. The most popular RTILs are those based on imidazolium cations [101,102], being 1-butyl-3-methyl imidazolium tetrafluoroborate (BMIMBF$_4$), 1-butyl-3-methyl imidazolium hexafluorophosphate (BMIMPF$_6$), 1-ethyl-3-methyl imidazolium tetrafluoroborate (EMIMBF$_4$) and 1-ethyl-3-methyl imidazolium hexafluorophosphate (EMIMPF$_6$) probably the first ones to be broadly commercialized. Their properties have been profusely described, especially including their complete miscibility with EtOH [103–110], the other main component of the reaction medium, used both as cosolvent and nucleophile. As these RTILs possess a very high boiling point, the use of their mixtures with EtOH allows using these binary mixtures at high reaction temperatures, without any noticeable EtOH evaporation. Hence, ethanolysis of (R, S)-**1** was tested at two different temperatures, 90 °C (similar to the maximum tested with *iso*octane) and 120 °C, a temperature higher than the boiling point of the organic solvent. The results are depicted in Figures 2–5. Table 2 summarizes the parameters obtained from the fitting of the corresponding progress curves.

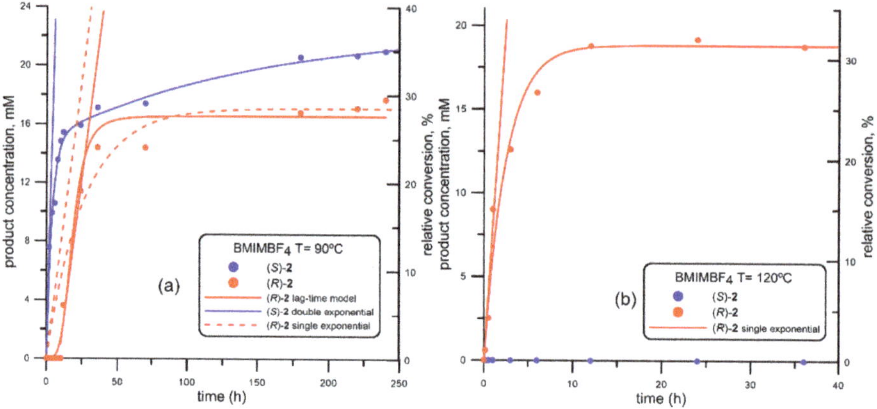

Figure 2. Progress curve of the BTL2-catalyzed production of both enantiomers of mandelic acid ((R)-acid, in red; (S)-acid, in blue) via ethanolysis of racemic 2-(butyryloxy)-2-phenylacetic acid (R, S)-**1**, using 1-butyl-3-methyl imidazolium tetrafluoroborate (BMIMBF$_4$) at different temperatures. (**a**) 90 °C; (**b**) 120 °C. Fitting parameters shown in Table 2, corresponding to entries **#4** (BMIMBF$_4$ at 90 °C) and **#5** (BMIMBF$_4$ at 120 °C).

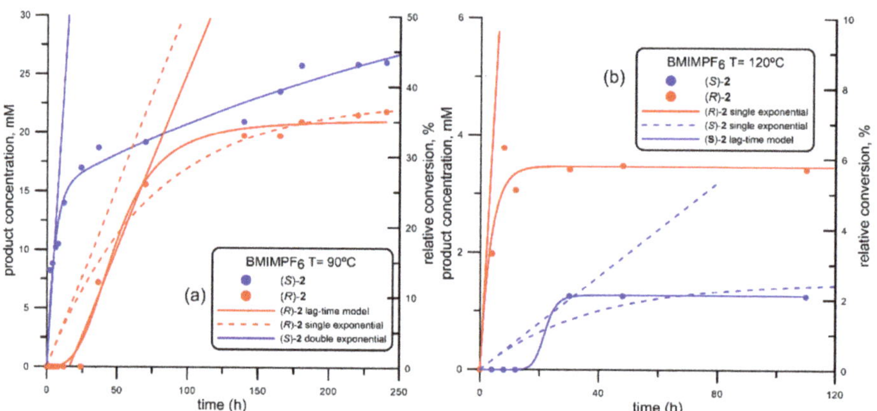

Figure 3. Progress curve of the BTL2-catalyzed production of both enantiomers of mandelic acid ((R)-acid, in red; (S)-acid, in blue) via ethanolysis of racemic 2-(butyryloxy)-2-phenylacetic acid (R, S)-**1**, using 1-butyl-3-methyl imidazolium hexafluorophosphate (BMIMPF$_6$) at different temperatures. (**a**) 90 °C; (**b**) 120 °C. Fitting parameters shown in Table 2, corresponding to entries **#6** (BMIMPF$_6$ at 90 °C) and **#7** (BMIMPF$_6$ at 120 °C).

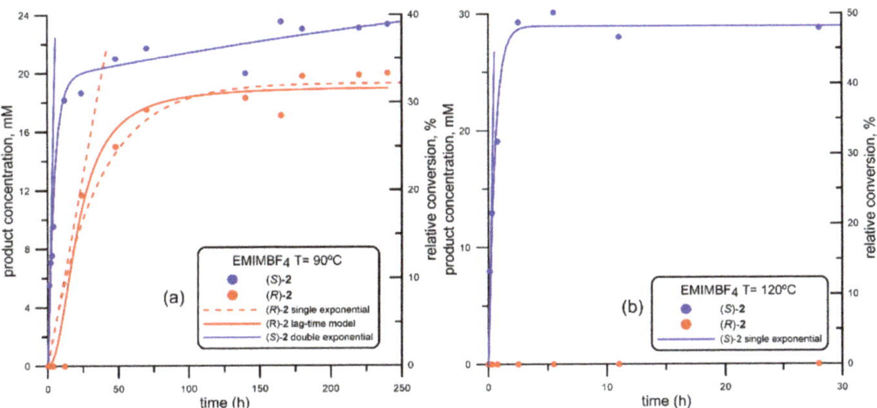

Figure 4. Progress curve of the BTL2-catalyzed production of both enantiomers of mandelic acid ((*R*)-acid, in red; (*S*)-acid, in blue) via ethanolysis of racemic 2-(butyryloxy)-2-phenylacetic acid (*R*, *S*)-**1**, using 1-ethyl-3-methyl imidazolium tetrafluoroborate (EMIMBF$_4$) at different temperatures. (**a**) 90 °C; (**b**) 120 °C. Fitting parameters shown in Table 2, corresponding to entries **#8** (EMIMBF$_4$ at 90 °C) and **#9** (EMIMBF$_4$ at 120 °C).

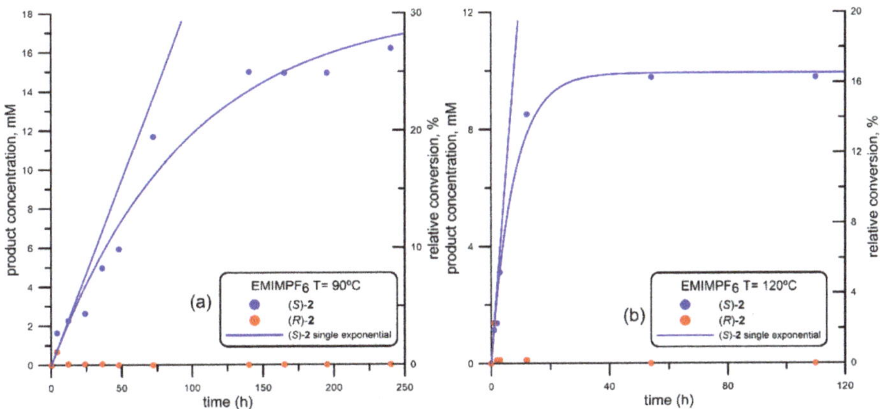

Figure 5. Progress curve of the BTL2-catalyzed production of both enantiomers of mandelic acid ((*R*)-acid, in red; (*S*)-acid, in blue) via ethanolysis of racemic 2-(butyryloxy)-2-phenylacetic acid (*R*, *S*)-**1**, using 1-ethyl-3-methyl imidazolium hexafluorophosphate (EMIMPF$_6$) at different temperatures. (**a**) 90 °C; (**b**) 120 °C. Fitting parameters shown in Table 2, corresponding to entries **#9** (EMIMPF$_6$ at 90 °C) and **#10** (EMIMPF$_6$ at 120 °C).

Table 2. Quantitative assessment of the ethanolysis of racemic 2-(butyryloxy)-2-phenylacetic acid (R,S)-**1** catalyzed by BTL2 using different room-temperature ionic liquids (RTILs) at different temperatures.

	RTIL	T (°C)	V_S [1]	V_R [1]	V_{MAX}/V_{min} [2]	t_{MAX} [3]	[C] max [4]	P [5]	[(S)-2] [6]	[(R)-2] [6]	E [6]
#4	BMIM BF$_4$	90	3.85	0.76	5.1	10	13.6	1.36	15.42	13.6	1.1
#5	BMIM BF$_4$	120	-	8.14	-	36	18.8	0.52	0	18.8	>200
#6	BMIM-PF$_6$	90	2.0	0.31	6.4	24	17.0	0.71	14.0	0	>200
#7	BMIM-PF$_6$	120	0.04	0.96	24	8	6.3	0.79	0	3.06	>200
#8	EMIM-BF$_4$	90	4.08	0.52	7.8	12	18.2	3.03	18.2	0	>200
#9	EMIM-BF$_4$	120	53.4	0	-	2.5	29.2	11.7	28.1	0	>200
#10	EMIM-PF$_6$	90	1.86	-	-	240	16.2	0.81	2.28	0.06	51.3
#11	EMIM-PF$_6$	120	1.3	-	-	54	9.8	0.18	8.52	0	>200

[1] initial rate (mM/h); [2] VS/V_R in all cases except for entries #5 and #7, when it should be V_R/VS; [3] higher reaction time (h) at which only one enantiomer is detected; [4] concentration (mM) of the only isomer detected at that higher reaction time; [5] productivity (mM acid/h) at the higher reaction time; [6] enantiomeric ratio, calculated at 12 h.

2.2.1. Ethanolysis of (R, S)-1 Using RTILs Based on 1-Butyl-3-methyl Imidazolium (BMIM) as Solvent

When the tetrafluoroborate (BMIMBF$_4$) solvent was used (Figure 2), it can be observed how using the lower reaction temperature (90 °C, Figure 2a) led to a different kinetic behavior in the generation of both enantiomers of mandelic acid. In fact, as the (S)-acid (in blue) was detected from the earlier reaction stages, the correspondent (R)-acid (in red) was not produced until a lag-time of around 12 h had been overpassed, experimenting a rapid increase in its production leading to an overall sigmoid curve (Figure 2a, red solid line); remarkably, a similar initial rate (VS = 3.85 mM/h, Table 2) was calculated considering a single exponential model (Figure 2a, red dotted line) or the sigmoid lag-time model, red solid line). In this case, (S)-acid (in blue) was the best-recognized enantiomer, as also observed using *iso*octane (Figure 1), although for BMIMBF$_4$ the initial rate was 35 times higher than that observed for isooctane (VS = 0.11 mM/h, Table 1). Furthermore, the reaction in this RTIL was not only faster, but also more enantioselective than in *iso*octane, as the lag-time observed for the generation of the (R)-acid allowed the production of exclusively (S)-**2** at the first stages of the reaction (t_{MAX} 10 h, [(S)-**2**] $_{MAX}$ 13.6%, corresponding to 22.7% conversion).

Another interesting aspect to be taken into account is that, while the generation (R)-**2** remained constant after a certain reaction time (around 30% after 50 h, according to the sigmoid fitting), the production of (S)-**2** was slowly increasing after this time, although at a lower reaction rate than that observed at the early stages; that is the reason the overall (S)-**2** production followed a double exponential fitting. This slower second reaction rate could be caused by an inhibition promoted by the increasing amounts of (R)-**2** present in the reaction media, as the slope change in the (S)-**2** production is observed only after a certain accumulation of (R)-**2**.

However, when performing the ethanolysis at 120 °C (Figure 2b), the observed reaction pattern was radically different from that obtained at 90 °C, as only the (R) enantiomer of mandelic acid was produced through a fast single exponential fit, displaying an initial rate (V_R = 8.14 mM/h, Table 2) twice that one obtained for the preferentially recognized (S)-**2** at 90 °C. Indeed, the reaction did not progress after 40 h, and no traces of (S)-**2** were detected under these conditions, so that the enantioselectivity was complete, leading to an overall conversion of around 30%.

An inversion in the stereobias of BTL2 in the recognition of both enantiomers of mandelic acid had been previously reported, although in the hydrolysis of (R,S)-**1** and associated with the different methodology and support used for the immobilization of this lipase [45,47–49], as already mentioned in the Introduction. A simple explanation of this modification in the enantiodiscrimination of BTL2 upon increasing the temperature would demand a detailed computational study, which is out of the scope of this manuscript.

The results of the ethanolysis of (R,S)-**1** using BMIMPF$_6$ are shown in Figure 3.

As can be seen, the behavior was similar to that observed using BMIMBF$_4$; at 90 °C (Figure 3a), the (S) enantiomer of mandelic acid is quickly detected, once again following a double exponential fit, at a smaller initial rate than that one obtained using BMIMBF$_4$, but also higher if compared to that calculated using *iso*octane (Table 1). Similarly, the inversion in the stereobias at 120 °C is also detected, but now the reaction proceeded very poorly compared to that depicted in Figure 2b, as only a maximum of 6% conversion is detected for (*R*)-**2**.

2.2.2. Ethanolysis of (R, S)-1 Using RTILs Based on 1-Ethyl-3-methyl Imidazolium (EMIM) as Solvent

The results obtained using RTILs in which 1-ethyl-3-methyl imidazolium (EMIM) is the cation are shown in Figures 4 and 5. More concretely, Figure 4 shows the reaction progress curves using EMIMBF$_4$ at 90 °C (Figure 4a) and 120 °C (Figure 4b).

Comparing Figure 4a (EMIMBF$_4$) with Figure 2a (BMIMBF$_4$), a similar comportment is observed. As depicted in Figure 4a, the lag-time for the detection of (*R*)-**2** was slightly higher than that observed using BMIMPF$_4$ and also the initial rate (4,08 vs 3.85 mM h^{-1}, Table 2), so that it was possible to detect only (*S*)-**2** in the first 12 hours, reaching a concentration of 18.2 mM (around 30% conversion) with complete enantioselectivity Once again, as the reaction proceeded and the (*R*)-**2** enantiomer was being produced (lag-time and single exponential models almost similar), the rate of production of (*S*)-**2** was reduced, so that once again the overall behavior for (*R*)-**2** could be fitted to a double exponential curve.

When the reaction temperature is increased up to 120 °C (Figure 4b), a very fast kinetic resolution can be observed. In fact, the initial rate in the generation of (*S*)-**2** (VS = 53.4 mM h^{-1}, Table 2) was one order of magnitude higher than that obtained at 90 °C; moreover, no traces of (*R*)-**2** were detected during the whole reaction time, so that the shape of progress curve fits to a single exponential plot leading to almost 50% conversion (the maximum for a kinetic resolution) in only 2.5 hours.

Results obtained using EMIMPF$_6$ as solvents are depicted in Figure 5. As can be seen, at both temperatures only small traces of (*R*)-**2** were detected, while the generation of (*S*)-**2** followed single exponential kinetics. At 90 °C, the kinetic resolution observed using this solvent (Figure 5a) was slower (VS around one half, See Table 2, entries #8 vs #10) than that obtained with EMIMBF$_4$ (Figure 4a), although this fact was compensated with an absence of production of (*R*)-**2**, leading to 30% conversion after 250 h. When increasing the temperature to 120 °C (Figure 5b), the kinetic resolution was slightly slower, and the maximum conversion was half that obtained at 90 °C.

3. Discussion

It is generally accepted that enzymes in RTILs are more active and stable as the hydrophobicity of the RTIL increases (see the review from Liu and co-workers [75], as well as the references cited therein). This fact is commonly related to the higher preservation of the essential water layer surrounding the enzyme structure, resulting in a decrease of the protein–ion interactions with a concomitant reduction of enzyme denaturation [76]. Therefore, the use of water-immiscible RTILs, more hydrophobic than the corresponding water-soluble ones, has been recommended [75]. However, these rules were described for pure RTILs. In fact, mixtures of RTILs and organic solvents have been reported to increase the catalytic activity, stability and enantioselectivity of enzymes compared to the single RTIL, probably by reducing the viscosity of the RTIL and diminishing mass-transfer limitations [75]; on the other hand, the proportion and the nature of the organic cosolvent is usually crucial to reach good activity and selectivity (see [75] and references cited therein). Varela et al. [110] collected some data on theoretical studies of mixtures of RTILs and alcohols, concluding that molecular dynamics simulations of the solvation of alcohols in RTILs may resemble water behavior. Thus, the different regions of the bulk RTIL can interact with the analogous (polar or apolar) parts of the solute molecules, and this fact is pivotal for understanding both the mesomorphic structure of the mixtures as well as their dynamics, following a pattern termed "nanostructured solvation" [110]. However, theoretical studies describing the effect of RTIL/organic solvent mixtures on the structure of enzymes are still missing.

Particularly, the use of an apolar organic solvent (toluene) has been reported to increase BTL2 rigidity, as deduced from molecular dynamics simulations at temperatures as high as 450°K (176.85 °C) [72], without observing any tendency for the lid to open. These same authors described that either inserting a thin layer of water around the enzyme or promoting a single point mutation (G116P) would be required for retaining activity in acyl-transfer processes [72]. Additionally, another very recent study confirmed the reduced flexibility of BTL2 in nonpolar organic solvents, using molecular dynamics on toluene and cyclohexane, and thus confirming the enhancement of thermostability of BTL2 in the presence of these type of solvents [38]. These theoretical studies would support our results obtained using *iso*octane (Figure 1 and Table 1), where we observed how increasing the reaction temperature up to 90 °C promoted a moderate rise in the reaction rate, but with no enantioselectivity, as the lid should not open (the only water present in the medium would be that one retained in the enzymatic liophilizate, not enough to reach a proper concentration) and therefore not upholding the lid flexibility required for the proper enzymatic enantiodiscrimination [111,112].

It has been also described that polar solvents (water and short-chain alcohols) lead to enhanced fluctuation of BTL2's lid at low temperatures, but surprisingly the open conformation turned out to be more stable in EtOH than in water or MeOH [38]. Thus, considering the beneficial effect of EtOH on BTL2, we decide to check the catalytic performance in mixtures between different RTILs and EtOH. As pointed out in Section 4.2, EtOH was used in a high molar excess compared to the starting substrate (3.42 M EtOH versus 60 mM for (*R*,*S*)-**1**), but if we consider the molar fraction of the RTIL and EtOH, the situation is different, as indicated in Table 3.

Table 3. Composition of the different reaction mixtures EtOH/RTIL (1 mL/4 mL, V/V).

RTIL	MW [1] (g/mol)	Density [1] (g/mL)	[RTIL], M	[EtOH], M	Molar Fraction X_{RTIL}
BMIMBF$_4$	226.02	1.21	4.28	3.42	0.56
BMIMPF$_6$	284.19	1.38	3.89	3.42	0.53
EMIMBF$_4$	197.97	1.29	5.22	3.42	0.60
EMIMPF$_6$	256.13	1.48	4.62	3.42	0.57

[1] data from SciFinder database.

As can be seen from Table 3, the molar composition of the reaction mixture is not exactly the same, because of the differences in the molecular weight and density of the four RTILs, although an average value of (0.56 ± 0.04) for X_{RTIL} can be taken as an average.

The most hydrophobic RTIL, water-insoluble BMIMPF$_6$, is definitively not the best option for the ethanolysis of (*R*,*S*)-**1**, as shown in Figure 3, neither at 90 nor at 120 °C; in fact, the initial rate in the generation of (*S*)-**2** at 90 °C is not the highest, although enantioselectivity is perfect (E > 200) up to 24 h. At 120 °C, an inversion in the stereobias at 120 °C was observed, but only a maximum of 6% conversion is detected for (*R*)-**2**. Looking at the literature, (*R*)-**1** is the recognized enantiomer in the hydrolysis of (*R*,*S*)-**1** by wild-type BTL2 [56], as well as by some immobilized preparation of this enzyme [45], so that the "canonical" recognition, as predicted by the well-known Kazlauskas' rule (Scheme 2a), based on the relative size of substituents around the stereocenter [113] would be that one depicted in Scheme 2b.

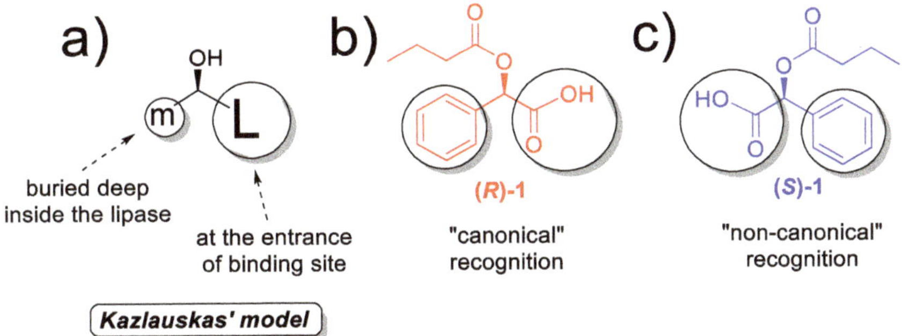

Scheme 2. (**a**) Kazlauskas' rule; (**b**) "canonical" recognition of (*R*)-**1** enantiomer; (**c**) "noncanonical" recognition of (*S*)-**1** enantiomer.

Actually, the large (L) binding pockets is located at the entrance of binding site, while the other medium (m) pocket is buried deep inside the lipase; thus, this would mean that the phenyl ring of the (*R*)-substrate would be the one interacting with the cavity inside the 3D structure of BTL2 in the canonical recognition pattern. This interaction could be caused by a π-stacking of the phenyl moiety of (*R*)-**1** with Phe17, a residue which changes its conformation in the open structure and allows the access of the substrate to the catalytic serine [37]; a similar interaction has been proposed for other aromatic substrates with lipases [114,115]. However, using BMIMPF$_6$ at 90 °C, the non-Kazlauskas recognition (Scheme 2c) is majoritarian, while increasing temperature up to 120 °C, an alteration of the enantioselectivity is observed, but this could be attributed to an enzyme inactivation, as long as the activity dropped dramatically.

Changing from BMIMPF$_6$ (Figure 3) to BMIMBF$_4$ (Figure 2), the resolution become faster at 90 °C, (Table 2), but even better at 120 °C, when the inversion of the canonical recognition is absolute, up to the point that no (*S*)-**2** is detected at all, and the (*S*)-selection is maintained until the reaction ends (after 12 h, Figure 2b). Filice et al. [116] described that the BF$_4$ does not cause negative effects on BTL2 in aqueous media, as it happens with other lipases. However, the change of the canonical (Scheme 2b) recognition to the non-Kazlauskas pattern (Scheme 2c) upon heating at 120 °C may be caused by many possible factors. Indeed, a different hydrogen bonding arrangement is one of the reported reason for changing the enantioselection of lipases [117,118], and it is well known that the anionic component of a RTIL is the main responsible of establishing the hydrogen–bonding network with the enzyme [75]; anyhow, this assumption would demand a systematic molecular dynamics study, out of the scope of this paper. In this sense, some preliminary data (not published) obtained in our group have revealed that the half-life of an hydrogen bond between BTL2 and BMIMBF$_4$ at 120 °C is twice that one in water, partially reinforcing our hypotheses.

On the other hand, the use of more hydrophilic 1-ethyl-3-methyl imidazolium (EMIM) cation has been recommended for BTL2 [116]. Similarly, for other lipases it has been also shown how the shorter the alkyl chain in the cationic imidazolium, the higher the activity [119], while Filice et al. recommended the use of hexafluorophosphate (PF$_6^-$) combined with EMIM [116]. In fact, by looking at the progress curves obtained in the ethanolysis of (*R,S*)-**1** using EtOH/EMIMPF$_6$ and depicted in Figure 5a (90 °C) and Figure 5b (120 °C), the kinetic resolutions are quite good, although better at 90 °C. In any case, it is noteworthy to observe how BTL2 is totally enantioselective under these reaction conditions, as no traces of the "canonical" (*R*)-**2** isomer were detected.

When using EtOH/EMIMBF$_4$ at 90 °C (Figure 4a), the situation is quite similar to that obtained with BMIMBF$_4$ (Figure 2a) or BMIMPF$_6$ (Figure 3a); nevertheless, the kinetic resolution is much better when using EtOH/EMIMBF$_4$ at 120° (Figure 4b), as in only 2.5 h a total 50% conversion in (*S*)-**2** is obtained, and again no traces of the (*R*)-antipode are formed. To our knowledge, this is the best and faster lipase-catalyzed kinetic resolution ever reported for enantiomers of mandelic acid.

Thus, the best results, both in terms of reaction rate and enantioselectivity, are those obtained using this RTIL formed by the most hydrophilic cation and the most hydrophilic anion, at 120 °C. Once again, we cannot propose a certain reason to explain this behavior, as that was not the purpose of this manuscript. According to the theoretical studies from Shehata et al. [53], the presence of EtOH in the reaction medium would allow the fluctuation of the lid to allow the active site to get exposed and is very compatible with BTL2 stability, as for sure $EMIMBF_4$ has also demonstrated to be.

4. Materials

Lipase from *Geobacillus thermocatenulatus* was a kind gift of the Biochemistry, Genetics and Immunology Department of the Universidad de Vigo, Spain. All solvents of the highest purity commercially available and used without purification were purchased from Sigma-Aldrich-Fluka, (Barcelona, Spain). All other chemicals and RTILs (1-butyl-3-methyl imidazolium tetrafluoroborate ($BMIMBF_4$), 1-butyl-3-methyl imidazolium hexafluorophosphate ($BMIMPF_6$), 1-ethyl-3-methyl imidazolium tetrafluoroborate ($EMIMBF_4$) and 1-ethyl-3-methyl imidazolium hexafluorophosphate ($EMIMPF_6$) were purchased also from Sigma-Aldrich-Fluka (Barcelona, Spain).

4.1. Synthesis of (R,S) 2-(Butyryloxy)-2-phenylacetic Acid

Twenty mmol of mandelic acid in 200 mL of diethyl ether were added to 2.88 mL triethylamine (20 mmol). Subsequently, a solution of 2.131 mL (20 mmol) butyryl chloride in 100 mL ethyl ether was dropped. The reaction was carried out in a flask at 25 °C for approximately four hours, giving a yield of 50%. Reaction progress was followed by HPLC, ^{1}H-NMR and ^{13}C-NMR (Bruker AC-250 (^{1}H), 63 MHz (^{13}C), Bruker Corporation, Billerica, MA, USA.

Simply by adding water, the unreacted acid could be separated from the ester, which remained in the organic phase. Acid was removed performing successive washing with water. Subsequently, organic phase was dried with anhydrous sodium sulfate and the remaining ether was removed using rotary evaporator. Successive extraction with diethyl ether, allowed the isolation of 7 g of yellowish oil as a final product ((R, S) 2-butanoiloxy phenyl acetic). spectroscopic data were according to those previously reported in literature [120].

4.2. Resolution of 2-(Butyryloxy)-2-phenylacetic Acid by Alcoholysis Reaction

Reactions were carried out in closed glass vials and the temperature of the experiments varied between 40 and 120 °C. In the case of *iso*octane, temperatures of 40, 70 and 90 °C were used, whereas ionic liquids were tested at 90 and 120 °C. To maintain the temperature fixed, a thermostat-equipped bath oil was used for several days. The reaction mixture included: an organic solvent or RTIL (4 mL), 2-(butyryloxy)-2-phenylacetic acid (60 mM) and ethanol (1 mL), assuring a molar excess of alcohol versus the organic acid. Then, the enzyme (4 mg solid/mL) was added. In order to ensure that reaction is due only to the lipase, a reaction test without lipase was carried out. As a consequence of the employment of high temperatures, the vials were rapidly cooled down for sampling previously to the opening to avoid any evaporation of the alcohol; thus, aliquots of 100 μL were taken at different times. As RTILs cannot be directly injected into the HPLC (Constrametric 4100 pump, UV detector Spectromonitor 5000, LDC Analytical, Spain), the samples were extracted with 1 mL of diethyl ether and the solvent evaporated at room temperature; subsequently, they were re-diluted with 400 μL hexane/2-propanol (1:1) (*v/v*) and filtered using syringe filters (Millex-GV (PVDF), 0.22 μm pore size). HPLC analysis was performed using a Chiracel OD (20 μm (250 × 4.6 mm) chiral column, a mobile phase composed by *n*-hexane/2-propanol/trifluoroacetic acid (90/9/1), a 0.8 mL/min flux and a wavelength of 254 nm. Peak assignation was determined using pure compounds as standards.

5. Conclusions

Although thermophilic enzymes can efficiently work at very high temperature, they are more stereoselective when used at temperature below their optimal one, as they are more rigid under those reaction conditions. Consequently, finding a thermophilic enzyme capable to retain its activity and stereodiscrimination capability at high temperature would lead to very fast and stereoselective processes. In this paper, we have shown how binary mixtures of EtOH and room temperature ionic liquids (RTILs) are an excellent reaction media for the enantioselective kinetic resolution (KR) of racemic 2-(butyryloxy)-2-phenylacetic acid via ethanolysis catalyzed by lipase-form *Geobacillus thermocatenulatus* (BTL2) at very high reaction temperatures (120°). Thus, the KR carried out using an EtOH/BMIMBF$_4$ at 120 °C furnished (*R*)-mandelic acid as the only reaction product (E > 200) at very short reaction time (12 h), while by changing the composition of the cationic moiety of the RTIL, then using EtOH/EMIMBF$_4$ at the same temperature of 120 °C, an ever faster (2.5 h reaction time) and completely enantioselective KR led to enantiopure (*S*)-mandelic acid. This is the faster described KR for this substrate. The previously reported beneficial effect of EtOH on BTL2, stabilizing its open active conformation even at extreme temperatures, as deduced from molecular dynamics, is supporting our results; on the other hand, the establishment of very strong hydrogen bonds between the enzyme and the RTIL could be the responsible for the observed thermostability.

Author Contributions: Experimental data, J.R.-M., O.K.; writing—original draft preparation, J.R.-M., O.K.; writing—review and editing, A.R.A., J.M.S.-M. Both corresponding authors have similarly contributed to the final version of the manuscript. All authors have read and agreed to the published version of the manuscript.

Funding: This research was partially funded by Projects CTQ2017-86170-R (MINECO-Spanish Government) and PR87/19-22676 (Banco de Santander-Complutense Research Projects).

Conflicts of Interest: The authors declare no conflict of interest.

References

1. Wiltschi, B.; Cernava, T.; Dennig, A.; Casas, M.G.; Geier, M.; Gruber, S.; Haberbauer, M.; Heidinger, P.; Acero, E.H.; Kratzer, R.; et al. Enzymes revolutionize the bioproduction of value-added compounds: From enzyme discovery to special applications. *Biotechnol. Adv.* **2020**, *40*, 51. [CrossRef] [PubMed]
2. Woodley, J.M. New frontiers in biocatalysis for sustainable synthesis. *Curr. Opin. Green Sustain. Chem.* **2020**, *21*, 22–26. [CrossRef]
3. Huang, X.; Cao, M.; Zhao, H. Integrating biocatalysis with chemocatalysis for selective transformations. *Curr. Opin. Chem. Biol.* **2020**, *55*, 161–170. [CrossRef] [PubMed]
4. Li, J.; Amatuni, A.; Renata, H. Recent advances in the chemoenzymatic synthesis of bioactive natural products. *Curr. Opin. Chem. Biol.* **2020**, *55*, 111–118. [CrossRef] [PubMed]
5. Sheldon, R.A.A.; Brady, D.; Bode, M.L.L. The Hitchhiker's guide to biocatalysis: Recent advances in the use of enzymes in organic synthesis. *Chem. Sci.* **2020**, *11*, 2587–2605. [CrossRef]
6. Sandoval, B.A.; Hyster, T.K. Emerging strategies for expanding the toolbox of enzymes in biocatalysis. *Curr. Opin. Chem. Biol.* **2020**, *55*, 45–51. [CrossRef]
7. Chapman, J.; Ismail, A.; Dinu, C. Industrial Applications of Enzymes: Recent Advances, Techniques, and Outlooks. *Catalysts* **2018**, *8*, 238. [CrossRef]
8. de Gonzalo, G.; Domínguez de María, P. *Biocatalysis: An Industrial Perspective*; Royal Society of Chemistry: London, UK, 2018. [CrossRef]
9. Woodley, J.M. Accelerating the implementation of biocatalysis in industry. *Appl. Microbiol. Biotechnol.* **2019**, *103*, 4733–4739. [CrossRef]
10. Hughes, G.; Lewis, J.C. Introduction: Biocatalysis in Industry. *Chem. Rev.* **2018**, *118*, 1–3. [CrossRef]
11. Domínguez de María, P.; de Gonzalo, G.; Alcántara, A.R. Biocatalysis as useful tool in asymmetric synthesis: An assessment of recently granted patents (2014–2019). *Catalysts* **2019**, *9*, 802. [CrossRef]
12. Alcántara, A.R. Biotransformations in Drug Synthesis: A Green and Powerful Tool for Medicinal Chemistry. *J. Med. Chem.Drug. Des.* **2018**, *1*, 1–7. [CrossRef]
13. Hoyos, P.; Pace, V.; Alcántara, A.R. Chiral Building Blocks for Drugs Synthesis via Biotransformations. In *Asymmetric Synthesis of Drugs and Natural Products*; Nag, A., Ed.; CRC Press: Boca Raton, FL, USA, 2018; pp. 346–448.
14. Alcántara, A.R. Biocatalysis and Pharmaceuticals: A Smart Tool for Sustainable Development. *Catalysts* **2019**, *9*, 792. [CrossRef]

15. Truppo, M.D. Biocatalysis in the Pharmaceutical Industry: The Need for Speed. *ACS Med. Chem. Lett.* **2017**, *8*, 476–480. [CrossRef] [PubMed]

16. Rosenthal, K.; Lutz, S. Recent developments and challenges of biocatalytic processes in the pharmaceutical industry. *Curr. Opin. Green Sustain. Chem.* **2018**, *11*, 58–64. [CrossRef]

17. Lalor, F.; Fitzpatrick, J.; Sage, C.; Byrne, E. Sustainability in the biopharmaceutical industry: Seeking a holistic perspective. *Biotechnol. Adv.* **2019**, *37*, 698–703. [CrossRef] [PubMed]

18. Sheldon, R.A.; Woodley, J.M. Role of Biocatalysis in Sustainable Chemistry. *Chem. Rev.* **2018**, *118*, 801–838. [CrossRef] [PubMed]

19. Sheldon, R.A.; Brady, D. Broadening the Scope of Biocatalysis in Sustainable Organic Synthesis. *ChemSusChem* **2019**, *12*, 2859–2881. [CrossRef]

20. Kumar, S.; Dangi, A.K.; Shukla, P.; Baishya, D.; Khare, S.K. Thermozymes: Adaptive strategies and tools for their biotechnological applications. *Bioresour. Technol.* **2019**, *278*, 372–382. [CrossRef]

21. Han, H.W.; Ling, Z.M.; Khan, A.; Virk, A.K.; Kulshrestha, S.; Li, X.K. Improvements of thermophilic enzymes: From genetic modifications to applications. *Bioresour. Technol.* **2019**, *279*, 350–361. [CrossRef]

22. González-Siso, M.-I. Editorial for the Special Issue: Thermophiles and Thermozymes. *Microorganisms* **2019**, *7*, 62. [CrossRef]

23. Atalah, J.; Caceres-Moreno, P.; Espina, G.; Blamey, J.M. Thermophiles and the applications of their enzymes as new biocatalysts. *Bioresour. Technol.* **2019**, *280*, 478–488. [CrossRef] [PubMed]

24. Hait, S.; Mallik, S.; Basu, S.; Kundu, S. Finding the generalized molecular principles of protein thermal stability. *Proteins* **2020**, *88*, 788–808. [CrossRef] [PubMed]

25. Liszka, M.J.; Clark, M.E.; Schneider, E.; Clark, D.S. Nature Versus Nurture: Developing Enzymes That Function Under Extreme Conditions. In *Annual Review of Chemical and Biomolecular Engineering*; Prausnitz, J.M., Ed.; Annual Reviews: Palo Alto, CA, USA, 2012; Volume 3, pp. 77–102.

26. Stepankova, V.; Bidmanova, S.; Koudelakova, T.; Prokop, Z.; Chaloupkova, R.; Damborsky, J. Strategies for Stabilization of Enzymes in Organic Solvents. *ACS Catal.* **2013**, *3*, 2823–2836. [CrossRef]

27. Dumorne, K.; Cordova, D.C.; Astorga-Elo, M.; Renganathan, P. Extremozymes: A Potential Source for Industrial Applications. *J. Microbiol. Biotechnol.* **2017**, *27*, 649–659. [CrossRef] [PubMed]

28. Filho, D.G.; Silva, A.G.; Guidini, C.Z. Lipases: Sources, immobilization methods, and industrial applications. *Appl. Microb. Biotechnol.* **2019**, *103*, 7399–7423. [CrossRef] [PubMed]

29. Daiha, K.D.; Angeli, R.; de Oliveira, S.D.; Almeida, R.V. Are lipases still important biocatalysts? A study of scientific publications and patents for technological forecasting. *PLoS ONE* **2019**, *10*, 20. [CrossRef] [PubMed]

30. Dwivedee, B.P.; Soni, S.; Sharma, M.; Bhaumik, J.; Laha, J.K.; Banerjee, U.C. Promiscuity of lipase-catalyzed reactions for organic synthesis: A recent update. *ChemistrySelect* **2018**, *3*, 2441–2466. [CrossRef]

31. Sarmah, N.; Revathi, D.; Sheelu, G.; Rani, K.Y.; Sridhar, S.; Mehtab, V.; Sumana, C. Recent advances on sources and industrial applications of lipases. *Biotechnol. Prog.* **2018**, *34*, 5–28. [CrossRef]

32. Priyanka, P.; Tan, Y.Q.; Kinsella, G.K.; Henehan, G.T.; Ryan, B.J. Solvent stable microbial lipases: Current understanding and biotechnological applications. *Biotechnol. Lett.* **2019**, *41*, 203–220. [CrossRef] [PubMed]

33. Kumar, A.; Dhar, K.; Kanwar, S.S.; Arora, P.K. Lipase catalysis in organic solvents: Advantages and applications. *Biol. Proced. Online* **2016**, *18*, 2. [CrossRef] [PubMed]

34. Mohtashami, M.; Fooladi, J.; Haddad-Mashadrizeh, A.; Housaindokht, M.R.; Monhemi, H. Molecular mechanism of enzyme tolerance against organic solvents: Insights from molecular dynamics simulation. *Int. J. Biol. Macromol.* **2019**, *122*, 914–923. [CrossRef] [PubMed]

35. Elleuche, S.; Schroder, C.; Antranikian, G. Lipolytic extremozymes from psychro- and (hyper-)thermophilic prokaryotes and their potential for industrial applications. In *Biotechnology of Extremophiles: Advances and Challenges*; Rampelotto, P.H., Ed.; Springer International Publishing Ag: Cham, Switzerland, 2016; Volume 1, pp. 351–374.

36. Lajis, A.F.B. Realm of Thermoalkaline Lipases in Bioprocess Commodities. *J. Lipids* **2018**, *2018*, 5659683. [CrossRef]

37. Carrasco-Lopez, C.; Godoy, C.; de las Rivas, B.; Fernandez-Lorente, G.; Palomo, J.M.; Guisan, J.M.; Fernandez-Lafuente, R.; Martinez-Ripoll, M.; Hermoso, J.A. Activation of bacterial thermoalkalophilic lipases is spurred by dramatic structural rearrangements. *J. Biol. Chem.* **2009**, *284*, 4365–4372. [CrossRef] [PubMed]

38. Shehata, M.; Timucin, E.; Venturini, A.; Sezerman, O.U. Understanding thermal and organic solvent stability of thermoalkalophilic lipases: Insights from computational predictions and experiments. *J. Mol. Model.* **2020**, *26*, 1–12. [CrossRef]

39. Nazina, T.N.; Tourova, T.P.; Poltaraus, A.B.; Novikova, E.V.; Grigoryan, A.A.; Ivanova, A.E.; Lysenko, A.M.; Petrunyaka, V.V.; Osipov, G.A.; Belyaev, S.S.; et al. Taxonomic study of aerobic thermophilic bacilli: Descriptions of *Geobacillus subterraneus* gen. nov., sp nov and *Geobacillus uzenensis* sp nov from petroleum reservoirs and transfer of *Bacillus stearothermophilus*, *Bacillus thermocatenulatus*, *Bacillus thermoleovorans*, *Bacillus kaustophilus*, *Bacillus thermoglucosidasius* and *Bacillus thermodenitrificans* to *Geobacillus* as the new combinations *G. stearothermophilus*, *G. thermocatenulatus*, *G. thermoleovorans*, *G. kaustophilus*, *G. thermoglucosidasius* and *G. thermodenitrificans*. *Int. J. Syst. Evol. Microbiol.* **2001**, *51*, 433–446. [CrossRef]

40. Schmidt-Dannert, C.; Sztajer, H.; Stocklein, W.; Menge, U.; Schmid, R.D. Screening, purification and properties of a thermophilic lipase from *Bacillus thermocatenulatus*. *Biochim. Biophys. Acta Lipids Lipid Metab.* **1994**, *1214*, 43–53. [CrossRef]

41. Schmidt-Dannert, C.; Rua, M.L.; Schmid, R.D. *Bacillus thermocatenulatus* lipase: A thermoalkalophilic lipase with interesting properties. *Biochem. Soc. Trans.* **1997**, *25*, 178–182. [CrossRef]

42. Schmidt-Dannert, C.; Rua, M.L.; Atomi, H.; Schmid, R.D. Thermoalkalophilic lipase of *Bacillus thermocatenulatus*. 1. Molecular cloning, nucleotide sequence, purification and some properties. *Biochim. Biophys. Acta Lipids Lipid Metab.* **1996**, *1301*, 105–114. [CrossRef]

43. Carrasco-Lopez, C.; Godoy, C.; de las Rivas, B.; Fernandez-Lorente, G.; Palomo, J.M.; Guisan, J.M.; Fernandez-Lafuente, R.; Martinez-Ripoll, M.; Hermoso, J.A. Crystallization and preliminary X-ray diffraction studies of the BTL2 lipase from the extremophilic microorganism *Bacillus thermocatenulatus*. *Acta Crystallogr. F Struct. Biol. Commun.* **2008**, *64*, 1043–1045. [CrossRef]

44. Liu, A.M.F.; Somers, N.A.; Kazlauskas, R.J.; Brush, T.S.; Zocher, F.; Enzelberger, M.M.; Bornscheuer, U.T.; Horsman, G.P.; Mezzetti, A.; Schmidt-Dannert, C.; et al. Mapping the substrate selectivity of new hydrolases using colorimetric screening: Lipases from Bacillus thermocatenulatus and Ophiostoma piliferum, esterases from Pseudomonas fluorescens and Streptomyces diastatochromogenes. *Tetrahedron Asymmetry* **2001**, *12*, 545–556. [CrossRef]

45. Palomo, J.M.; Fernandez-Lorente, G.; Rua, M.L.; Guisan, J.M.; Fernandez-Lafuente, R. Evaluation of the lipase from Bacillus thermocatenulatus as an enantioselective biocatalyst. *Tetrahedron Asymmetry* **2003**, *14*, 3679–3687. [CrossRef]

46. Palomo, J.M.; Ortiz, C.; Fuentes, M.; Fernandez-Lorente, G.; Guisan, J.M.; Fernandez-Lafuente, R. Use of immobilized lipases for lipase purification via specific lipase-lipase interactions. *J. Chromatogr. A* **2004**, *1038*, 267–273. [CrossRef] [PubMed]

47. Palomo, J.M.; Segura, R.L.; Fernandez-Lorente, G.; Pernas, M.; Rua, M.L.; Guisan, J.M.; Fernandez-Lafuente, R. Purification, immobilization, and stabilization of a lipase from Bacillus thermocatenulatus by interfacial adsorption on hydrophobic supports. *Biotechnol. Prog.* **2004**, *20*, 630–635. [CrossRef]

48. Fernandez-Lorente, G.; Cabrera, Z.; Godoy, C.; Fernandez-Lafuente, R.; Palomo, J.M.; Guisan, J.M. Interfacially activated lipases against hydrophobic supports: Effect of the support nature on the biocatalytic properties. *Process. Biochem.* **2008**, *43*, 1061–1067. [CrossRef]

49. Fernandez-Lorente, G.; Godoy, C.A.; Mendes, A.A.; Lopez-Gallego, F.; Grazu, V.; de las Rivas, B.; Palomo, J.M.; Hermoso, J.; Fernandez-Lafuente, R.; Guisan, J.M. Solid-phase chemical amination of a lipase from *Bacillus thermocatenulatus* to improve its stabilization via covalent immobilization on highly activated glyoxyl-agarose. *Biomacromolecules* **2008**, *9*, 2553–2561. [CrossRef]

50. Bolivar, J.M.; Mateo, C.; Godoy, C.; Pessela, B.C.C.; Rodrigues, D.S.; Giordano, R.L.C.; Fernandez-Lafuente, R.; Guisan, J.M. The co-operative effect of physical and covalent protein adsorption on heterofunctional supports. *Process. Biochem.* **2009**, *44*, 757–763. [CrossRef]

51. Godoy, C.A.; de las Rivas, B.; Filice, M.; Fernandez-Lorente, G.; Guisan, J.M.; Palomo, J.M. Enhanced activity of an immobilized lipase promoted by site-directed chemical modification with polymers. *Process. Biochem.* **2010**, *45*, 534–541. [CrossRef]

52. Godoy, C.A.; de las Rivas, B.; Bezbradica, D.; Bolivar, J.M.; Lopez-Gallego, F.; Fernandez-Lorente, G.; Guisan, J.M. Reactivation of a thermostable lipase by solid phase unfolding/refolding Effect of cysteine residues on refolding efficiency. *Enzyme Microb. Technol.* **2011**, *49*, 388–394. [CrossRef] [PubMed]

53. Godoy, C.A.; de las Rivas, B.; Grazu, V.; Montes, T.; Manuel Guisan, J.; Lopez-Gallego, F. Glyoxyl-Disulfide Agarose: A Tailor-Made Support for Site-Directed Rigidification of Proteins. *Biomacromolecules* **2011**, *12*, 1800–1809. [CrossRef] [PubMed]

54. Godoy, C.A.; Fernandez-Lorente, G.; de las Rivas, B.; Filice, M.; Guisan, J.M.; Palomo, J.M. Medium engineering on modified Geobacillus thermocatenulatus lipase to prepare highly active catalysts. *J. Mol. Catal. B Enzym.* **2011**, *70*, 144–148. [CrossRef]

55. Lopez-Gallego, F.; Abian, O.; Manuel Guisan, J. Altering the Interfacial Activation Mechanism of a Lipase by Solid-Phase Selective Chemical Modification. *Biochemistry* **2012**, *51*, 7028–7036. [CrossRef] [PubMed]

56. Godoy, C.A.; Romero, O.; de la Rivas, B.; Mateo, C.; Fernandez-Lorente, G.; Guisan, J.M.; Palomo, J.M. Changes on enantioselectivity of a genetically modified thermophilic lipase by site-directed oriented immobilization. *J. Mol. Catal. B Enzym.* **2013**, *87*, 121–127. [CrossRef]

57. Marciello, M.; Bolivar, J.M.; Filice, M.; Mateo, C.; Guisan, J.M. Preparation of Lipase-Coated, Stabilized, Hydrophobic Magnetic Particles for Reversible Conjugation of Biomacromolecules. *Biomacromolecules* **2013**, *14*, 602–607. [CrossRef]

58. Bautista-Barrufet, A.; Lopez-Gallego, F.; Rojas-Cervellera, V.; Rovira, C.; Pericas, M.A.; Guisan, J.M.; Gorostiza, P. Optical Control of Enzyme Enantioselectivity in Solid Phase. *ACS Catal.* **2014**, *4*, 1004–1009. [CrossRef]

59. Mendes, A.A.; Oliveira, P.C.; Velez, A.M.; Giordano, R.C.; Giordano, R.d.L.C.; de Castro, H.F. Evaluation of immobilized lipases on poly-hydroxybutyrate beads to catalyze biodiesel synthesis. *Int. J. Biol. Macromol.* **2012**, *50*, 503–511. [CrossRef]

60. Herranz, S.; Marciello, M.; Olea, D.; Hernandez, M.; Domingo, C.; Velez, M.; Gheber, L.A.; Guisan, J.M.; Moreno-Bondi, M.C. Dextran-Lipase Conjugates as Tools for Low Molecular Weight Ligand Immobilization in Microarray Development. *Anal. Chem.* **2013**, *85*, 7060–7068. [CrossRef]

61. Guajardo, N.; Bernal, C.; Wilson, L.; Cabrera, Z. Asymmetric hydrolysis of dimethyl-3-phenylglutarate in sequential batch reactor operation catalysed by immobilized Geobacillus thermocatenulatus lipase. *Catal. Today* **2015**, *255*, 21–26. [CrossRef]

62. Godoy, C.A. New Strategy for the Immobilization of Lipases on Glyoxyl-Agarose Supports: Production of Robust Biocatalysts for Natural Oil Transformation. *Int. J. Mol. Sci.* **2017**, *18*, 2130. [CrossRef]

63. Lopez-Tejedor, D.; de las Rivas, B.; Palomo, J.M. Ultra-Small Pd(0) Nanoparticles into a Designed Semisynthetic Lipase: An Efficient and Recyclable Heterogeneous Biohybrid Catalyst for the Heck Reaction under Mild Conditions. *Molecules* **2018**, *23*, 2358. [CrossRef]

64. Romero, O.; de las Rivas, B.; Lopez-Tejedor, D.; Palomo, J.M. Effect of Site-Specific Peptide-Tag Labeling on the Biocatalytic Properties of Thermoalkalophilic Lipase from Geobacillus thermocatenulatus. *ChemBioChem* **2018**, *19*, 369–378. [CrossRef] [PubMed]

65. Moreno-Perez, S.; Fernandez-Lorente, G.; Romero, O.; Guisan, J.M.; Lopez-Gallego, F. Fabrication of heterogeneous biocatalyst tethering artificial prosthetic groups to obtain omega-3-fatty acids by selective hydrolysis of fish oils. *RSC Adv.* **2016**, *6*, 97659–97663. [CrossRef]

66. Cowan, D.A.; Fernandez-Lafuente, R. Enhancing the functional properties of thermophilic enzymes by chemical modification and immobilization. *Enzyme Microb. Technol.* **2011**, *49*, 326–346. [CrossRef]

67. Godoy, C.A.; de las Rivas, B.; Guisan, J.M. Site-directing an intense multipoint covalent attachment (MCA) of mutants of the *Geobacillus thermocatenulatus* lipase 2 (BTL2): Genetic and chemical amination plus immobilization on a tailor-made support. *Process. Biochem.* **2014**, *49*, 1324–1331. [CrossRef]

68. Kajiwara, S.; Yamada, R.; Matsumoto, T.; Ogino, H. N-linked glycosylation of thermostable lipase from Bacillus thermocatenulatus to improve organic solvent stability. *Enzyme Microb. Technol.* **2020**, *132*. [CrossRef] [PubMed]

69. Godoy, C.A.; Klett, J.; Di Geronimo, B.; Hermoso, J.A.; Guisan, J.M.; Carrasco-Lopez, C. Disulfide Engineered Lipase to Enhance the Catalytic Activity: A Structure-Based Approach on BTL2. *Int. J. Mol. Sci.* **2019**, *20*, 5245. [CrossRef] [PubMed]

70. Karimi, E.; Karkhane, A.A.; Yakhchali, B.; Shamsara, M.; Aminzadeh, S.; Torktaz, I.; Hosseini, M.; Safari, Z. Study of the effect of F17A mutation on characteristics of *Bacillus thermocatenulatus* lipase expressed in *Pichia pastoris* using *in silico* and experimental methods. *Biotech. Appl. Biochem.* **2014**, *61*, 264–273. [CrossRef]

71. Goodarzi, N.; Karkhane, A.A.; Mirlohi, A.; Tabandeh, F.; Torktas, I.; Aminzadeh, S.; Yakhchali, B.; Shamsara, M.; Ghafouri, M.A.-S. Protein engineering of *Bacillus thermocatenulatus* lipase via deletion of the alpha 5 helix. *Appl. Biochem. Biotech.* **2014**, *174*, 339–351. [CrossRef]

72. Yenenler, A.; Venturini, A.; Burduroglu, H.C.; Sezerman, O.U. Investigating the structural properties of the active conformation BTL2 of a lipase from *Geobacillus thermocatenulatus* in toluene using molecular dynamic simulations and engineering BTL2 via in-silico mutation. *J. Mol. Model.* **2018**, *24*, 13. [CrossRef]

73. Yukselen, O.; Timucin, E.; Sezerman, U. Predicting the impact of mutations on the specific activity of *Bacillus thermocatenulatus* lipase using a combined approach of docking and molecular dynamics. *J. Mol. Recognit.* **2016**, *29*, 466–475. [CrossRef]

74. Khaleghinejad, S.H.; Motalleb, G.; Karkhane, A.A.; Aminzadeh, S.; Yakhchali, B. Study the effect of F17S mutation on the chimeric Bacillus thermocatenulatus lipase. *J. Genet. Eng. Biotechnol.* **2016**, *14*, 83–89. [CrossRef]

75. Wang, S.H.; Meng, X.H.; Zhou, H.; Liu, Y.; Secundo, F.; Liu, Y. Enzyme stability and activity in non-aqueous reaction systems: A mini review. *Catalysts* **2016**, *6*, 32. [CrossRef]

76. Elgharbawy, A.A.; Riyadi, F.A.; Alam, M.Z.; Moniruzzaman, M. Ionic liquids as a potential solvent for lipase-catalysed reactions: A review. *J. Mol. Liq.* **2018**, *251*, 150–166. [CrossRef]

77. Elgharbawy, A.A.M.; Moniruzzaman, M.; Goto, M. Recent advances of enzymatic reactions in ionic liquids: Part II. *Biochem. Eng. J.* **2020**, *154*, 23. [CrossRef]

78. Itoh, T. Ionic Liquids as Tool to Improve Enzymatic Organic Synthesis. *Chem. Rev.* **2017**, *117*, 10567–10607. [CrossRef]

79. de los Rios, A.P.; Irabien, A.; Hollmann, F.; Fernandez, F.J.H. Ionic Liquids: Green Solvents for Chemical Processing. *J. Chem.* **2013**, *2013*, 402172. [CrossRef]

80. Clarke, C.J.; Tu, W.C.; Levers, O.; Brohl, A.; Hallett, J.P. Green and Sustainable Solvents in Chemical Processes. *Chem. Rev.* **2018**, *118*, 747–800. [CrossRef]

81. Terreni, M.; Pagani, G.; Ubiali, D.; Fernandez-Lafuente, R.; Mateo, C.; Guisan, J.M. Modulation of penicillin acylase properties via immobilization techniques: One-pot chemoenzymatic synthesis of cephamandole from cephalosporin C. *Bioorg. Med. Chem. Lett.* **2001**, *11*, 2429–2432. [CrossRef]

82. Furlenmeier, A.; Quitt, P.; Vogler, K.; Lanz, P. 6-Acyl Derivatives of Aminopenicillanic Acid. U.S. Patent US3957758A, 18 May 1976.

83. Su, X.P.; Bhongle, N.N.; Pflum, D.; Butler, H.; Wald, S.A.; Bakale, R.P.; Senanayake, C.H. A large-scale asymmetric synthesis of (*S*)-cyclohexylphenyl glycolic acid. *Tetrahedron Asymmetry* **2003**, *14*, 3593–3600. [CrossRef]

84. Glushkov, R.G.; Ovcharova, I.M.; Muratov, M.A.; Kaminka, M.E.; Mashkovsky, M.D. Synthesis and pharmacological activity of new oxyaminoalkylxanthines and dialkylaminoalkylxanthines 1,4-diazepino and pyrazino[1,2,3-*g*,*h*] purine derivatives. *Khimiko-Farmatsevticheskii Zhurnal* **1977**, *11*, 30–35.

85. Bousquet, A.; Musolino, A. Hydroxyacetic Ester Derivatives, Namely (R)-methyl 2-(sulfonyloxy)-2-(chlorophenyl)acetates, Preparation Method, and Use as Synthesis Intermediates for Clopidogrel. WO9918110A1, 15 April 1999.

86. Bast, A.; Leurs, R.; Timmerman, H. Cyclandelate as a calcium modulating agent in rat cerebral-cortex. *Drugs* **1987**, *33*, 67–74. [CrossRef]

87. Gokce, M.; Utku, S.; Gur, S.; Ozkul, A.; Gumus, F. Synthesis, in vitro cytotoxic and antiviral activity of cis-[Pt(*R*(-) and *S*(+)-2-alpha,-hydroxybenzylbenzimidazole)(2)Cl-2] complexes. *Eur. J. Med. Chem.* **2005**, *40*, 135–141. [CrossRef]

88. Yamamoto, K.; Fujimatsu, I.; Komatsu, K. Purification and characterization of the nitrilase from *Alcaligenes faecalis* ATCC-8750 responsible for enantioselective hydrolysis of mandelonitrile. *J. Ferment. Bioeng.* **1992**, *73*, 425–430. [CrossRef]

89. Yamamoto, K.; Oishi, K.; Fujimatsu, I.; Komatsu, K.I. Production of *R*-(-)-mandelic acid from mandelonitrile by *Alcaligenes faecalis* ATCC-8750. *Appl. Environ. Microbiol.* **1991**, *57*, 3028–3032. [CrossRef]

90. Poterala, M.; Dranka, M.; Borowiecki, P. Chemoenzymatic Preparation of Enantiomerically Enriched (*R*)-(-)-Mandelic Acid Derivatives: Application in the Synthesis of the Active Agent Pemoline. *Eur. J. Org. Chem.* **2017**, *2017*, 2290–2304. [CrossRef]

91. Arroyo, M.; de la Mata, I.; Garcia, J.L.; Barredo, J.L. *Biocatalysis for Industrial Production of Active Pharmaceutical Ingredients (APIs)*; Academic Press Ltd.: Cambridge, MA, USA; Elsevier Science Ltd.: London, UK, 2017; pp. 451–473. [CrossRef]

92. Martin, J.R.; Nus, M.; Gago, J.V.S.; Sanchez-Montero, J.M. Selective esterification of phthalic acids in two ionic liquids at high temperatures using a thermostable lipase of *Bacillus thermocatenulatus*: A comparative study. *J. Mol. Catal. B Enzym.* **2008**, *52–53*, 162–167. [CrossRef]

93. Pizzilli, A.; Zoppi, R.; Hoyos, P.; Gómez, S.; Gatti, F.G.; Hernáiz, M.J.; Alcántara, A.R. First stereoselective acylation of a primary diol possessing a prochiral quaternary center mediated by lipase TL from Pseudomonas stutzeri. *Tetrahedron* **2015**, *71*, 9172–9176. [CrossRef]

94. Chamorro, S.; Sanchez-Montero, J.M.; Alcantara, A.R.; Sinisterra, J.V. Treatment of *Candida rugosa* lipase with short-chain polar organic solvents enhances its hydrolytic and synthetic activities. *Biotechnol. Lett.* **1998**, *20*, 499–505. [CrossRef]

95. Borreguero, I.; Carvalho, C.M.L.; Cabral, J.M.S.; Sinisterra, J.V.; Alcantara, A.R. Enantioselective properties of *Fusarium solani pisi* cutinase on transesterification of acyclic diols: Activity and stability evaluation. *J. Mol. Catal. B Enzym.* **2001**, *11*, 613–622. [CrossRef]

96. Chamorro, S.; Alcántara, A.R.; de la Casa, R.M.; Sinisterra, J.V.; Sánchez-Montero, J.M. Small water amounts increase the catalytic behaviour of polar organic solvents pre-treated *Candida rugosa* lipase. *J. Mol. Catal. B Enzym.* **2001**, *11*, 939–947. [CrossRef]

97. Domínguez de María, P.; Martínez-Alzamora, F.; Moreno, S.P.; Valero, F.; Rúa, M.L.; Sánchez-Montero, J.M.; Sinisterra, J.V.; Alcántara, A.R. Heptyl oleate synthesis as useful tool to discriminate between lipases, proteases and other hydrolases in crude preparations. *Enzyme Microb, Technol.* **2002**, *31*, 283–288. [CrossRef]

98. Alcántara, A.R.; De María, P.D.; Fernández, M.; Hernáiz, M.J.; Sánchez-Montero, J.M.; Sinisterra, J.V. Resolution of racemic acids, esters and amines by *Candida rugosa* lipase in slightly hydrated organic media. *Food Technol. Biotechnol.* **2004**, *42*, 343–354.

99. Alfonsi, K.; Colberg, J.; Dunn, P.J.; Fevig, T.; Jennings, S.; Johnson, T.A.; Kleine, H.P.; Knight, C.; Nagy, M.A.; Perry, D.A.; et al. Green chemistry tools to influence a medicinal chemistry and research chemistry based organisation. *Green Chem.* **2008**, *10*, 31–36. [CrossRef]

100. Bardsley, W.G. *SIMFIT—A Computer Package for Simulation, Curve-Fitting and Statistical-Analysis using Life-Science Models*; Plenum Press Div. Plenum Publishing Corp.: New York, NY, USA, 1993; pp. 455–458.

101. Zheng, D.X.; Dong, L.; Huang, W.J.; Wu, X.H.; Nie, N. A review of imidazolium ionic liquids research and development towards working pair of absorption cycle. *Renew. Sust. Energ. Rev.* **2014**, *37*, 47–68. [CrossRef]

102. Green, M.D.; Long, T.E. Designing Imidazole-Based Ionic Liquids and Ionic Liquid Monomers for Emerging Technologies. *Polym. Rev.* **2009**, *49*, 291–314. [CrossRef]

103. Domanska, U.; Marciniak, A. Solubility of ionic liquid [emim] [PF_6] in alcohols. *J. Phys. Chem. B* **2004**, *108*, 2376–2382. [CrossRef]

104. Heintz, A. Recent developments in thermodynamics and thermophysics of non-aqueous mixtures containing ionic liquids. A review. *J. Chem. Thermodyn.* **2005**, *37*, 525–535. [CrossRef]

105. Sahandzhieva, K.; Tuma, D.; Breyer, S.; Kamps, A.P.S.; Maurer, G. Liquid-liquid equilibrium in mixtures of the ionic liquid 1-n-butyl-3-methylimidazolium hexafluorophosphate and an alkanol. *J. Chem. Eng. Data* **2006**, *51*, 1516–1525. [CrossRef]

106. Pereiro, A.B.; Rodriguez, A. Study on the phase behaviour and thermodynamic properties of ionic liquids containing imidazolium cation with ethanol at several temperatures. *J. Chem. Thermodyn.* **2007**, *39*, 978–989. [CrossRef]

107. Abdulagatov, I.M.; Tekin, A.; Safarov, J.; Shahverdiyev, A.; Hassel, E. Densities and excess, apparent, and partial molar volumes of binary mixtures of $BMIMBF_4$ plus ethanol as a function of temperature, pressure, and concentration. *Int. J. Thermophys.* **2008**, *29*, 505–533. [CrossRef]

108. Domanska, U. Solubilities and thermophysical properties of ionic liquids. *Pure Appl. Chem.* **2005**, *77*, 543–557. [CrossRef]

109. Guo, Y.M.; Wang, X.; Tao, X.Y.; Shen, W.G. Liquid-liquid equilibrium and heat capacity measurements of the binary solution {ethanol+1-butyl-3-methylimidazolium hexafluorophosphate}. *J. Chem. Thermodyn.* **2017**, *115*, 342–351. [CrossRef]

110. Varela, L.M.; Mendez-Morales, T.; Carrete, J.; Gomez-Gonzalez, V.; Docampo-Alvarez, B.; Gallego, L.J.; Cabeza, O.; Russina, O. Solvation of molecular cosolvents and inorganic salts in ionic liquids: A review of molecular dynamics simulations. *J. Mol. Liq.* **2015**, *210*, 178–188. [CrossRef]

111. Secundo, F.; Carrea, G.; Tarabiono, C.; Gatti-Lafranconi, P.; Brocca, S.; Lotti, M.; Jaeger, K.E.; Puls, M.; Eggert, T. The lid is a structural and functional determinant of lipase activity and selectivity. *J. Mol. Catal. B Enzym.* **2006**, *39*, 166–170. [CrossRef]

112. Khan, F.I.; Lan, D.; Durrani, R.; Huan, W.; Zhao, Z.; Wang, Y. The lid domain in lipases: Structural and functional determinant of enzymatic properties. *Front. Bioeng. Biotechnol.* **2017**, *5*, 16. [CrossRef]

113. Kazlauskas, R.J.; Weissfloch, A.N.E.; Rappaport, A.T.; Cuccia, L.A. A rule to predict which enantiomer of a secondary alcohol reacts faster in reactions catalysed by cholesterol esterase, lipase from *Pseudomonas cepacia*, and lipase from *Candida rugosa*. *J. Org. Chem.* **1991**, *56*, 2656–2665. [CrossRef]

114. Borreguero, I.; Sánchez-Montero, J.M.; Sinisterra, J.V.; Rumbero, A.; Hermoso, J.A.; Alcántara, A.R. Regioselective resolution of 1,n-diols catalysed by lipases: A rational explanation of the enzymatic selectivity. *J. Mol. Catal. B Enzym.* **2001**, *11*, 1013–1024. [CrossRef]

115. Borreguero, I.; Sinisterra, J.V.; Rumbero, A.; Hermoso, J.A.; Martínez-Ripoll, M.; Alcántara, A.R. Acyclic phenylalkanediols as substrates for the study of enzyme recognition. Regioselective acylation by porcine pancreatic lipase: A structural hypothesis for the enzymatic selectivity. *Tetrahedron* **1999**, *55*, 14961–14974. [CrossRef]

116. Filice, M.; Romero, O.; Abian, O.; de las Rivas, B.; Palomo, J.M. Low ionic liquid concentration in water: A green and simple approach to improve activity and selectivity of lipases. *RSC Adv.* **2014**, *4*, 49115–49122. [CrossRef]

117. Maraite, A.; Hoyos, P.; Carballeira, J.D.; Cabrera, Á.C.; Ansorge-Schumacher, M.B.; Alcántara, A.R. Lipase from *Pseudomonas stutzeri*: Purification, homology modelling and rational explanation of the substrate binding mode. *J. Mol. Catal. B Enzym.* **2013**, *87*, 88–98. [CrossRef]

118. Min, B.; Park, J.; Sim, Y.K.; Jung, S.; Kim, S.H.; Song, J.K.; Kim, B.T.; Park, S.Y.; Yun, J.; Park, S.; et al. Hydrogen-bonding-driven enantioselective resolution against the Kazlaukas rule to afford gamma-amino alcohols by *Candida rugosa* lipase. *ChemBioChem* **2015**, *16*, 77–82. [CrossRef]

119. Nascimento, P.A.M.; Pereira, J.F.B.; Santos-Ebinuma, V.D. Insights into the effect of imidazolium-based ionic liquids on chemical structure and hydrolytic activity of microbial lipase. *Bioprocess. Biosyst. Eng.* **2019**, *42*, 1235–1246. [CrossRef] [PubMed]

120. Palomo, J.M.; Fernandez-Lorente, G.; Guisan, J.M.; Fernandez-Lafuente, R. Modulation of immobilized lipase enantioselectivity via chemical amination. *Adv. Synth. Catal.* **2007**, *349*, 1119–1127. [CrossRef]

 *catalysts*

Article

Immobilization of *Arabidopsis thaliana* Hydroxynitrile Lyase (*At*HNL) on EziG Opal

José Coloma [1,2], Tim Lugtenburg [1], Muhammad Afendi [1], Mattia Lazzarotto [1,3], Paula Bracco [1], Peter-Leon Hagedoorn [1], Lucia Gardossi [3] and Ulf Hanefeld [1,*]

[1] Biokatalyse, Afdeling Biotechnologie, Technische Universiteit Delft, Van der Maasweg 9, 2629 HZ Delft, The Netherlands; J.L.ColomaHurel@tudelft.nl (J.C.); t.g.lugtenburg@tudelft.nl (T.L.); MuhammadFarhanbinMuhammadAfendi@student.tudelft.nl (M.A.); mattia.lazzarotto@uni-graz.at (M.L.); paulabracco@gmail.com (P.B.); P.L.Hagedoorn@tudelft.nl (P.-L.H.)

[2] Facultad de Ciencias Agropecuarias, Ingeniería Agroindustrial, Universidad Laica Eloy Alfaro de Manabí, Avenida Circunvalación s/n, Manta 13-05-2732, Ecuador

[3] Dipartimento di Scienze Chimiche e Farmaceutiche, Università degli Studi di Trieste, Via Licio Giorgieri 1, 34127 Trieste, Italy; gardossi@units.it

* Correspondence: u.hanefeld@tudelft.nl; Tel.: +31-15-278-9304

Received: 17 July 2020; Accepted: 5 August 2020; Published: 8 August 2020

Abstract: *Arabidopsis thaliana* hydroxynitrile lyase (*At*HNL) catalyzes the selective synthesis of (*R*)-cyanohydrins. This enzyme is unstable under acidic conditions, therefore its immobilization is necessary for the synthesis of enantiopure cyanohydrins. EziG Opal is a controlled porosity glass material for the immobilization of His-tagged enzymes. The immobilization of His$_6$-tagged *At*HNL on EziG Opal was optimized for higher enzyme stability and tested for the synthesis of (*R*)-mandelonitrile in batch and continuous flow systems. *At*HNL-EziG Opal achieved 95% of conversion after 30 min of reaction time in batch and it was recycled up to eight times with a final conversion of 80% and excellent enantioselectivity. The EziG Opal carrier catalyzed the racemic background reaction; however, the high enantioselectivity observed in the recycling study demonstrated that this was efficiently suppressed by using citrate/phosphate buffer saturated methyl-tert-butylether (MTBE) pH 5 as reaction medium. The continuous flow system achieved 96% of conversion and excellent enantioselectivity at 0.1 mL min^{-1}. Lower conversion and enantioselectivity were observed at higher flow rates. The specific rate of *At*HNL-EziG Opal in flow was 0.26 mol h^{-1} g$_{enzyme}$$^{-1}$ at 0.1 mL min^{-1} and 96% of conversion whereas in batch, the immobilized enzyme displayed a specific rate of 0.51 mol h^{-1} g$_{enzyme}$$^{-1}$ after 30 min of reaction time at a similar level of conversion. However, in terms of productivity the continuous flow system proved to be almost four times more productive than the batch approach, displaying a space-time-yield (STY) of 690 mol$_{product}$ h^{-1} L^{-1} g$_{enzyme}$$^{-1}$ compared to 187 mol$_{product}$ h^{-1} L^{-1} g$_{enzyme}$$^{-1}$ achieved with the batch system.

Keywords: *Arabidopsis thaliana*; hydroxynitrile lyase; oxynitrilase; His-tag; immobilization; batch; continuous flow

---

1. Introduction

Hydroxynitrile lyases (HNLs) are enzymes that catalyze the synthesis of enantiopure cyanohydrins (Scheme 1), known building blocks for the production of fine chemicals, pharmaceuticals and cosmetics [1–4]. HNL catalyzed reactions are faced with two problems, the chemical formation of racemic cyanohydrins and product racemization due to the reaction equilibrium [5]. These limitations can be overcome by performing the reactions in buffer saturated organic solvent and adjusting the pH to the lower limit accepted for HNLs [5,6]. These conditions are not the natural environment of HNLs, as they have to be stabilized for instance by immobilization on a suitable carrier.

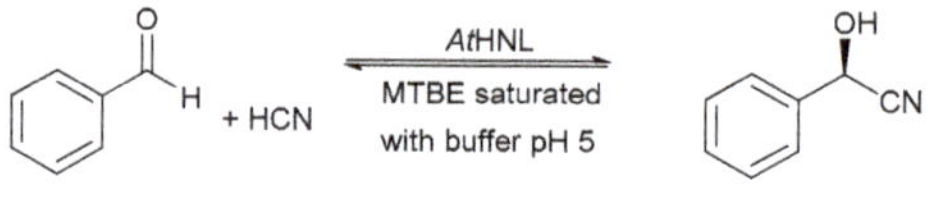

Scheme 1. *At*HNL catalyzed hydrocyanation of benzaldehyde yielding (*R*)-mandelonitrile.

Improved stability, activity and selectivity of immobilized enzymes have been reported earlier [7,8]. In addition, immobilization enables the increase of enzyme loading and facilitates recycling and downstream processing. To achieve the beneficial aspects mentioned before, the characteristics of enzyme and carrier must be considered. However, there is not a general method to immobilize an enzyme and its feasibility must be determined experimentally [9,10].

Immobilized metal ion chromatography (IMAC) is a widely used technique for the purification and immobilization of His-tagged enzymes. The enzyme immobilization is based on the affinity of divalent metal ions such as Zn^{2+}, Cu^{2+}, Ni^{2+} or Co^{2+} to the imidazole ring of histidines. Chelated Ni^{2+} on nitrilotriacetic acid (Ni-NTA) has turned out to be the most effective combination for the purification of His-tagged proteins [11]. However, nickel induced genotoxicity, carcinogenicity and immunotoxicity has been reported [12]. Hence, the development of a carrier with a non-toxic metal ion is highly desirable.

A new set of carriers (EziG, provided by EnginZyme AB, Stockholm, Sweden) containing non-toxic Fe^{3+} (>10 μmol g^{-1}) on its surface has been developed for the immobilization of His-tagged enzymes. These materials have a core made of controlled porosity glass (CPG) particles facilitating mass transfer from reactants and products to the material due to its interconnecting pore structure (circa 1.8 mL g^{-1}). In addition, its non-compressible non-swelling nature is an advantage compared to NTA agarose. The porous surface can be coated with an organic polymer to tailor carriers with different hydrophobic characteristics such as EziG Opal (hydrophilic), EziG Coral (hydrophobic) and EziG Amber (semi-hydrophobic). Given the hydrophilic surface of His_6-tagged *At*HNL [13], its immobilization was performed on EziG Opal. Moreover, EziG Opal has been found to be suitable for reactions in organic solvents [14], a crucial property to enable the synthesis of enantiopure cyanohydrins together with low pH required in the reaction medium [5,15–19].

Some successful studies with different immobilized enzymes on EziG carriers have been reported earlier. An ω-transaminase was active in methyl-tert-butylether (MTBE) and a Baeyer–Villiger monooxygenase (BVMO) together with two cofactor-regenerating enzymes displayed increased stability [14]. An ω-transaminase from *Arthrobacter* sp. (AsR-ωTA) on EziG Amber was highly stable in batch and continuous flow systems [20]. On the other hand, when an old yellow enzyme (OYE) was immobilized on EziG Opal, poor recyclability was observed, and the initial conversion dropped to 56% after two reaction cycles [21]. The loss of activity of OYE was assumed to be due to enzyme leaching and/or deactivation of the enzyme. Likewise, the enzyme arylmalonate decarboxylase (AMDase) presented a significant loss of activity during recycling studies on all EziG carriers [22]. The loss of activity of AMDase activity on EziG carriers was explained to be due to enzyme leaching because the of the lower strength of the coordinate bond and due to local pH changes by the acidic product of the reaction.

Enzyme catalyzed reactions in flow are gaining attention due to improved productivity, easier downstream processing and efficiency of scale-up compared to batch systems. Reduced reaction times and enhanced selectivity are reported benefits of performing reactions in flow [23–26]. On top of all these benefits mentioned before, continuous flow reactions allow to optimize resource utilization, reduce reaction volumes and consequently achieve waste reductions and lower energy consumption [27]. Furthermore, they allow handling of toxic and reactive reagents such as cyanide [28] in a safer manner.

The aim of this study is to immobilize *At*HNL on EziG Opal based on the His-tag/Fe^{3+} affinity and compare its performance for the synthesis of (*R*)-mandelonitrile with the earlier reported successful immobilization of *At*HNL on Celite by adsorption [15]. Important parameters such as stability, specific rate and productivity were investigated in batch and flow systems.

2. Results and Discussion

*At*HNL was recombinantly produced with a His$_6$-tag to enable its purification and immobilization by metal-ion affinity. It was successfully overexpressed in *E. coli* BL21(DE3) and purified displaying a specific activity of 136.5 ± 3.2 U mg^{-1}. *At*HNL was purified prior to its immobilization to avoid binding of other proteins with affinity to the EziG Opal carrier.

2.1. Batch Reactions

All batch reactions were performed with *At*HNL-EziG Opal tightly packed into tea bags. Earlier research revealed that while the material of the bags had no influence on the conversion and enantioselectivity, it was essential to pack the bags tightly [16,18,29]. A magnetic stirrer was attached to the tea bag to enable the rotation of the immobilized enzyme and stirrer simultaneously. This set up avoids mechanical attrition of the carrier caused by the stirrer and facilitates the manipulation of the immobilized enzyme for recyclability studies without any loss of enzyme material. A leaching assay showed that *At*HNL did not leach from EziG Opal carrier to the reaction medium (Figure S1). Similarly, no leaching had been reported previously for hydrocyanation reactions catalyzed by immobilized HNLs on siliceous carriers in general and *At*HNL specifically [15,16,18,29].

Once it was established that EziG Opal is a suitable carrier for the immobilization of *At*HNL (Figure S1), preliminary time studies using different enzyme loadings of *At*HNL-EziG Opal for the synthesis of (*R*)-mandelonitrile were performed (Figure 1). The different enzyme loadings showed a huge difference in conversion and enantiopurity during four hours of reaction time. In these preliminary experiments, *At*HNL was immobilized on EziG Opal by incubating an enzyme solution with the carrier in an orbital shaker (see Section 3.7 for details). The rotation enabled the enzyme to bind to the carrier but some precipitation was observed and it might explain the results in Figure 1. On the other hand, an earlier report [15] showed that *At*HNL on Celite R-633 displayed near complete conversion and excellent enantioselectivity after 45 min using 5 mg mL^{-1} (circa 400 U) and MTBE saturated with citrate/phosphate buffer pH 5.5. The enzyme was immobilized by adsorption in that study. In addition, the successful immobilizations of *Prunus amydalus* HNL (*Pa*HNL) [16], *Manihot esculenta* HNL (*Me*HNL) [18] and *Granulicella tundricola* HNL (*Gt*HNL) [29] on Celite were also performed by adsorbing all the enzyme solution into the carrier until saturation, which means that the enzyme solution is completely adsorbed into the carrier, just like in the case of Celite. All these results suggest that the immobilization of *At*HNL on EziG Opal had to be optimized.

In order to optimize the immobilization of *At*HNL on EziG Opal, the enzyme was immobilized by either incubating an enzyme solution in an orbital shaker or by adding it dropwise to EziG Opal carrier in such a way that the carrier absorbs the enzyme solution completely, as in the case of Celite [15,16,18,29]. Additionally, the effect of drying *At*HNL-EziG Opal, which might influence the enzyme performance, was investigated. For this, *At*HNL-EziG Opal was used either immediately after its immobilization (wet *At*HNL-EziG Opal) or after 16 h of drying under vacuum in a desiccator over silica gel. Figure 2 shows the effect of drying *At*HNL-EziG Opal and the immobilization method on the bioconversions. The immobilization of *At*HNL in an orbital shaker with subsequent drying had a large negative impact on the conversion and enantioselectivity for the synthesis of (*R*)-mandelonitrile (Figure 2, dotted line and diamonds). The reaction catalyzed by wet *At*HNL-EziG Opal (Figure 2, dashed line and triangles) proceeded faster and with improved enantioselectivity (92% of conversion and 92% of enantioselectivity) as compared to the dried *At*HNL-EziG Opal (43% of conversion and 63% of enantioselectivity). As drying proved to have a negative impact on the enzyme activity and enantioselectivity, two reactions with wet *At*HNL-EziG Opal immobilized by either incubation in an orbital shaker or adsorption were performed. A faster reaction was observed when the enzyme was immobilized by adsorption (Figure 2, continuous line and crosses), comparable to the results with Celite [15]. Similar conversion (circa 95% in both cases) was obtained after 4 h of reaction time, but enantioselectivity was slightly better for the enzyme immobilized by adsorption (96% *ee*) as compared to the enzyme immobilized by incubation (92% *ee*). Surprisingly, the effect of drying on the reaction rate was negligible for *At*HNL immobilized by adsorption on EziG Opal (Figure 2, dashed line and dots). Conversions of circa 96% were obtained for both dried and non-dried *At*HNL-EziG Opal immobilized by adsorption within 30 min and little change was observed in the following 3.5 h. A possible explanation is that the enzyme is immediately stabilized right after its adsorption into the pores of the carrier, thus it is capable to endure the mechanical stress caused by the orbital shaker as well as the drying.

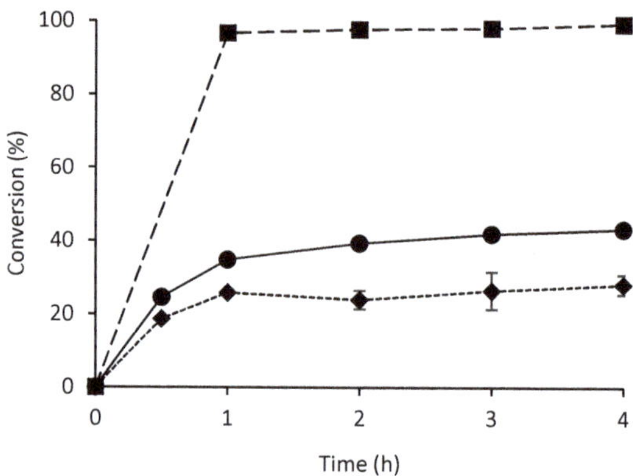

Figure 1. Synthesis of (*R*)-mandelonitrile using different enzyme loadings. Immobilization was performed by incubating enzyme and carrier in an orbital shaker and subsequent drying. Dashed line and squares (20 U mg^{-1}, final *ee* = 99%), solid line and dots (10 U mg^{-1}, final *ee* = 63%) and dotted line and diamonds (5 U mg^{-1}, final *ee* = 23%). Conditions: Ratio benzaldehyde:HCN in citrate/phosphate buffered MTBE, pH 5, 1:4, benzaldehyde (100 µL, 1 mmol), 2 mL HCN solution in citrate/phosphate buffered MTBE (1.5–2 M) pH 5, 27.5 µL (0.1 mmol) 1,3,5-triisopropylbenzene as internal standard (I.S.) and a teabag filled with *At*HNL immobilized on 60 mg EziG Opal. The reaction was stirred at 900 rpm at room temperature. Error bars correspond to the standard deviation of duplicates (*n* = 2) HPLC samples of the single experiment.

The lower conversions observed when the enzyme was immobilized by incubation might be explained by the loss of the *At*HNL dimeric structure caused by the rotation in an orbital shaker. In fact, some precipitation was observed after the incubation time. Earlier reports have shown that the enzyme stability is enhanced by higher oligomeric states [30]. Indeed, the ability of *Me*HNL to form tetramers in solution whereas *At*HNL forms dimers, explained the superior stability to higher temperatures and lower pH-values observed for *Me*HNL as compared to *At*HNL [31]. Similarly, *Me*HNL proved to be more stable than dimeric *Hevea brasiliensis* HNL (*Hb*HNL) for the synthesis of (*S*)-mandelonitrile in a monolith microreactor [17]. The formation of *Me*HNL tetrameric structures was attributed as the most likely reason for its higher stability.

Earlier studies [15] demonstrate a significant influence of the water content on enzyme activity of immobilized *At*HNL on Celite R-633, indicating that the minimal water content of *At*HNL-Celite for retaining enzymatic activity is 10% (*w/w*) of the immobilized enzyme. Additionally, the stability of *Me*HNL as CLEA or immobilized on Celite R-633 is highly dependent on the water entrapped in the carrier [18,32]. This effect can be ruled out here as buffer saturated MTBE was used as solvent.

Silica carriers, such as EziG Opal, are known to catalyze the chemical racemic background reaction [15–19,33,34] (Figures S2 and S3). However, the enantioselectivities achieved here in batch reactions under the optimized immobilization condition demonstrated the efficient suppression of this undesired reaction.

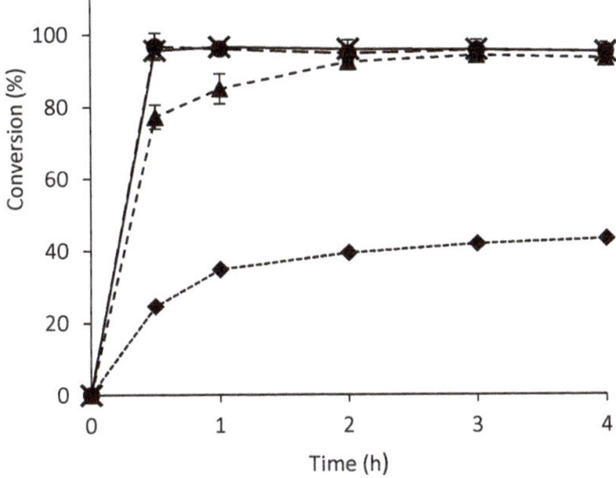

Figure 2. Effect of immobilization method and drying on the synthesis of (*R*)-mandelonitrile. Continuous line and crosses is the reaction with wet *At*HNL-EziG Opal (adsorption), final *ee* = 96.2%; dashed line and dots is the reaction with dried *At*HNL-EziG Opal (adsorption) final *ee* = 93.8%; dashed line and triangles is the reaction with wet *At*HNL-EziG Opal (incubation in orbital shaker), final *ee* = 92.3% and dotted line and diamonds is the reaction with dried *At*HNL-EziG Opal (incubation in orbital shaker), final *ee* = 63.3%. Conditions: Ratio benzaldehyde:HCN in citrate/phosphate buffered MTBE, pH 5, 1:4, benzaldehyde (100 μL, 1 mmol), 2 mL HCN solution in citrate/phosphate buffered MTBE (1.5–2 M) pH 5, 27.5 μL (0.1 mmol) 1,3,5-triisopropylbenzene as I.S. and a teabag filled with *At*HNL immobilized on 60 mg EziG Opal. All reactions were performed with enzyme loading of 10 U mg^{-1} and the mol ratio of monomeric *At*HNL:Fe^{3+} was 1:5. The reaction was stirred at 900 rpm at room temperature. Error bars of wet *At*HNL- EziG Opal (adsorption), wet *At*HNL-EziG Opal (incubation) and dried *At*HNL-EziG Opal (incubation) correspond to the standard deviation of duplicate (*n* = 2) HPLC samples of the single experiment. Error bars of the reaction with dried *At*HNL-EziG Opal (adsorption) correspond to the standard deviation of triplicate (*n* = 3) HPLC samples of the single experiment.

In addition to enzymatic activity and enantioselectivity, the stability of immobilized enzymes is crucial in biocatalytic applications. Indeed, one of the main objectives of enzyme immobilization is the potential for recycling the biocatalyst [4,6,35]. Since the best results for the synthesis of (*R*)-mandelonitrile were obtained with wet *At*HNL-EziG Opal (10 U mg^{-1}) immobilized by adsorption, a recyclability study was performed under these conditions (Figure 3). In order to avoid enzyme overloading on the carrier which might lead to misinterpretations in the recyclability study, the mol ratio of monomeric *At*HNL:Fe^{3+} used was 1:5, thus ensuring any enzyme deactivation is visible during the reaction cycles. Near complete conversion and excellent enantioselectivity (>99%) were observed during 7 cycles. When 10 U mg^{-1} of the enzyme were immobilized by incubation in an orbital shaker and subsequent drying, the recycling was unsuccessful (data not shown), whereas an enzyme loading of 20 U mg^{-1} led to a biocatalyst that could be recycled five cycles (Figure S4).

Figure 3 shows that EziG Opal enables to recycle *At*HNL over several cycles with good conversion and enantioselectivity under the conditions of this study; accomplishing one of the main objectives of enzyme immobilization. Similarly, *At*HNL on Celite displayed good conversion (>95%) and excellent enantioselectivity (>98% *ee*) during five consecutive reaction cycles [15]. Also, the successful recyclability of ω-transaminase from *Arthrobacter* sp. (AsR-ωTA) immobilized on EziG Amber (semi hydrophobic polymer surface) has been reported [20]. The immobilized AsR-ωTA (10 mg, 10% enzyme loading, w w^{-1}) was used for the kinetic resolution of *rac*-α-methylbenzylamine during 16 consecutive reaction cycles with excellent conversion and enantioselectivity. On the other hand, poor recyclability was recently reported [21] for the bioreduction of α-methyl-trans-cinnamaldehyde with a co-immobilized preparation of old yellow enzyme 3 (OYE3) and glucose dehydrogenase (GDH) on EziG Opal (OYE3/GDH- EziG Opal). The conversion dropped to 56% after only two reaction cycles. However, it is worthy to mention that after 11 cycles (almost no conversion) the addition of GDH increased the conversion up to 30% suggesting that GDH was gradually deactivated or leached from the carrier over the reaction cycles. Also, the synthesis of enantiopure (S)-arylpropinate using arylmalonate decarboxylase (AMDase) immobilized on three EziG carriers with different surface hydrophobicity has been reported [22]. The best activity was obtained with the hydrophilic carrier (EziG Opal). Unfortunately, the enzyme was nearly fully deactivated after the second reaction cycle for the three EziG carriers. This significant loss of enzymatic activity was attributed to enzyme leaching or local pH shifts inside the porous carriers.

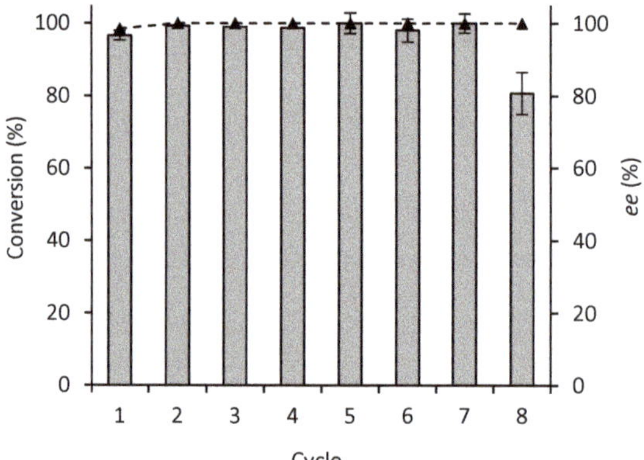

Figure 3. Recycling of wet *At*HNL-EziG Opal (10 U mg^{-1}) in eight successive cycles. Immobilization was performed by adsorption. Conversion (bars), enantiomeric excess (dotted line and triangles). Conditions: Ratio benzaldehyde:HCN in citrate/phosphate buffered MTBE, pH 5, 1:4, benzaldehyde (100 μL, 1 mmol), 2 mL HCN solution in citrate/phosphate buffered MTBE (1.5–2 M) pH 5, 27.5 μL (0.1 mmol) 1,3,5-triisopropylbenzene as I.S. and a teabag filled with *At*HNL immobilized on 60 mg EziG Opal. Mol ratio of monomeric *At*HNL:Fe^{3+} was 1:5. The reaction was stirred at 900 rpm at room temperature. The enzyme was washed for 1 min with 100 mM citrate/phosphate buffer saturated MTBE pH 5 after each cycle. Reaction time: 1 h. Error bars correspond to the standard deviation of duplicates (*n* = 2).

2.2. Continuous Flow Reactions

Immobilization enables the use of enzyme catalyzed synthesis reactions in continuous flow. Several advantages have been reported for this approach such as increased productivity, enhanced stability, reduced enzyme inhibition and easier downstream processing [26,36,37]. In addition, the reduced risk of manipulation of hazardous reagents, such as hydrogen cyanide, due to the smaller reaction volume used for the biocatalytic reactions is advantageous [28].

*At*HNL was immobilized on EziG Opal in accordance with the optimized method (adsorption + wet *At*HNL-EziG Opal) developed for batch reactions and tested in a continuous flow reactor (CFR). Figure 4 shows the synthesis of (R)-mandelonitrile at different flow rates. Near complete conversion and excellent enantioselectivity were achieved at 0.1 mL min^{-1}. An important decrease in enantioselectivity was observed at flow rates above 0.2 mL min^{-1} suggesting that *At*HNL on EziG Opal suffers from the competing chemical background reaction catalyzed by the carrier. Indeed, the reduced enantioselectivity could be explained as the result of the carrier catalyzed chemical reaction (Figure S3). Water content and pH had a major impact on the synthesis of *rac*-mandelonitrile. Pure EziG Opal formed circa 5% of *rac*-mandelonitrile due to the chemical background reaction whereas the addition of 306 μL of phosphate buffer pH 5 (same volume used for the enzyme immobilization) increased its formation up to 26% at 0.1 mL min^{-1} (Figure S3).

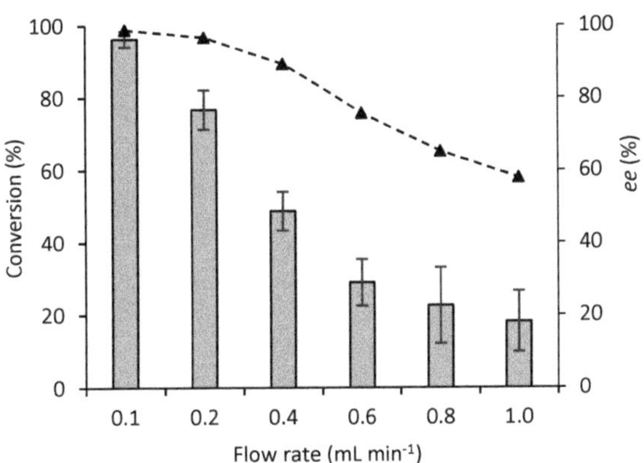

Figure 4. Synthesis of (*R*)-mandelonitrile using wet *At*HNL-EziG Opal (10 U mg^{-1}) in a CFR. Immobilization was performed by adsorption. Conversion (bars), enantiomeric excess (dotted line and triangles). Conditions: benzaldehyde (0.5 M), HCN solution in citrate/phosphate buffered MTBE (1.5–2 M) pH 5, 1,3,5-triisopropylbenzene (50 mM, I.S.), a CFR with *At*HNL immobilized on 150 mg EziG Opal. Mol ratio of monomeric *At*HNL:Fe^{3+} was 1:4. Reactions were performed at room temperature. Error bars correspond to the standard deviation of triplicates (*n* = 3).

The stability of *At*HNL-EziG Opal was evaluated in the synthesis of (*R*)-mandelonitrile at 0.1 mL min^{-1} during 12 h on continuous operation. At this flow rate near complete conversion was achieved with a mol ratio of monomeric *At*HNL:Fe^{3+} of 1:4, thus the robustness of the reaction system could be evaluated. Overall, *At*HNL-EziG Opal displayed good conversion and high enantioselectivity during the stability study (Figure 5). Conversion of 74% and enantioselectivity of 89% respectively were achieved after 12 h of continuous operation. The decreased conversion and enantioselectivity after 12 h might be explained by enzyme deactivation due to the low pH 5 and the chemical reaction catalyzed by the carrier.

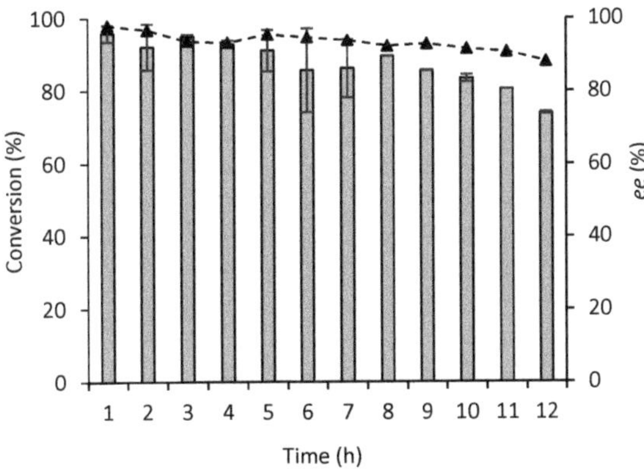

Figure 5. Stability of wet *At*HNL-EziG Opal (10 U mg^{-1}) in continuous flow at 0.1 mL min^{-1}. Immobilization was performed by adsorption. Conversion (bars), enantiomeric excess (dotted line and triangles). Conditions: benzaldehyde (0.5 M), HCN solution in citrate/phosphate buffered MTBE (1.5–2 M) pH 5, 1,3,5-triisopropylbenzene (50 mM, I.S.), a CFR with *At*HNL immobilized on 150 mg EziG Opal. Mol ratio of monomeric *At*HNL:Fe^{3+} was 1:4. Reactions were performed at room temperature. Error bars correspond to the standard deviation of duplicates (*n* = 2) during the first 7 h. From hour 8, error bars correspond to the standard deviation of duplicates (*n* = 2) HPLC samples of the single experiment.

2.3. Comparison between Batch and Continuous Flow Systems

The comparison of the performance of the batch and continuous flow systems cannot be made based on conversions due to the different set ups used. To establish a clear comparison regarding the performance of *At*HNL-EziG Opal in batch and continuous flow, the specific rate and space-time-yield (STY) at a similar level of conversion were calculated.

The specific rate at 0.1 mL min^{-1} (96% of conversion) was 0.26 mol h^{-1} g$_{enzyme}$$^{-1}$, surprisingly, it is half of the specific rate calculated in batch. At 0.4 mL min^{-1}, a similar rate (0.53 mol h^{-1} g$_{enzyme}$$^{-1}$) was observed as compared to the batch system. However, the reduced conversion (49%) might make downstream processing more problematic. Higher flow rates did not further improve the specific rate. *At*HNL-EziG Opal in batch displayed higher specific rate (0.51 mol h^{-1} g$_{enzyme}$$^{-1}$) compared to *At*HNL-Celite [15] (0.20 mol h^{-1} g$_{enzyme}$$^{-1}$) at 96% of conversion. In both cases the reported conversion was achieved after 30 min reaction time.

Previously, *At*HNL was immobilized on Celite R-633 [19] and tested for the synthesis of (*R*)-mandelonitrile, the reaction in flow using a packed bed reactor did not enhance the rate of the reaction as compared to the batch system; the best specific rate calculated for the continuous flow system was 0.04 mol h^{-1} g$_{enzyme}$$^{-1}$ at 0.04 mL min^{-1} (85% of conversion), whereas the batch system showed 0.07 mol h^{-1} g$_{enzyme}$$^{-1}$ (circa 91% of conversion). These results are circa six and seven-fold lower as compared to the best specific rate in flow (0.1 mL) and batch respectively reported in this study. In another study [17], the continuous flow synthesis of (*S*)-mandelonitrile with immobilized *Hb*HNL on a siliceous monolith microreactor was 8 times faster as compared to the batch system and displayed a specific rate of 0.50 mmol min^{-1} g$_{enzyme}$$^{-1}$ at circa 95% conversion and 0.2 mL min^{-1}. This result is twice the specific rate observed for *At*HNL-EziG Opal at a similar level of conversion and might be explained by diffusion limitation due to the partial blockage of the pores of EziG Opal during the enzyme immobilization by adsorption. The irregular structure of the microchannels and mesopores found in monolith microreactors overcome this limitation.

The space-time-yield (STY) is a parameter commonly used to compare the productivity of reactors with different size. Batch systems often require rapid stirring to reduce mass transfer limitations that may shorten the lifetime of the immobilized enzyme. On the other hand, stirring is not required in flow, thus this problem is avoided, and better productivities can be achieved [17,25,29]. Indeed, the *At*HNL on EziG Opal catalyzed synthesis of (*R*)-mandelonitrile in flow displayed a STY of 690 mol$_{product}$ h^{-1} L^{-1} g$_{enzyme}$$^{-1}$, whereas the batch approach led to only 187 mol$_{product}$ h^{-1} L^{-1} g$_{enzyme}$$^{-1}$ showing that the flow system greatly enhanced productivity. In batch, a similar productivity has been achieved previously with *At*HNL-Celite (150 mol$_{product}$ h^{-1} L^{-1} g$_{enzyme}$$^{-1}$) [15].

Comparing the results for *At*HNL of this study with the literature reports again demonstrates the advantages of flow chemistry. The synthesis of (*S*)-mandelonitrile in a siliceous monolith microreactor using either *Hevea brasiliensis* HNL (11.3 mg total protein; 1120 U per monolith) or *Manihot esculenta* HNL (17.4 mg total protein; 1310 U per monolith) showed STYs of 555 and 405 mol$_{product}$ h^{-1} L^{-1} g$_{enzyme}$$^{-1}$ [17]. Recently, a ω-transaminase from *Arthrobacter* sp. (*AsR*-ωTA) was immobilized on EziG Amber (semi-hydrophobic carrier) for the kinetic resolution of *rac*-α-methylbenzylamine (*rac*-α-MBA) [20]. The enzyme was shown to be highly stable on this carrier and was able to perform the kinetic resolution of *rac*-α-MBA during 96 consecutive hours with excellent enantioselectivity (49% conversion and 99% *ee*). This flow system achieved a space time yield of 184 mol$_{product}$ h^{-1} L^{-1} g$_{enzyme}$$^{-1}$. The productivities reported in this study are comparable with the productivities reported for immobilized HNLs on siliceous carriers and other enzymes on EziG carriers.

3. Materials and Methods

3.1. Chemicals

Except when reported otherwise all chemicals were bought from Sigma Aldrich (Schnelldorf, Germany). Isopropanol and heptane were of HPLC grade (≥99%) and used as HPLC solvents.

1,3,5-triisopropylbenzene (97%) was from Fluka Chemie (Buchs, Switzerland). Potassium cyanide (KCN, 97%) from J.T. Baker (Deventer, The Netherlands) was used as cyanide source in the HCN solution. (±)-Mandelonitrile from Acros Organics (Geel, Belgium) was purified by flash chromatography (PE/MTBE 9:1/3:7).

3.2. Heterologous Expression of Arabidopsis Thaliana HNL (AtHNL)

pET28a-*At*HNL expression plasmid containing the *At*HNL gene (GenBank accession number AAN13041, EC:4.1.2.10) codon optimized for *E. coli* and with a polyhistidine tag (His$_6$-tag) (see Table S1) was obtained from the group of Martina Pohl (Institute of Bio-and Geosciences, Jülich, Germany). *E. coli* BL21(DE3) was transformed with the expression plasmid for the production of the His-tagged *At*HNL. A preculture was prepared by inoculating one single colony of *E. coli* BL21(DE3)-pET28a*At*HNL in 10 mL of Lysogeny Broth (LB) medium with kanamycin (40 μg mL^{-1}) and incubated overnight (Eppendorf/New Brunswick Scientific Incubator Shaker Excella E24 Series, Nijmegen, The Netherlands) at 37 °C, 180 rpm. Subsequently, this preculture was used for the inoculation of 1 L of Terrific Broth (TB) medium containing kanamycin (40 μg mL^{-1}) and incubated at 37 °C, 120 rpm. When the OD$_{600}$ reached 0.7–0.9 the protein expression was induced by adding 1 mL of 0.1 M isopropyl β-D-thiogalactoside (IPTG) per liter of culture (0.1 mM IPTG final concentration) and cultivation was continued at 25 °C, 180 rpm for 20 h.

Cells were harvested by centrifugation at 4 °C, 3600× *g* rpm during 20 min (Sorvall RC6, Thermo Fisher Scientific, Landsmeer, The Netherlands). The supernatant was discarded and the pellets were washed with 30 mL of 10 mM sodium phosphate buffer pH 7, frozen in liquid nitrogen and stored at −80 °C.

3.3. Enzyme Purification

The pellets containing *At*HNL were resuspended in lysis buffer (10 mM potassium phosphate buffer pH 7.4 + DNase) and lysed in a cell disruptor (Constant Systems Ltd., Daventry, United Kingdom) at 1.5 kBar and 4 °C to avoid protein denaturation. The cell free extracts were collected by centrifugation at 48,000× *g*, 1 h, 4 °C (Sorvall RC6, Thermo Fisher Scientific, Landsmeer, The Netherlands). The enzyme was purified by using a NGC Chromatography system (Bio-Rad, Veenendaal, The Netherlands) by immobilized metal ion chromatography (IMAC) with chelating Ni^{2+} Sepharose (HiTrap Chelating HP 5 mL, GE Healthcare) according to the manufacturer [38]. 20 mM sodium phosphate + 0.5 M NaCl + 20 mM imidazole pH 7.4 was used for the enzyme binding and 20 mM sodium phosphate + 0.5 M NaCl + 0.5 M imidazole pH 7.4 was used for the enzyme elution.

The purified *At*HNL was concentrated with a 10 kDa MWCO Amicon filter (Millipore, Amsterdam-Zuidoost, The Netherlands) and desalted with a PD-10 desalting column (Cytiva, Medemblik, The Netherlands) according to the supplier instructions [39].

3.4. Enzymatic Activity Aassay

*At*HNL activity was determined spectrophotometrically (Agilent Technologies Cary 60 UV-VIS, Amstelveen, The Netherlands) according to the literature [15] with slight modifications. The cleavage of *rac*-mandelonitrile into benzaldehyde and hydrogen cyanide was followed at 280 nm and 25 °C in 1 cm quartz glass cuvettes. Briefly, 1400 μL of 50 mM citrate/phosphate buffer pH 5 and 200 μL of enzyme solution (in 5 mM phosphate buffer pH 6.5) were mixed and incubated for 30 s at 25 °C. The reaction was started by adding 400 μL of 60 mM *rac*-mandelonitrile solution (80 μL of *rac*-mandelonitrile in 10 mL 3 mM citrate/phosphate buffer, pH 3.5). The enzymatic activity was calculated with the molar extinction coefficient of benzaldehyde (ε_{280} = 1.376 mM^{-1} cm^{-1}) and the background reaction (performed without enzyme) was subtracted in the final calculation.

One unit of *At*HNL activity is the amount of micromoles of *rac*-mandelonitrile converted per minute in citrate/phosphate buffer pH 5 at 25 °C.

3.5. Synthesis of Hydrogen Cyanide (HCN) Solution in MTBE

A HCN solution in MTBE was made according to earlier studies [15–19,29]. 25 mL MTBE and 10 mL MilliQ water were mixed in a 100 mL Erlenmeyer and kept at 0 °C. 0.1 mol potassium cyanide (6.51 g) was dissolved in the mixture and magnetically stirred for 15 min. 10 mL of 30% (*v/v*) HCl solution was added slowly and stirring was continued for 2 min. The HCN solution was allowed to reach room temperature (circa 20 °C). The organic and aqueous phases were separated using a separation funnel and the organic layer containing HCN was collected. The separation was performed twice more after adding 7 mL of MTBE each time. Finally, 5 mL of 50 mM citrate/phosphate buffer pH 5 was added to the organic fraction collected and it was stored in a dark bottle at 4 °C.

The HCN concentration was determined in accordance to the literature [40]. 1 mL of HCN solution was added to 5 mL of 2 M NaOH in a 25 mL Erlenmeyer. The mixture was stirred for 2 min. Potassium chromate was added as indicator. The solution was titrated with 0.1 M silver nitrate. The cyanide reacts 1:1 with the silver and precipitates.

3.6. Immobilization of AtHNL on EziG Opal by Adsorption

The immobilization of *At*HNL on EziG Opal by adsorption was performed as described previously [16,18,29]. Given volumes of *At*HNL solution were concentrated in Amicon filters with a 10 kDa MW cut-off, and subsequently added dropwise to 60 mg of EziG Opal (batch) or 150 mg of EziG Opal (flow). For batch reactions, *At*HNL-EziG Opal was tightly packed in a regular paper tea bag [16,29] and either directly used for biocatalytic reactions or dried 16 h under vacuum in a desiccator over silica gel before packing. A magnetic stirrer was attached to the tea bags as reported earlier [16,29] to ensure tight packing and rapid stirring without mechanical attrition of the carrier. Reactions in flow were performed with wet *At*HNL-EziG Opal (without drying and packing). The ratio of enzyme solution to carrier (µL:mg) was 2:1 in all cases to ensure that the enzyme solution was completely absorbed by the carrier. The immobilization of different enzyme units was achieved by determining the enzyme activity and adjusting the amount of enzyme solution before its concentration.

3.7. Immobilization of AtHNL on EziG Opal by Incubation

The immobilization of *At*HNL on EziG Opal by incubation was performed according to the manufacturer (see the instruction manual in the supplementary information). 2 mL of enzyme solution with the required units to be immobilized was incubated with 60 mg of carrier in an orbital shaker (model RM:2M) at 30 rpm during 2 h at room temperature. The binding of the enzyme to the carrier was monitored by determining the protein concentration of the supernatant after immobilization using the BCA protein determination (Pierce BCA protein assay kit, Thermo Fisher Scientific, Landsmeer, The Netherlands) in accordance with the manufacturer instructions [41].

3.8. Synthesis of (R)-Mandelonitrile in Batch

Several (*R*)-mandelonitrile syntheses were performed with 60 mg of immobilized *At*HNL-EziG Opal. The reaction conditions were as follows: 100 µL benzaldehyde (1 mmol), 27.5 µL 1,3,5-triisopropylbenzene (internal standard (I.S.)), 2 mL HCN in 50 mM citrate/phosphate buffered MTBE pH 5 (1.5–2 M), tea bag filled with 60 mg immobilized enzyme, 900 rpm and room temperature. The ratio benzaldehyde to HCN solution was 1:4. The mole ratio *At*HNL:Fe^{3+} was 1:5.

3.9. Enzyme Recyclability in Batch

The enzyme recyclability was determined by several cycles of (*R*)-mandelonitrile synthesis according to [15,16,18,29]. The reaction conditions were as follows: benzaldehyde (100 µL, 1 mmol), 27.5 µL 1,3,5-triisopropylbenzene (I.S.), 2 mL HCN in 50 mM citrate/phosphate buffered MTBE pH 5 (1.5–2 M), tea bag filled with 60 mg immobilized enzyme, 900 rpm and room temperature. The ratio benzaldehyde to HCN solution was 1:4. The mol ratio *At*HNL:Fe^{3+} was 1:5. Between each cycle the immobilized enzyme was washed for 1 min with 50 mM citrate/phosphate buffered MTBE, pH 5, and stored at 4 °C in fresh citrate/phosphate buffered MTBE, pH 5.

3.10. Synthesis of (R)-Mandelonitrile in Continuous Flow

One milliliter stainless steel flow reactor (6.4 cm length, 0.45 cm inner diameter) was used for the continuous synthesis of (R)-mandelonitrile with 150 mg of immobilized *At*HNL on EziG Opal (10 U mg^{-1}). The reactor was filled with 100 mg of non-porous glass beads (1 mm diameter) and 150 mg of EziG Opal containing immobilized enzyme (final reaction volume = 0.3 mL). 10 cm of polytetrafluoroethylene (PTFE) tubing with 1.5 mm inner diameter was used to connect a syringe pump (Knauer, Germany) with the reactor. Initial conditions were as follows: 0.5 M benzaldehyde, 1.5–2 M HCN in 100 mM citrate/phosphate buffered MTBE, pH 5 and 50 mM 1,3,5-triisopropylbenzene asI.S.. The synthesis of (R)-mandelonitrile was evaluated at different flow rates (from 0.1 to 1 mL min^{-1}) by chiral HPLC. The mole ratio *At*HNL:Fe^{3+} was 1:4. The flow rate was checked at each sampling time by the difference of weight. Reactions were performed at room temperature. No significant pressure drop or increase was observed within the timeframe of the experiments.

3.11. Enzyme Stability in Continuous Flow

The stability of immobilized *At*HNL on EziG Opal (10 U mg^{-1}) was evaluated by performing a synthesis reaction during 12 h at 0.1 mL min^{-1} on stream at room temperature. The mol ratio *At*HNL:Fe^{3+} was 25%. Samples were drawn at regular intervals and analyzed by chiral HPLC.

3.12. Analysis

Samples (10 µL) were taken at different times during the reaction run and added to 990 µL of heptane:2-propanol 95:5 in 1.5 mL Eppendorf tubes. A small amount of anhydrous magnesium sulphate (MgSO$_4$) was used to remove the water from the solution and the Eppendorf tubes were centrifuged at 13,000× g rpm for 1 min. 850 µL of the supernatant was transferred to a 4 mL HPLC vial and 10 µL was injected into the HPLC (Chiralpak AD-H column, column size: 0.46 cm I.D × 25 cm). Heptane and 2-propanol were used as mobile phase with a flow rate of 1 mL min^{-1} and the UV detector was set at 216 nm. The column temperature was set at 40 °C. The samples in the autosampler were maintained at 4 °C.

4. Conclusions

*At*HNL was successfully immobilized on EziG Opal by an optimized methodology. *At*HNL-EziG Opal was recycled up to seven times in batch with nearly complete conversion and excellent enantioselectivity. The continuous flow system displayed excellent conversion and enantioselectivity at 0.1 mL min^{-1} and allowed to increase four times the productivity for the synthesis of (R)-mandelonitrile as compared to the batch system.

Supplementary Materials: The following information is available online at http://www.mdpi.com/2073-4344/10/8/899/s1, Figure S1: Leaching assay of *At*HNL-EziG Opal, Figure S2: Blank reaction in batch, Figure S3: Blank reaction in flow, Figure S4. Recycling of *At*HNL-EziG Opal (20 U mg^{-1}) in eight successive cycles. Figure S5. HPLC detection of benzaldehyde and 1,3,5-triisopropylbenzene during 8 h of incubation, Table S1: *At*HNL gene and Amino acid sequences.

Author Contributions: U.H. conceptualized and supervised the study; J.C., T.L. and M.A. performed the experiments; M.L. and P.B. developed the *At*HNL overexpression and purification method; P.-L.H., L.G. and U.H. reviewed the manuscript and supervised the study. J.C. wrote and edited the manuscript. All authors have read and agreed to the published version of the manuscript.

Funding: This research was funded by the Secretary of Higher Education, Science, Technology and Innovation of Ecuador (Senescyt) and Universidad Laica Eloy Alfaro de Manabí (ULEAM) and by BE-BASIC, grant number FES-0905 to P.B.

Acknowledgments: We thank Martina Pohl (Institute of Bio- and Geosciences—Jülich, Germany) for providing the pET28a-*At*HNL expression plasmid. EnginZyme AB is also acknowledged for the kindly provision of EziG Opal used in this study. Mattia Lazzarotto is grateful to the European Commission and to the University of Trieste for an ERASMUS+ fellowship.

Conflicts of Interest: The authors declare no conflict of interest. The funders had no role in the design of the study; in the collection, analyses, or interpretation of data; in the writing of the manuscript, or in the decision to publish the results.

References

1. Dadashipour, M.; Asano, Y. Hydroxynitrile lyases: Insights into biochemistry, discovery, and engineering. *ACS Catal.* **2011**, *1*, 1121–1149. [CrossRef]
2. Steiner, K.; Glieder, A.; Gruber-Khadjawi, M. Cyanohydrin formation/Henry reaction. In *Science of Synthesis: Biocatalysis in Organic Synthesis*; Georg Thieme Verlag: Stuttgart, Germany, 2015; Volume 2, pp. 1–30.
3. Lanfranchi, E.; Steiner, K.; Glieder, A.; Hajnal, I.; Sheldon, R.A.; van Pelt, S.; Winkler, M. Mini-Review: Recent developments in hydroxynitrile lyases for industrial biotechnology. *Recent Pat. Biotechnol.* **2013**, *7*, 197–206. [CrossRef]
4. Bracco, P.; Busch, H.; von Langermann, J.; Hanefeld, U. Enantioselective synthesis of cyanohydrins catalysed by hydroxynitrile lyases—A review. *Org. Biomol. Chem.* **2016**, *14*, 6375–6389. [CrossRef]
5. Faber, K. *Biotransformations in Organic Chemistry*, 7th ed.; Springer Nature: Basel, Switzerland, 2018; pp. 224–233. [CrossRef]
6. Hanefeld, U. Immobilisation of hydroxynitrile lyases. *Chem. Soc. Rev.* **2013**, *42*, 6308–6321. [CrossRef]
7. Zhu, Y.; Chen, Q.; Shao, L.; Jia, Y.; Zhang, X. Microfluidic immobilized enzyme reactors for continuous biocatalysis. *React. Chem. Eng.* **2020**, *5*, 9–32. [CrossRef]
8. Cantone, S.; Ferrario, V.; Corici, L.; Ebert, C.; Fattor, D.; Spizzo, P.; Gardossi, L. Efficient immobilisation of industrial biocatalysts: Criteria and constraints for the selection of organic polymeric carriers and immobilisation methods. *Chem. Soc. Rev.* **2013**, *42*, 6262–6276. [CrossRef] [PubMed]
9. Liese, A.; Hilterhaus, L. Evaluation of immobilized enzymes for industrial applications. *Chem. Soc. Rev.* **2013**, *42*, 6236–6249. [CrossRef] [PubMed]
10. Hanefeld, U.; Gardossi, L.; Magner, E. Understanding enzyme immobilisation. *Chem. Soc. Rev.* **2009**, *38*, 453–468. [CrossRef] [PubMed]
11. Block, H.; Maertens, B.; Spriestersbach, A.; Brinker, N.; Kubicek, J.; Fabis, R.; Labahn, J.; Schäfer, F. *Immobilized-Metal Affinity Chromatography (IMAC) in Methods in Enzymology*; Academic Press: Cambridge, MA, USA, 2009; pp. 439–473. [CrossRef]
12. Das, K.K.; Reddy, R.C.; Bagoji, I.B.; Das, S.; Bagali, S.; Mullur, L.; Khodnapur, J.P.; Biradar, M.S. Primary concept of nickel toxicity—An overview. *J. Basic Clin. Physiol. Pharmacol.* **2018**, *30*, 141–152. [CrossRef] [PubMed]
13. Andexer, J.N.; Staunig, N.; Eggert, T.; Kratky, C.; Pohl, M.; Gruber, K. Hydroxynitrile lyases with α/β-hydrolase fold: Two enzymes with almost identical 3D structures but opposite enantioselectivities and different reaction mechanisms. *ChemBioChem* **2012**, *13*, 1932–1939. [CrossRef] [PubMed]
14. Cassimjee, K.E.; Kadow, M.; Wikmark, Y.; Humble, M.S.; Rothstein, M.L.; Rothstein, D.M.; Bäckvall, J.-E. A general protein purification and immobilization method on controlled porosity glass: Biocatalytic applications. *Chem. Commun.* **2014**, *50*, 9134–9137. [CrossRef] [PubMed]
15. Okrob, D.; Paravidino, M.; Orru, R.V.A.; Wiechert, W.; Hanefeld, U.; Pohl, M. Hydroxynitrile lyase from *Arabidopsis thaliana*: Identification of reaction parameters for enantiopure cyanohydrin synthesis by pure and immobilized catalyst. *Adv. Synth. Catal.* **2011**, *353*, 2399–2408. [CrossRef]
16. Bracco, P.; Torrelo, G.; Noordam, S.; de Jong, G.; Hanefeld, U. Immobilization of *Prunus amygdalus* hydroxynitrile lyase on Celite. *Catalysts* **2018**, *8*, 287. [CrossRef]
17. van der Helm, M.P.; Bracco, P.; Busch, H.; Szymańska, K.; Jarzębski, A.; Hanefeld, U. Hydroxynitrile lyases covalently immobilized in continuous flow microreactors. *Catal. Sci. Technol.* **2019**, *9*, 1189–1200. [CrossRef]
18. Torrelo, G.; van Midden, N.; Stloukal, R.; Hanefeld, U. Immobilized hydroxynitrile lyase: A comparative study of recyclability. *ChemCatChem* **2014**, *6*, 1096–1102. [CrossRef]

19. Brahma, A.; Musio, B.; Ismayilova, U.; Nikbin, N.; Kamptmann, S.; Siegert, P.; Jeromin, G.E.; Ley, S.V.; Pohl, M. An orthogonal biocatalytic approach for the safe generation and use of HCN in a multistep continuous preparation of chiral *O*-Acetylcyanohydrins. *Synlett* **2015**, *27*, 262–266. [CrossRef]
20. Böhmer, W.; Knaus, T.; Volkov, A.; Slot, T.K.; Shiju, N.R.; Cassimjee, K.E.; Mutti, F.G. Highly efficient production of chiral amines in batch and continuous flow by immobilized ω-transaminases on controlled porosity glass metal-ion affinity carrier. *J. Biotechnol.* **2019**, *291*, 52–60. [CrossRef]
21. Tentori, F.; Bavaro, T.; Brenna, E.; Colombo, D.; Monti, D.; Semproli, R.; Ubiali, D. Immobilisation of old yellow enzymes via covalent or coordination bonds. *Catalysts* **2020**, *10*, 260. [CrossRef]
22. Aßmann, M.; Mügge, C.; Gaßmeyer, S.K.; Enoki, J.; Hilterhaus, L.; Kourist, R.; Liese, A.; Kara, S. Improvement of the process stability of Arylmalonate decarboxylase by immobilization for biocatalytic profen synthesis. *Front. Microbiol.* **2017**, *8*, 448. [CrossRef]
23. Yu, T.; Ding, Z.; Nie, W.; Jiao, J.; Zhang, H.; Zhang, Q.; Xue, C.; Duan, D.; Yamada, Y.M.A.; Li, P. Recent advances in continuous-flow enantioselective catalysis. *Chem. Eur. J.* **2020**, *26*, 5729–5747. [CrossRef]
24. Yoo, W.J.; Ishitani, H.; Saito, Y.; Laroche, B.; Kobayashi, S. Reworking organic synthesis for the modern age: Synthetic strategies based on continuous-flow addition and condensation reactions with heterogeneous catalysts. *J. Org. Chem.* **2020**, *85*, 5132–5145. [CrossRef] [PubMed]
25. Thompson, M.P.; Peñafiel, I.; Cosgrove, S.C.; Turner, N.J. Biocatalysis using immobilized enzymes in continuous flow for the synthesis of fine chemicals. *Org. Process Res. Dev.* **2019**, *23*, 9–18. [CrossRef]
26. Akwi, F.M.; Watts, P. Continuous flow chemistry: Where are we now? Recent applications, challenges and limitations. *Chem. Commun.* **2018**, *54*, 13894–13928. [CrossRef] [PubMed]
27. Sheldon, R.A.; Woodley, J.M. Role of biocatalysis in sustainable chemistry. *Chem. Rev.* **2018**, *118*, 801–838. [CrossRef]
28. Movsisyan, M.; Delbeke, E.I.P.; Berton, J.K.E.T.; Battilocchio, C.; Ley, S.V.; Stevens, C.V. Taming hazardous chemistry by continuous flow technology. *Chem. Soc. Rev.* **2016**, *45*, 4892–4928. [CrossRef]
29. Coloma, J.; Guiavarc'h, Y.; Hagedoorn, P.L.; Hanefeld, U. Probing batch and continuous flow reactions in organic solvents: *Granulicella tundricola* hydroxynitrile lyase (*Gt*HNL). *Catal. Sci. Technol.* **2020**, *10*, 3613–3621. [CrossRef]
30. Vieille, C.; Zeikus, G.J. Hyperthermophilic enzymes: Sources, uses, and molecular mechanisms for thermostability. *Microbiol. Mol. Biol. Rev.* **2001**, *65*, 1–43. [CrossRef]
31. Guterl, J.K.; Andexer, J.N.; Sehl, T.; von Langermann, J.; Frindi-Wosch, I.; Rosenkranz, T.; Fitter, J.; Gruber, K.; Kragl, U.; Eggert, T.; et al. Uneven twins: Comparison of two enantiocomplementary hydroxynitrile lyases with α/β-hydrolase fold. *J. Biotechnol.* **2009**, 166–173. [CrossRef]
32. Paravidino, M.; Sorgedrager, M.J.; Orru, R.V.; Hanefeld, U. Activity and enantioselectivity of the hydroxynitrile lyase *Me*HNL in dry organic solvents. *Chem. Eur. J.* **2010**, *16*, 7596–7604. [CrossRef]
33. Effenberger, F.; Eichhorn, J.; Roos, J. Enzyme catalyzed addition of hydrocyanic acid to substituted pivalaldehydes—A novel synthesis of (*R*)-pantolactone. *Tetrahedron Asymmetry* **1995**, *6*, 271–282. [CrossRef]
34. Costes, D.; Wehtje, E.; Adlercreutz, P. Hydroxynitrile lyase-catalyzed synthesis of cyanohydrins in organic solvents: Parameters influencing activity and enantiospecificity. *Enzym. Microb. Technol.* **1999**, *25*, 384–391. [CrossRef]
35. Abdelraheem, E.M.M.; Busch, H.; Hanefeld, U.; Tonin, F. Biocatalysis explained: From pharmaceutical to bulk chemical production. *React. Chem. Eng.* **2019**, *4*, 1878–1894. [CrossRef]
36. Britton, J.; Majumdar, S.; Weiss, G.A. Continuous flow biocatalysis. *Chem. Soc. Rev.* **2018**, *47*, 5891–5918. [CrossRef] [PubMed]
37. Tamborini, L.; Fernandes, P.; Paradisi, F.; Molinari, F. Flow bioreactors as complementary tools for biocatalytic process intensification. *Trends Biotechnol.* **2018**, *36*, 73–88. [CrossRef]
38. Affinity Chromatography: Principles and Methods. Available online: www.sigmaaldrich.com/content/dam/sigma-aldrich/docs/Sigma-Aldrich/General_Information/1/ge-affinity-chromatography.pdf (accessed on 14 October 2019).
39. PD-10 Desalting Columns. Available online: http://wwwuser.gwdg.de/~{}jgrossh/protocols/protein-purification/PD10.pdf (accessed on 14 October 2019).

40. Van Langen, L.M.; van Rantwijk, F.; Sheldon, R.A. Enzymatic hydrocyanation of a sterically hindered aldehyde. Optimization of a chemoenzymatic procedure for (*R*)-2-Chloromandelic Acid. *Org. Process. Res. Dev.* **2003**, *7*, 828–831. [CrossRef]
41. User Guide: Pierce BCA Protein Assay Kit. Available online: https://www.thermofisher.com/document-connect/document-connect.html?url=https%3A%2F%2Fassets.thermofisher.com%2FTFS-Assets%2FLSG%2Fmanuals%2FMAN0011430_Pierce_BCA_Protein_Asy_UG.pdf&title=VXNlciBHdWlkZTogUGllcmNlIEJDQSBQcm90ZWluIEFzc2F5IEtpdpdA (accessed on 15 October 2019).

Article

Enzymatic Synthesis of Estolides from Castor Oil

Amine Arslan, Anders Rancke-Madsen and Jesper Brask *

Novozymes A/S, Biologiens Vej 2, 2800 Kgs. Lyngby, Denmark; aminekarslan@gmail.com (A.A.);
arm@novozymes.com (A.R.-M.)
* Correspondence: jebk@novozymes.com

Received: 22 June 2020; Accepted: 14 July 2020; Published: 24 July 2020

Abstract: Estolides are fatty acid polyesters with applications in both industry and consumer products. Recently, reports have emerged detailing lipase-catalyzed synthesis of estolides from free hydroxy fatty acids. In this paper, we describe a simple alternative enzymatic process, in which castor oil is directly converted to an estolide mixture by *Candida antarctica* lipase A (CALA) catalyzed transesterification. The reaction mixture is analyzed by NMR to determine the estolide number (EN) and MALDI MS to identify individual components, in addition to titration to determine the acid value (AV). Estolide trimers and tetramers (EN 2–3) were formed over 24 h in a system with 2:1 (*v/v*) castor oil–water. Further, utilizing different lipase specificities, addition of *Thermomyces lanuginosus* lipase (TLL), allowed the CALA product mixture to be cleaned up by hydrolyzing attached glycerol. In addition, a three-enzyme process is suggested, in which a simple alcohol is added and *Candida antarctica* lipase B (CALB) is used to esterify carboxylic acids to lower AV.

Keywords: estolides; castor oil; lipase; candida antarctica lipase A

1. Introduction

Estolides are a class of fatty acid polyesters with interesting biological properties [1]. Further, synthetic estolides have recently received increasing attention. Attractive physical properties, including improved hydrolytic stability compared to glycerides, while being biodegradable and originating from renewable resources, has spurred interest in diverse industrial uses, like as biolubricants [2–4], plasticizers [5], and applications to provide emulsification or a viscosity increase [6].

Estolides can be prepared from hydroxy fatty acids by esterification (Figure 1), or from unsaturated fatty acids by the acid-catalyzed addition to the double bond [7–9]. Harsh conditions can however result in colored products and undesired side reactions that would require extra costs for purification. Enzymatic synthesis of estolides using lipases has therefore been investigated as an alternative to overcome this [2,3,10–15]. The enzyme-mediated reactions promise mild reaction conditions such as ambient temperature and pressure and neutral pH. Together with high selectivity this results in very clean reactions. Like the acid-catalyzed routes, control of polymerization degree (chain length) however continues to be a challenge. A chemical alternative with utilization of protecting groups seeks to address this but appears cumbersome and costly in large scale [16].

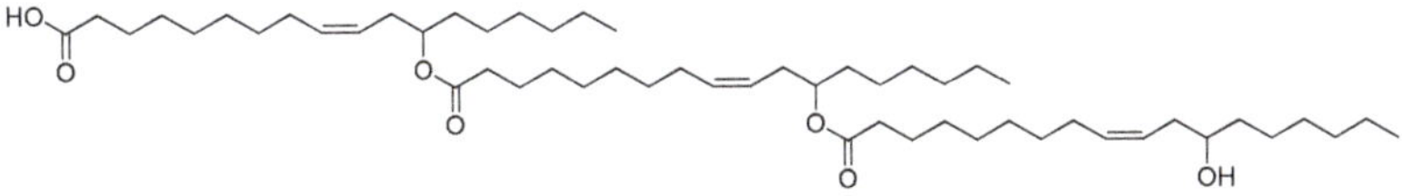

Figure 1. An estolide trimer based on ricinoleic acid.

Castor oil contains 90% ricinoleic acid (12-hydroxy-9-*cis*-octadecenoic acid) and is therefore a natural starting point for estolide synthesis. The reported enzymatic approaches however typically go through hydrolysis of castor oil to produce free ricinoleic acid, or start from commercial ricinoleic acid, which is then condensed to estolides [2,3,9,10,14,15]. What we report here is the lipase-catalyzed synthesis of estolides directly from castor oil by transesterification in a green process using only water as solvent and producing only glycerol as byproduct.

2. Results

2.1. Lipase Screening

2.1.1. Liquid Enzymes

Commercial production of estolides from ricinoleic acid would first involve fat splitting of castor oil. As an alternative to this, lipases may be able to catalyze the direct transesterification of castor oil into estolides. To investigate this, a range of Novozymes' lipolytic enzymes were assayed. This included the *Candida antarctica* lipase A (CALA), the *Candida antarctica* lipase B (CALB), as well as lipases from *Thermomyces lanuginosus* (TLL), *Rhizomucor miehei* (RML), and *Geotrichum candidum* (GCL), and finally a genetically engineered variant of *Humicola insolens* cutinase (HIC). To enable a high throughput for screening different conditions, reactions were conducted in parallel in a sealed 24-well round-bottom deep-well plate, with efficient magnetic stirring in all wells using a custom-designed stirrer-hotplate. The lipases were assayed in a system with castor oil–water (2:1) at 40 °C for 24 h, after which estolide formation was determined quantitatively by ^{1}H NMR, and hydrolysis was quantified by titration of free carboxylic acids. In Table 1, this is reported as the average estolide number (EN) and the acid value (AV), respectively.

Table 1. Lipase screening.

Lipase	EN	AV
Blank	0.0	2
CALA	2.45	25
CALB	0.02	25
TLL	0.05	79
RML	0.09	110
GCL	0.52	43
HIC	0.45	88

The EN is the number of ricinoleic acid units added to a base unit. The NMR analysis relied on the esterified ricinoleic acid C*H*-OR signal at 4.8–4.9 ppm and the non-esterified ricinoleic C*H*-OH signal at 3.6–3.7 ppm. The EN was directly found as the ratio of these two integrals. Based on this NMR quantification, CALA clearly resulted in the most significant estolide formation while keeping a relatively low degree of hydrolysis. This demonstrates a significant enzymatic preference for a lipophilic nucleophilic donor, to catalyze the transesterification over hydrolysis, even in a system with abundant water.

Since the NMR analysis only describes average properties, MS-methods were investigated to identify and characterize individual components in the reaction mixtures. LC-MS analyses provided some chromatographic separation on a C18 RP-column, whereas the coupling to ESI-MS with application of MS^2 and MS^3 methods, allowed isobaric substances to be studied further. Specifically, loss of the oleate radical m/z 280, from fragmentation of the secondary -CH-OR, was frequently identified. For providing an overview of the species, MALDI MS was however preferred. As expected, such analysis revealed a mixture of estolide oligomers, as well as glycerides and glyceride/estolide hybrids (glycerol-containing estolides). Several isomers have the same molecular mass. In Figure 2, these are shown schematically and grouped in structural classes. Pure estolides are referred to as E# and glycerides are called G#. The number (#) denotes the estolide number. MALDI analysis of the CALA reaction identified estolides up to E4 and glycerides up to G6. The other lipases, also showing less estolide formation by NMR, formed only the smaller components (Table 2).

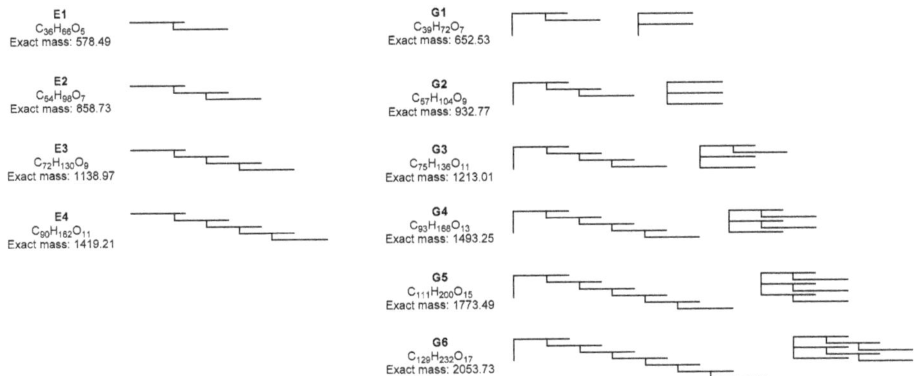

Figure 2. Structures of estolides (E) and glycerides (G). Within each glyceride group (G1–G6) the different isomers have identical chemical formulas and masses (not all isomers are shown).

Table 2. MALDI MS analysis of product mixtures from lipase screening, showing presence (+) and absence (-) of ricinoleic acid (R), estolides (E1–E4), and glycerides (G1–G6).

Lipase	R	E1	E2	E3	E4	G1	G2	G3	G4	G5	G6
CALA	+	+	+	+	+	+	+	+	+	+	+
CALB	+	-	-	-	-	-	+	-	-	-	-
TLL	+	-	-	-	-	+	+	-	-	-	-
RML	+	-	-	-	-	+	+	-	-	-	-
GCL	+	+	+	-	-	+	+	+	-	-	-

2.1.2. Immobilized Enzymes

Since water concentration is expected to play a significant role in the reaction, a few experiments were also conducted with immobilized lipases. These included both the already tested lipases, as well as the *Candida rugosa* lipase. All were immobilized on polymethacrylate resin beads. Reactions were conducted both in the absence of added water, and with a 2:1 oil–water ratio. Interestingly, none of the enzymes performed as well as seen with the liquid enzymes. Immobilized CALA only provided a modest average EN = 0.18. The significant difference between reactions catalyzed by free and immobilized enzymes could be related to transport limitations in the viscous oil and 3-phase system with oil, water, and the porous support material.

2.2. Optimizing the CALA Reaction

2.2.1. Temperature

Performing the CALA reaction at 25 °C, 40 °C, and 65 °C, resulted in EN of 1.38, 2.13, and 2.03, respectively. MALDI MS characterization showed identical results for the three temperatures, with ricinoleic acid (R), E1–E4, and G1–G5 being identified. Based on this, it was decided to fix the temperature at 40 °C in the following experiments.

2.2.2. Water Content

To determine optimal water content, volume of the water phase was varied (Figure 3). No added water resulted in fluctuating results, whereas a plateau was obtained around 10–40% water with EN =

2–2.5. More water resulted in lower conversions. The estolides and glycerides identified with MALDI MS followed the conversion, i.e., up to E4 and G6 were found in the reactions with high EN, whereas only the lower species were found in reactions with lower conversion (MALDI MS data shown in Supplementary Material, Table S1). For subsequent experiments, it was decided to maintain the 33% water phase (2:1 oil–water ratio).

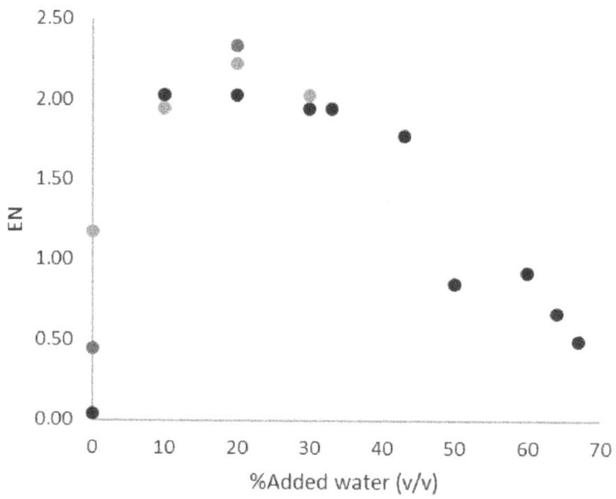

Figure 3. EN measured by NMR as a function of added water to the reaction (% of total reaction volume). Some reactions were performed in duplicates or triplicates.

2.2.3. Time Course

Using 40 °C and 2:1 oil–water, EN was followed over time. As shown in Figure 4, estolide formation initially increased rapidly, then leveled off, reaching EN = 2.40 after 24 h and EN = 2.67 after 48 h. The MALDI MS data further showed that the longer oligomers are only formed slowly. Only in the 24 h reaction mixture (and later) can E3 and E4 be identified.

2.2.4. pH-Effect

Results so far indicate that polymerization stops after E4. It was speculated, that this could be explained by enzyme inactivation resulting from a pH-drop when free fatty acids are released. To examine this, phosphate buffer pH 7.0 was tested as the aqueous phase. Whereas the unbuffered system resulted in a pH-drop to 5, 100 mM phosphate resulted in pH 6, and 500 mM phosphate maintained pH 7. However, upholding pH 7 did not result in higher EN, indicating that other factors are limiting further reaction.

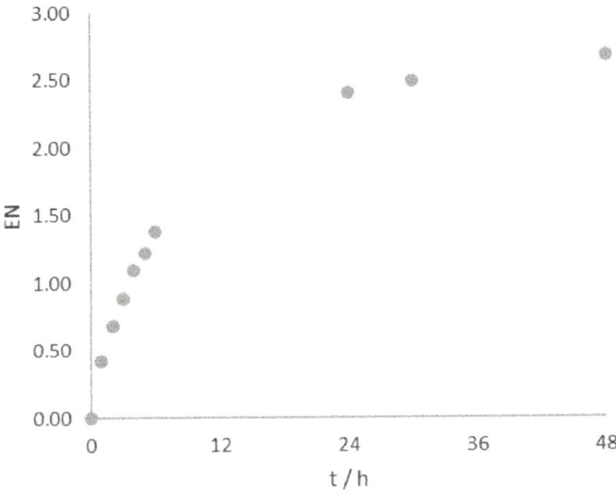

Figure 4. EN measured by NMR as a function of reaction time.

2.3. Two-Enzyme Process

2.3.1. Reaction Sequence

The MALDI MS results clearly show that a mixture of both estolides and glycerides in different lengths are formed. Removing the glycerol from G1–G6 by selective hydrolysis would convert them to estolides E1–E6, thereby effectively cleaning up the reaction mixture. Classical lipases such as TLL are known for their ability to hydrolyze glycerides. On this background, two-enzyme reactions were evaluated, combining the specificities of CALA and TLL. This involved both reactions with (i) CALA being added first to produce the estolides mixture, followed by glyceride hydrolysis by TLL; and (ii) TLL added first to procedure ricinoleic acid, followed by condensation to estolides by CALA; and (iii) both enzymes added together. In the two former approaches, the water phase was removed after step one, before adding the 2nd enzyme. However, only option (i) the two-step reaction with CALA first, produced good conversion to estolide (Table 3). Remarkably, due to the specificity of TLL toward glycerides, only very little estolide hydrolysis was observed. The MALDI MS analysis confirmed that the TLL treatment in "process (i) step 2" indeed does clean up the reaction mixture by converting glyceride oligomers to estolides (Table 4).

Table 3. EN measured by NMR for two-enzyme processes.

	Step 1	Step 2
(i) CALA, TLL	2.70	2.70
(ii) TLL, CALA	0.05	0.56
(iii) CALA+TLL	0.61	-

Table 4. MALDI MS analysis of product mixtures from two-enzyme processes.

Process	R	E1	E2	E3	E4	G1	G2	G3	G4	G5	G6
(i) step 1	+	+	+	+	+	+	+	+	+	+	+
(i) step 2	+	+	+	+	+	-	-	-	-	-	-
(ii) step 1	+	-	-	-	-	+	+	-	-	-	-
(ii) step 2	+	+	+	-	-	+	+	+	-	-	-
(iii)	+	+	+	-	-	+	-	-	-	-	-

2.3.2. Scale up

The two-enzyme process was performed in larger scale to provide material for further studies and to determine the isolated product yield. Hence, reaction setup was round-bottom flasks with magnetic stirring in a thermostat bath. Starting out with 50 g castor oil, the oil phase was reduced to 43 g after step 1 and further to 35 g after step 2. Assuming castor oil (MW 933 g/mol) is all converted to the estolide trimer E2 (MW 859 g/mol), this corresponds to a yield of 76%. The loss seems to be due to inefficient phase separation when water phases are removed, which may be improved with better centrifugation. NMR analysis resulted in EN = 2.45 after step 2. The glyceride to estolide conversion was again confirmed by MALDI MS (Figure 5).

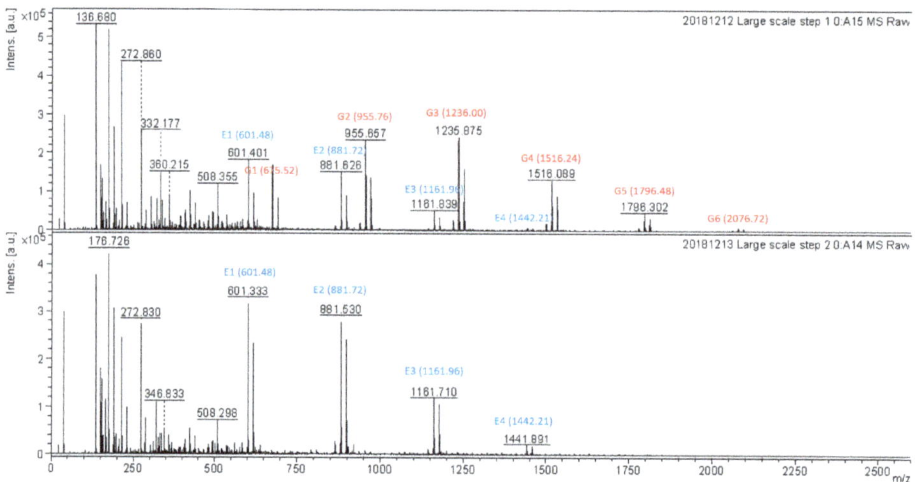

Figure 5. MALDI MS analysis of two-enzyme reaction, step 1 (top) and step 2 (bottom). E1–E4 are marked with blue, while glycerides G1–G6 are marked with red (with theoretical m/z for [M+Na]$^+$).

2.4. Three-Enzyme Process

It was speculated that further modification of the reaction mixture from the two-enzyme process, specifically esterification of the free carboxylic acids, could result in attractive low-AV estolides. This is especially important for biolubricants, and other applications involving contact with machinery. Therefore, a three-enzyme process was preliminarily investigated, in which the product from the two-step process was further reacted with 1-hexanol, catalyzed by immobilized CALB (Novozym 435). After removing the immobilized enzyme by filtration and evaporating excess 1-hexanol, the estolide mixture was analyzed by MALDI MS. This clearly showed significant formation of hexyl esters of all estolides (+84 Da, see MALDI MS spectrum S2 in Supplementary Material). NMR showed a triplet signal at 4.05 ppm resulting from the -CH$_2$O- of hexyl esters (see NMR spectrum S3 in Supplementary Material). From the NMR integrals it appears that 49% of the estolide oligomers are hexyl esters. Further reaction can likely be obtained through optimization of conditions, e.g., removing water to drive the equilibrium.

3. Discussion

CALA is known for its ability to accommodate bulky secondary alcohols [17]. Here, the lipase further demonstrates a remarkable preference for lipophilic acyl transfer reactions. Hence, despite very high water concentrations, transesterification of castor oil is strongly preferred over hydrolysis. This is unique for CALA relative to the other tested lipases and forms the basis for a one-step synthesis of estolides directly from castor oil. Most published lipase-catalyzed routes to estolides start out with producing hydroxy fatty acids in a separate step [2,3,12], or use costly commercially available ricinoleic acid [10,14,15]. Condensation to estolides is then typically achieved with *Candida rugosa* lipase (CRL) in either free or immobilized forms. With the immobilized form, the system can be dried to very low water activity, which is reported to drive the reaction. In such a system, Freire and coworkers reached EN = 7, as determined by ^{13}C NMR spectroscopy [2]. The estolide numbers obtained in the present study are lower (approx. 2.5, corresponding to trimers and tetramers on average), but what that means for potential applications is uncertain.

An inherent issue when producing estolides by polymerization reactions is how to control the distribution of products obtained. Producing estolides directly from castor oil is simple and cost-efficient, but also adds new components to the product mixture, namely the glycerides. Interestingly, we found that combining different lipase specificities in multi-stage enzymatic reactions have the potential to modify the products to target specific properties. Hence, glycerol is quantitatively removed by TLL-catalyzed hydrolysis of the glycerides, without affecting the estolides. Further, preliminary results indicate the potential of CALB in esterification of the terminal carboxylic acids to produce low-AV estolides, e.g., for biolubricants applications.

Estolides have been analyzed by gel permeation chromatography [18], HPLC, GC, and a battery of other analytical methods [19]. In the present study, ^{1}H NMR was found to be a fast and quantitative method for providing average EN-numbers directly from the integrals, as well as information about purity and degree of esterification in the 3-enzyme process. It is complemented by an also rapid and simple MALDI MS method, to provide information about individual species in the product mixtures. MS analysis of such a complex blend of isomers is obviously challenged by the fact that several species have the exact same molecular weight (see Figure 2). Preliminary results however demonstrated the potential of MS^n methods to differentiate between such isobaric substances.

For biotechnological innovation to move from laboratory scale proof-of-concept to industrial production scale, the new process has to offer unique advantages, e.g., properties of the product, or cost-savings compared to established methods. Factors such as safety and sustainability also become increasingly important. In fact, application of lipolytic enzymes for modification of vegetable oils is well-established in the industry. Building on this, the present work has outlined a potential simple and inexpensive procedure for converting castor oil directly into estolides. The process may be conducted at ambient conditions in stirred tank reactors with known commercially available liquid formulated enzyme. Immobilized enzymes, although easy to separate and reuse, are significantly more expensive, often forming a barrier for implementing enzymatic processes. The use of liquid enzyme(s) opens up for a single-use process, although it may also be recycled in the aqueous phase.

4. Materials and Methods

4.1. General Procedures

All chemicals were purchased from Merck-Sigma-Aldrich (St. Louis, MO, USA), while castor oil was provided by Jayant Agro-Organics Ltd. (Mumbai, India). All enzymes were provided by Novozymes A/S (Bagsvaerd, Denmark). Liquid formulated enzymes included NovoCor AD L (CALA), Lipozyme CALB L (CALB), Lipolase 100 L (TLL), Palatase 20,000 L (RML), Stickaway (HIC), and experimental lipase from *Geotrichum candidum* (GCL). Immobilized enzymes included Novozym 435 (CALB) as well as the above mentioned liquid enzymes immobilized to a loading of 100 mg/g by adsorption to Lewatit VP OC 1600 resin beads. Small-scale experiments were performed in parallel in a sealed 24-well round-bottom deep-well plate, using a custom-designed stirrer-hotplate. Mixing was ensured by crosshead magnetic stir bars in each well. Standard conditions were castor oil (2 g), water (1 mL), lipase (0.1 mg enzyme protein, EP, per g of castor oil), 40 °C, 500 rpm for 24 h. Prior to analysis, samples were inactivated at 99 °C, 600 rpm for 15 min on an ThermoMixer (Eppendorf, Hamburg, Germany). To separate the oil phase, small scale samples were centrifuged with an MiniSpin microcentrifuge (Eppendorf, Hamburg, Germany), whereas large scale samples were centrifuged with Multifuge 3 S-R (Heraeus, Hanau, Germany) large bench centrifuge.

4.2. Analytical Procedures

NMR samples were prepared by mixing oil sample (50 µL) with chloroform-d (700 µL) and analyzed on a Bruker Avance III HD 400 MHz instrument (Bruker Biospin, Fällanden, Switzerland). The EN was calculated from the integrals of the ricinoleic -CH-OH signal at 3.65 ppm and the esterified -CH-OR signal at 4.85 ppm as:

$$EN = I_{4.85}/I_{3.65}.$$

In the three-enzyme process, percent of carboxylic acid end-groups esterified with 1-hexanol was calculated from the hexyl ester -CH_2O- signal at 4.05 ppm, assuming there is one carboxylic acid end-group for every one free ricinoleic hydroxyl:

$$\%\text{hexyl ester} = 100\% \times I_{4.05}/2/I_{3.65}$$

MALDI MS analyses were performed on a Bruker UltrafleXtreme MALDI-TOF (Bruker Daltonik, Bremen, Germany) mass spectrometer. Samples were diluted to a final concentration of 1 mg/mL in MeCN-H_2O (1:1) and spotted on the target plate with Super-DHB matrix. A Bruker amaZon SL instrument was used for LC-MS analyses, eluting a C18 RP-column with a gradient of 50% to 100% methanol (solvent B) vs. 0.1 mM aqueous NaCl (solvent A), and detecting with ESI MS in positive mode. Titrations were performed by dissolving oil (ca. 0.5 mL, no water phase) in 2-propanol (10 mL). After the addition of 1% phenolphthalein (0.5 mL), 0.1 M NaOH was added dropwise until a permanent pink color was observed. For the blank sample, oil (1 mL) was titrated with 0.025 M NaOH. Acid value (AV) (mg/g) was calculated as:

$$AV = c_{NaOH} \times V_{NaOH} \times M_{KOH}/m_{sample}$$

4.3. Multi-Enzyme Processes

For the upscaled two-enzyme process, castor oil (50 g) was mixed with water (25 mL) and added NovoCor AD L (0.73 mL, 12.5 mg EP). The mixture was incubated in a round-bottom flask under rapid magnetic stirring (500 rpm) at 40 °C for 24 h. Subsequently, the sample was inactivated at 99 °C for 15 min and centrifuged for 15 min. The oil phase (43 g) was then again mixed with water (21 mL) and added Lipolase 100 L (0.64 mL, 10.8 mg enzyme protein). The reaction was stirred for another 24 h at 40 °C. Inactivation and centrifugation of the second step was performed in a similar matter as step one, resulting in 35 g estolide mixture. For the three-enzyme process, estolide mixture from the two-step process (4 g) was mixed with 1-hexanol (2 mL) and Novozym 435 (10 mg) in a small round-bottom flask and incubated for 24 h at 40 °C. The reaction was stopped by enzyme inactivation. No phase separation was observed. Immobilized enzyme was removed by filtration and excess 1-hexanol was removed by rotary evaporation.

Supplementary Materials: The following are available online at http://www.mdpi.com/2073-4344/10/8/835/s1, Figure S1: MS-spectrum of 3-enzyme process, Figure S2: NMR spectrum of 3-enzyme process. Table S1: MALDI MS analysis of product mixtures with different water content.

Author Contributions: Conceptualization, A.R.-M. and J.B.; Data curation, J.B.; Investigation, A.A.; Supervision, A.R.-M. and J.B.; Writing—original draft, A.A.; Writing—review & editing, J.B. All authors have read and agreed to the published version of the manuscript.

Funding: This research received no external funding.

Acknowledgments: The authors wish to thank Jayant Agro-Organics Ltd. (India) for donating castor oil and Laila Lo Leggio (University of Copenhagen) for fruitful discussions during the project.

Conflicts of Interest: A.R.-M. and J.B. are employed by Novozymes A/S, a leading manufacturer of industrial enzymes, including the ones used in this study.

References

1. Kolar, M.J.; Nelson, A.T.; Chang, T.N.; Ertunc, M.E.; Christy, M.P.; Ohlsson, L.; Harrod, M.; Kahn, B.B.; Siegel, D.; Saghatelian, A. Faster protocol for endogenous fatty acid esters of hydroxy fatty acid (FAHFA) measurements. *Anal. Chem.* **2018**, *90*, 5358–5365. [CrossRef] [PubMed]
2. Greco-Duarte, J.; Collaco, A.C.A.; Costa, A.M.M.; Silva, L.O.; Da Silva, J.A.C.; Torres, A.G.; Fernandez-Lafuente, R.; Freire, D.M.G. Understanding the degree of estolide enzymatic polymerization and the effects on its lubricant properties. *Fuel* **2019**, *245*, 286–293. [CrossRef]
3. Greco-Duarte, J.; Cavalcanti-Oliveira, E.D.; Da Silva, J.A.C.; Fernandez-Lafuente, R.; Freire, D.M.G. Two-step enzymatic production of environmentally friendly biolubricants using castor oil: Enzyme selection and product characterization. *Fuel* **2017**, *202*, 196–205. [CrossRef]
4. McNutt, J.; He, Q. Development of biolubricants from vegetable oils via chemical modification. *J. Ind. Eng. Chem.* **2016**, *36*, 1–12. [CrossRef]
5. Stolp, L.J.; Joseph, E.; Kodali, D.R. Synthesis and evaluation of soy fatty acid ester estolides as bioplasticizers in poly (vinyl chloride). *J. Am. Oil Chem. Soc.* **2019**, *96*, 1291–1302. [CrossRef]
6. Isbell, T.A. Chemistry and physical properties of estolides. *Grasas Aceites* **2011**, *62*, 8–20. [CrossRef]
7. Cermak, S.C.; Isbell, T.A. Synthesis of estolides from oleic and saturated fatty acids. *J. Am. Oil Chem. Soc.* **2001**, *78*, 557–565. [CrossRef]
8. Isbell, T.A.; Kleiman, R.; Plattner, B.A. Acid-catalyzed condensation of oleic-acid into estolides and polyestolides. *J. Am. Oil Chem. Soc.* **1994**, *71*, 169–174. [CrossRef]
9. Wang, G.S.; Sun, S.D. Synthesis of ricinoleic acid estolides by the esterification of ricinoleic acids using functional acid ionic liquids as catalysts. *J. Oleo Sci.* **2017**, *66*, ess17031. [CrossRef] [PubMed]
10. Bodalo-Santoyo, A.; Bastida-Rodriguez, J.; Maximo-Martin, M.F.; Montiel-Morte, M.C.; Murcia-Almagro, M.D. Enzymatic biosynthesis of ricinoleic acid estolides. *Biochem. Eng. J.* **2005**, *26*, 155–158. [CrossRef]
11. Hayes, D.G.; Kleiman, R. Lipase-catalyzed synthesis and properties of estolides and their esters. *J. Am. Oil Chem. Soc.* **1995**, *72*, 1309–1316. [CrossRef]
12. Martin-Arjol, I.; Isbell, T.A.; Manresa, A. Mono-estolide synthesis from trans-8-hydroxy-fatty acids by lipases in solvent-free media and their physical properties. *J. Am. Oil Chem. Soc.* **2015**, *92*, 1125–1141. [CrossRef]
13. Todea, A.; Otten, L.G.; Frissen, A.E.; Arends, I.; Peter, F.; Boeriu, C.G. Selectivity of lipases for estolides synthesis. *Pure Appl. Chem.* **2015**, *87*, 51–58. [CrossRef]
14. Bodalo, A.; Bastida, J.; Maximo, M.F.; Montiel, M.C.; Gomez, A.; Murcia, M.D. Production of ricinoleic acid estolide with free and immobilized lipase from Candida rugosa. *Biochem. Eng. J.* **2008**, *39*, 450–456. [CrossRef]
15. Bodalo, A.; Bastida, J.; Maximo, M.F.; Montiel, M.C.; Murcia, M.D.; Ortega, S. Influence of the operating conditions on lipase-catalysed synthesis of ricinoleic acid estolides in solvent-free systems. *Biochem. Eng. J.* **2009**, *44*, 214–219. [CrossRef]
16. Zerkowski, J.A.; Nunez, A.; Solaiman, D.K.Y. Structured estolides: Control of length and sequence. *J. Am. Oil Chem. Soc.* **2008**, *85*, 277–284. [CrossRef]
17. Naik, S.; Basu, A.; Saikia, R.; Madan, B.; Paul, P.; Chaterjee, R.; Brask, J.; Svendsen, A. Lipases for use in industrial biocatalysis: Specificity of selected structural groups of lipases. *J. Mol. Catal. B Enzym.* **2010**, *65*, 18–23. [CrossRef]
18. Bantchev, G.B.; Cermak, S.C.; Durham, A.L.; Price, N.P.J. Estolide molecular weight distribution via gel permeation chromatography. *J. Am. Oil Chem. Soc.* **2019**, *96*, 365–380. [CrossRef]
19. Isbell, T.A.; Kleiman, R. Characterization of estolides produced from the acid-catalyzed condensation of oleic-acid. *J. Am. Oil Chem. Soc.* **1994**, *71*, 379–383. [CrossRef]

Article

Penicillin Acylase from *Streptomyces lavendulae* and Aculeacin A Acylase from *Actinoplanes utahensis*: Two Versatile Enzymes as Useful Tools for Quorum Quenching Processes

Rodrigo Velasco-Bucheli [1,†], Daniel Hormigo [1,‡], Jesús Fernández-Lucas [1,‡],
Pedro Torres-Ayuso [1,§], Yohana Alfaro-Ureña [1], Ana I. Saborido [1], Lara Serrano-Aguirre [1],
José L. García [2], Fernando Ramón [1,||], Carmen Acebal [1], Antonio Santos [3], Miguel Arroyo [1,¶]
and Isabel de la Mata [1,*,¶]

1   Department of Biochemistry and Molecular Biology, Faculty of Biology, Complutense University (UCM),
    José Antonio Novais 12, 28040 Madrid, Spain; rvelasc@unal.edu.co (R.V.-B.);
    daniel.hormigo@universidadeuropea.es (D.H.); jesus.fernandez2@universidadeuropea.es (J.F.-L.);
    pedro.torres-ayuso@nih.gov (P.T.-A.); yohanaalfaro@gmail.com (Y.A.-U.); asaborid@ucm.es (A.I.S.);
    larase01@ucm.es (L.S.-A.); faramon@ucm.es (F.R.); cacebals@ucm.es (C.A.); marroyos@ucm.es (M.A.)
2   Department of Environmental Biology, Biological Research Center (CSIC), Ramiro de Maeztu 9,
    28040 Madrid, Spain; jlgarcia@cib.csic.es
3   Department of Genetics, Physiology and Microbiology, Faculty of Biology, UCM, José Antonio Novais 12,
    28040 Madrid, Spain; ansantos@ucm.es
*   Correspondence: idlmata@ucm.es; Tel.: +34-913-944-150
†   Present address: Roche Diagnostics GmbH, Nonnenwald 2, 82377 Penzberg, Germany.
‡   Present address: Applied Biotechnology Group, European University of Madrid, Tajo, s/n. Urb. El Bosque,
    28670 Villaviciosa de Odón, Madrid, Spain.
§   Present address: Laboratory of Cell and Developmental Signaling, NCI ATRF, Frederick, MD 21701, USA.
|| Present address: Screening and Compound Profiling Institut de Recherches Servier, 125 Chemin de Ronde,
    78290 Croissy-sur-Seine, France.
¶   These two researchers share the position of last author.

Received: 3 June 2020; Accepted: 26 June 2020; Published: 1 July 2020

Abstract: Many Gram-negative bacteria produce *N*-acyl-homoserine lactones (AHLs), quorum sensing (QS) molecules that can be enzymatically inactivated by quorum quenching (QQ) processes; this approach is considered an emerging antimicrobial alternative. In this study, kinetic parameters of several AHLs hydrolyzed by penicillin acylase from *Streptomyces lavendulae* (*Sl*PA) and aculeacin A acylase from *Actinoplanes utahensis* (*Au*AAC) have been determined. Both enzymes catalyze efficiently the amide bond hydrolysis in AHLs with different acyl chain moieties (with or without 3-oxo modification) and exhibit a clear preference for AHLs with long acyl chains (C12-HSL > C14-HSL > C10-HSL > C8-HSL for *Sl*PA, whereas C14-HSL > C12-HSL > C10-HSL > C8-HSL for *Au*AAC). Involvement of *Sl*PA and *Au*AAC in QQ processes was demonstrated by *Chromobacterium violaceum* CV026-based bioassays and inhibition of biofilm formation by *Pseudomonas aeruginosa*, a process controlled by QS molecules, suggesting the application of these multifunctional enzymes as quorum quenching agents, this being the first time that quorum quenching activity was shown by an aculeacin A acylase. In addition, a phylogenetic study suggests that *Sl*PA and *Au*AAC could be part of a new family of actinomycete acylases, with a preference for substrates with long aliphatic acyl chains, and likely involved in QQ processes.

Keywords: penicillin acylase; aculeacin acylase; *N*-acyl-homoserine lactone acylases; quorum quenching; biofouling

1. Introduction

Quorum sensing (QS) is a bacterial cell-to-cell communication mechanism that allows bacteria to regulate a high diversity of biological functions such as bioluminescence, production of virulence factors, antibiotics and other secondary metabolites, and biofilm formation by releasing, detecting and responding to small, diffusible signal molecules called autoinducers [1,2]. The disruption of QS signaling, a process known as quorum quenching (QQ), is a promising alternative for controlling bacterial infections in human, animals or plants and anti-biofouling in membrane bioreactor systems [3–5]. Among the best-characterized QS molecules, *N*-acyl-homoserine lactones (AHLs) are the most commonly used by Gram-negative proteobacteria [6–9]. AHLs each consist of an acyl chain linked to a homoserine lactone core (HSL) by an amide bond [6]. Generally, AHL inactivation is accomplished by lactonases or acylases, although the inactivation by oxidoreductases acting on the C3 substituent in AHL acyl chain has also been reported [10]. Lactonases inactivate AHLs by hydrolyzing their lactone rings, but lactonolysis is reversible at acidic pH, producing an active form [11,12], limiting the use of these enzymes, whereas acylases are more advantageous as QQ agents since they catalyze the irreversible hydrolysis of the amide bond [12]. Although numerous AHL acylases have been characterized, the link between QQ and the ability to gain a competitive advantage due to the production of these enzymes has not conclusively been demonstrated [13]. Recently, it has been described that certain beta-lactam antibiotic resistant bacteria show quorum quenching activity as well [14]. However, whether the molecules involved in both processes are the same is yet unknown [14]. Thus, due to the large number of putative bacterial enzymes that might be involved in QQ processes, further investigation is needed to gain insight into the roles of such enzymes in both environmental issues and biotechnological applications [5,15,16].

The Ntn-hydrolase superfamily contains many enzymes with diverse activities, including β-lactam acylases, AHL acylases and proteasomes, among others [17]. Many of these enzymes have been classified according to their respective first reported activities, although this classification is not always necessarily in agreement with their true biological role [18]. Over the years, AHL acylase activity has been described for some penicillin acylases from Gram-negative bacteria, such as the penicillin G acylase from *Kluyvera citrophila* [19] and the penicillin V acylases (PVAs) from *Pectobacterium atrosepticum* and *Agrobacterium tumefaciens* [20]. Nevertheless, AHL degradation by penicillin acylases from filamentous Gram-positive bacteria has not been fully demonstrated so far.

Penicillin acylase from *Streptomyces lavendulae* ATCC 13664 (*Sl*PA, formerly abbreviated as *Sl*PVA in our previous reports) (EC 3.5.1.11) and aculeacin A acylase from *Actinoplanes utahensis* NRRL 12052 (*Au*AAC) (EC 3.5.1.70) are the unique described penicillin and echinocandin acylases, respectively, capable of efficiently hydrolyzing phenoxymethyl penicillin (penicillin V), several natural aliphatic penicillins (such as penicillin K, penicillin F and penicillin dihydro F) and aculeacin A [21–24]. Astoundingly, both show very similar substrate specificity, with a marked preference for amides bearing long hydrophobic acyl moieties [21,22,24]. Moreover, these extracellular heterodimeric (αβ) Ntn-hydrolases present interesting properties to be applied in the industrial production of semi-synthetic antibiotic and antifungal compounds in enzymatic bioreactors [25–30]. It is interesting that these enzymes are structurally related to the acyl-homoserine lactone acylase from *Streptomyces* sp. M664 [22], the only one characterized AHL acylase from filamentous bacteria [18,31], in contrast to the majority of QQ enzymes that belong to unicellular bacteria.

In the present study, we report the newly discovered AHL acylase activities of both *Sl*PA and *Au*AAC and suggest their possible involvement in QS interference since they can inhibit formation of biofilms by *Pseudomonas aeruginosa*. In addition, high identities between both enzymes and AHL quorum quenching acylases were revealed by comparative sequence analysis.

2. Results and Discussion

2.1. Substrate specificity of SlPA and AuAAC towards different AHLs

The substrate specificities of *Sl*PA and *Au*AAC were analyzed by measuring initial rates and calculating kinetic parameters for the hydrolysis of several enantiopure L-AHLs (Tables 1 and 2, respectively).

Table 1. Kinetic parameters for SlPA using different AHLs *.

Substrate	K_m (mM)	k_{cat} (s^{-1})	k_{cat}/K_m (mM^{-1} s^{-1})
C_8-HSL	1.19 ± 0.16	22.9 ± 1.2	19.2
3-oxo-C_8-HSL	0.93 ± 0.14	7.34 ± 0.42	7.9
C_{10}-HSL	0.25 ± 0.04	21.9 ± 1.1	87.6
3-oxo-C_{10}-HSL	0.40 ± 0.07	14.8 ± 1.0	37.0
C_{12}-HSL	0.13 ± 0.02	19.3 ± 1.1	148.5
3-oxo-C_{12}-HSL	0.22 ± 0.03	20.4 ± 1.3	92.7
C_{14}-HSL	0.039 ± 0.007	3.80 ± 0.25	97.4
3-oxo-C_{14}-HSL	0.137 ± 0.032	10.8 ± 1.9	78.8

* Reaction conditions: phosphate pH 8.0 containing DMSO 20% (*v/v*) at 45 °C.

Table 2. Kinetic parameters for *Au*AAC using different AHLs *.

Substrate	K_m (mM)	k_{cat} (s^{-1})	k_{cat}/K_m (mM^{-1} s^{-1})
C_8-HSL	2.70 ± 0.76	24.7 ± 3.6	9.2
3-oxo-C_8-HSL	0.45 ± 0.14	2.05 ± 0.20	4.6
C_{10}-HSL	0.19 ± 0.03	13.6 ± 0.8	71.6
3-oxo-C_{10}-HSL	0.47 ± 0.10	4.45 ± 0.39	9.5
C_{12}-HSL	0.10 ± 0.02	8.15 ± 0.56	81.5
3-oxo-C_{12}-HSL	0.17 ± 0.03	7.95 ± 0.57	46.8
C_{14}-HSL	0.013 ± 0.003	1.19 ± 0.11	91.5
3-oxo-C_{14}-HSL	0.023 ± 0.006	1.79 ± 0.23	77.8

* Reaction conditions: phosphate pH 8.0 containing DMSO 20% (*v/v*) at 45 °C.

As shown in Table 1, *Sl*PA efficiently hydrolyzed several AHLs and their 3-oxo-derivatives. Although the highest catalytic constant (k_{cat}) was found with C_8-HSL (22.9 s^{-1}), the enzyme was more efficient with substrates containing longer acyl chains. In fact, its catalytic efficiency (k_{cat}/K_m) was 4-fold higher with C_{10}-HSL (87.6 mM^{-1} s^{-1}) or C_{14}-HSL (97.4 mM^{-1} s^{-1}) and almost 8-fold higher when C_{12}-HSL was used (148.5 mM^{-1} s^{-1}). The presence of a 3-oxo group had a negative impact on catalytic efficiency, but this effect was less pronounced as the length of the acyl chain increased. These data suggest that the amide group and the ketone at the C3' position of the acyl chain of AHLs could be important for enzyme-substrate interaction. Among the 3-oxo derivatives, the highest catalytic efficiency value corresponded to 3-oxo-C_{12}-HSL (92.7 mM^{-1} s^{-1}) > 3-oxo-C_{14}-HSL > 3-oxo-C_{10}-HSL > 3-oxo-C_8-HSL. *Sl*PA did not show detectable activity on C_4-HSL, and its activity against C_6-HSL was poor. Likewise, the kinetic parameters for the *Au*AAC-catalyzed hydrolysis of different AHLs and their 3-oxo derivatives were also determined (Table 2), whose values were quite similar to those obtained using *Sl*PA (Table 1).

Like *Sl*PA, *Au*AAC exhibited hydrolytic activity towards C_8-HSL > C_{10}-HSL > C_{12}-HSL > C_{14}-HSL, C_8-HSL being the best substrate according to the observed catalytic constant (24.7 s^{-1}). This preference for a C_8-acyl chain is in agreement with previous kinetic studies employing various aliphatic penicillins as substrates. In this sense, both enzymes were reported to display the highest catalytic constant with penicillin K (22.7 s^{-1} for *Sl*PA and 33.3 s^{-1} for *Au*AAC, respectively) [21,24], a penicillin derivative which bears a C_8-acyl chain attached to 6-amino penicillanic acid (6-APA). In addition, K_m values for AHLs with acyl chains longer than eight carbons were generally lower than those reported for aliphatic penicillins in both *Sl*PA and *Au*AAC [21,24], pointing at the possible role of these enzymes in QQ processes. Although the activity towards 3-oxo derivatives was also detected, the highest catalytic efficiency value for *Au*AAC corresponded to 3-oxo-C_{14}-HSL (77.8 mM^{-1} s^{-1}) > 3-oxo-C_{12}-HSL > 3-oxo-C_{10}-HSL > 3-oxo-C_8-HSL, showing lower values than those observed for *Sl*PA. Among all the substrates included in this study, the best catalytic efficiency was observed with C_{14}-HSL (91.5 mM^{-1} s^{-1}). This preference of *Sl*PA and *Au*AAC for unsubstituted AHLs in C3′ position has been also observed in AHL acylases from *Shewanella* sp. MIB015 (AaC) [32] and *Acinetobacter* sp. Ooi24 (AmiE) [33]. Broad substrate specificity towards different AHLs and β-lactam antibiotics has also been described for other enzymes, besides *Sl*PA and *Au*AAC. For instance, the extracellular AHL acylase from *Streptomyces* sp. M664 (AhlM) efficiently hydrolyzes C_8-HSL, C_{10}-HSL and 3-oxo-C_{12}-HSL, whereas its deacylation activity towards short-acyl chain AHLs, C_6-HSL and 3-oxo-C_6-HSL was relatively low, and was negligible using C_4-HSL as the substrate [31], as is the case for *Sl*PA and *Au*AAC. In addition, AhlM also shows acylase activity towards penicillin G, expanding its substrate specificity to different structures such as β-lactam antibiotics.

It was believed that AHL acylase from *Acidovorax* sp. MR-S7 (MacQ) shows the broadest substrate specificity, since it can deacylate various AHLs, ranging from C_6 to C_{14} in length, and different β-lactams antibiotics, including penicillin derivatives (penicillin G, ampicillin, amoxicillin, and carbenicillin) and cephalosporin derivatives (cephalexin and cefadroxil) [34]. Nevertheless, no echinocandin acylase assays were reported for this enzyme.

It is important to point out that there is no report of kinetic parameters for AHLs by using AhlM and MacQ acylases. The activities of AhlM and MacQ acylases were measured by an AHL inactivation bioassay instead of the fluorescence HSL-OPA assay, which allows the quantification of the HSL released during the enzymatic reaction. Said bioassay has traditionally been used to check out qualitatively only the degradation or non-degradation of various AHLs ranging from C_6 to C_{14} in length using AHL acylases from microorganisms like *Ralstonia* sp XJ12B (AiiD) [35], *Pseudomonas aeruginosa* PAO1 (PvdQ and QuiP) [36,37], *Pseudomonas syringae* B728a (HacA and HacB) [38], *Anabaena* sp. PCC7120 (AiiC) [39], *Shewanella* sp. MIB015 (AaC) [32], *Deinococcus radiodurans* R1 (QqaR) [40] and *K. citrophila* DSM 2660 (*Kc*PGA) [19].

To the best of our knowledge, the kinetics of enzyme reactions catalyzed by AHL acylases have not been studied in detail, and only a few reports indicate catalytic efficiency values for different AHLs in order to deep into substrate specificity of these enzymes. In this sense, the HSL-OPA method was used to carry out kinetic studies of AHL acylases from *P. aeruginosa* PA01 (PA0305 and PvdQ) [36,41] and *Kc*PGA [19], and penicillin V acylases form *Pectobacterium atrosepticum* (*Pa*PVA) and *Agrobacterium tumefaciens* (*At*PVA) [20], but employing few AHLs as substrates.

Using 3-oxo-C_{12}-HSL as the substrate, k_{cat}/K_m values for *Pa*PVA (135 mM^{-1} s^{-1}), *At*PVA (26.8 mM^{-1} s^{-1}), PA0305 (78 mM^{-1} s^{-1}) and PvdQ (5.8 mM^{-1}s^{-1}) could be compared to those observed for *Sl*PA (92.7 mM^{-1} s^{-1}) and *Au*AAC (46.8 mM^{-1} s^{-1}) (Tables 1 and 2), although different reaction conditions (such as pH, temperature, buffer concentration and DMSO concentration) were employed in every case. The kinetic data support the strong activities of both *Sl*PA and *Au*AAC towards 3-oxo-C_{12}-HSL. In the case of *Pa*PVA and *At*PVA, the plot of the reaction velocity (v) as a function of 3-oxo-C_{12}-HSL concentration followed a sigmoidal pattern, and an allosteric behavior was attributed to the low substrate solubility in the reaction medium that did not allow one to reach substrate saturation [20], even though DMSO at 0.8% (*v/v*) was used to enhance 3-oxo-C_{12}-HSL solubility. In contrast, a hyperbolic behavior was observed in the case of *Sl*PA and *Au*AAC, likely due to the presence of DMSO at 20% (*v/v*) in the reaction that allowed total solubilization of higher concentrations of 3-oxo-C_{12}-HSL during the kinetic study (Figure S1). It is worth mentioning that activities of *Sl*PA and *Au*AAC were not affected in the presence of 20% DMSO.

Finally, using C_{12}-HSL as the substrate, the k_{cat}/K_m value for PA0305 [41] at pH 7.5 and 30 °C in the absence of DMSO (1.4 mM^{-1} s^{-1}) was significantly lower than the ones described for *SIPA* (148.5 mM^{-1} s^{-1}) and *AuAAC* (81.5 mM^{-1} s^{-1}) at pH 8.0 and 45 °C in the presence of DMSO at 20% (*v/v*) (Tables 1 and 2). Although kinetic constants have been reported for *KcPGA* employing C_6-HSL and 3-oxo-C_6-HSL, enzymatic activity of *SIPA* and *AuAAC* with these substrates was too low to perform an adequate kinetic characterization. In fact, a low catalytic activity (k_{cat}) was also observed in the case of *KcPGA* using C_6-HSL (0.03 s^{-1}) and 3-oxo-C_6-HSL (0.06 s^{-1}) [19].

2.2. Quorum Quenching Role of SIPA and AuAAC

The AHL acylase activities of both *SIPA* and *AuAAC* suggests that these enzymes may be involved in QQ processes. To demonstrate this hypothesis two different bioassays were used. The first one exploits QS-reporting violacein production by *Chromobacterium violaceum* CV026, and the second one monitors biofilm formation by *P. aeruginosa*.

Results of the forward bioassay carried out with *C. violaceum* CV026 employing C_6-HSL and C_8-HSL as inducers are shown in Figure 1. *C. violaceum* CV026 produced violacein only when C_6-HSL (or C_8-HSL) was added to the agar plate (Figure 1B vs. Figure 1A) and such production was slightly inhibited in the presence of *SIPA* (Figure 1C). The inhibition of violacein production was more evident when the AHL was previously incubated with *SIPA* (for 24 h at 40 °C) and further added to the medium (Figure 1D). This inhibitory effect was higher with C_8-HSL as expected due the higher catalytic efficiency of *SIPA* on this compound (Table 1). Similar results were observed for *AuAAC* in the forward bioassay of C_8-HSL hydrolysis (data not shown). Furthermore, the effect of both enzymes in violacein production by enzymatic hydrolysis of long-chain AHLs was detected by reverse bioassays (data not shown).

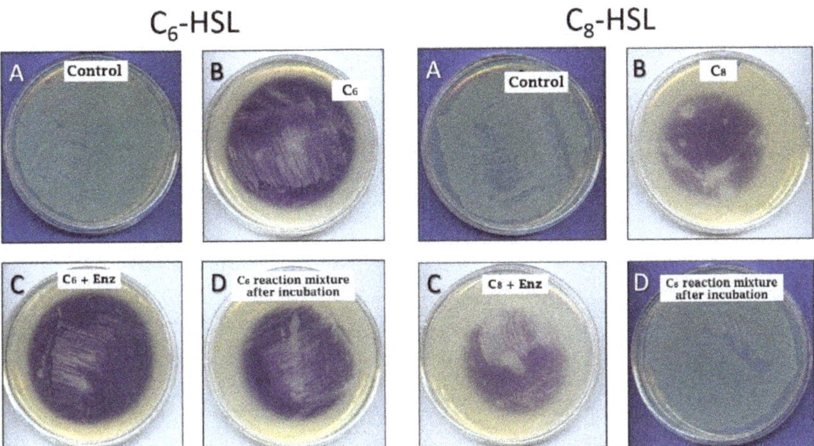

Figure 1. CV026-based forward bioassay of AHL hydrolysis by *SIPA*. C_6-HSL and C_8-HSL were used as inducers of violacein production. (**A**) *C. violaceum* CV026 without inductor; (**B**) *C. violaceum* CV026 grown in the presence of AHL; (**C**) *C. violaceum* CV026 grown in the presence of a mixture of AHL and *SIPA* (in vivo assay); (**D**) *C. violaceum* CV026 grown in the presence of a pre-incubated reaction mixture of AHL and *SIPA* (*in vitro* assay). C_6-HSL concentration was 20 mM, whereas C_8-HSL concentration was 5 mM.

QS interference by the wild-type strains producing *Sl*PA and *Au*AAC was confirmed using *C. violaceum* CV026-based forward bioassay (Figure 2B,C). Previously, *C. violaceum* CV026 was assessed to produce violacein when C_8-HSL was added to the agar plate (Figure 2A), and this production was not affected by the presence of wild-type *S. lividans* (Figure 2D). Finally, AHL acylase cleavage was also demonstrated employing the recombinant strains of *Streptomyces lividans* CECT 3376 and CECT 3377, expressing *Sl*PA and *Au*AAC respectively (Figure 2E,F), demonstrating that both enzymes can interfere in quorum sensing signaling in vivo.

On the other hand, *Sl*PA was tested for its ability to disrupt formation of biofilms by *P. aeruginosa*, a a QS controlled process (Figure 3). When *Sl*PA was added to the culture in its soluble form, disruption of biofilm formation was high (Figure 3C1), whereas it was only moderate when the enzyme was immobilized to silanized slides (Figure 3D1). In this sense, slide coating with *Sl*PA was confirmed by immunodetection (Figures 3B2 and 4D2). On the contrary, biofilm formation was not affected in the absence of *Sl*PA (Figure 3A1) or the presence of the heat-inactivated enzyme (Figure 3B1). Similar results were obtained when *Au*AAC was employed in the same experiment (data not shown).

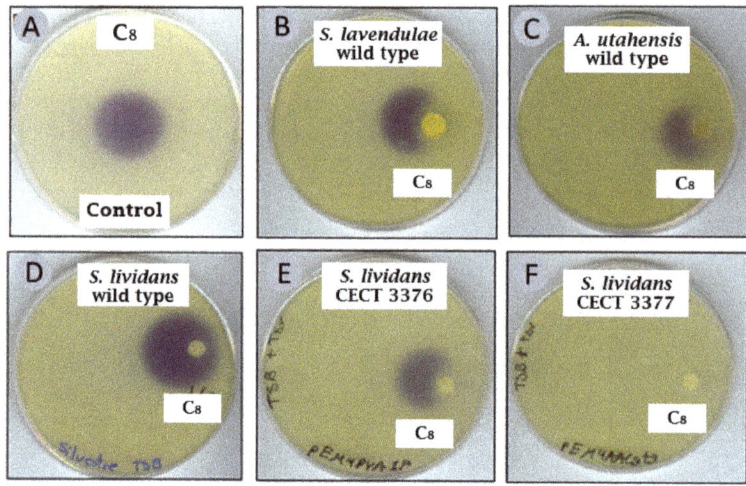

Figure 2. Detection of AHL degradation by recombinant *S. lividans* strains expressing *Sl*PA and *Au*AAC. CV026-based bioassay was used to monitor C_8-HSL cleavage. (**A**) *C violaceum* CV026 control (AHL non-degrading); (**B**) *Streptomyces lavendulae*; (**C**) *Actinoplanes utahensis*; (**D**) *Streptomyces lividans* (AHL non-degrading control); (**E**) recombinant *Streptomyces lividans* expressing *Sl*PA (CECT 3376); and (**F**) recombinant *Streptomyces lividans* expressing *Au*AAC (CECT 3377).

These results suggest that these acylases can be used to combat the formation of biofilms by *P. aeruginosa*. Corneal, lung and burn wound infections caused by *P. aeruginosa*, and the production of virulence factors (and the biofilm differentiation) of this opportunistic pathogen, are regulated by two QQ signals such as 3-oxo-C_{12}-HSL and C_4-HSL [1,42,43]. Besides, 3-oxo-C_8-HSL has demonstrated to increase cell-growth rate during the formation of *P. aeruginosa* biofilm on ultra-filtration membranes for advanced wastewater treatment [44]. In the present work, the presence of *Sl*PA and *Au*AAC has significantly reduced *P. aeruginosa* biofilm formation, likely due to hydrolysis of different AHLs and 3-oxo-AHLs with long acyl-chains, and this effect suggests potential clinical and environmental applications. In fact, the role of the AHL-degrading enzymes to prevent biofilm formation in wastewater treatment plants (WWTPs) is under study, since the inhibition of biofilms on ultrafiltration membranes of membrane bioreactors (MBR) could solve biofouling problems and increase the useful lifespan of filtration membranes minimizing operational costs [5]. Further experiments are warranted to gain a better understanding on the practicality of this advanced technology for biofouling control in MBR systems. Additional studies on the *QQ* acylase activity of *SI*PA and *Au*AAC are currently in progress in order to assess their inhibitory capacity of biofilm formation by bacteria inhabiting WWTPs.

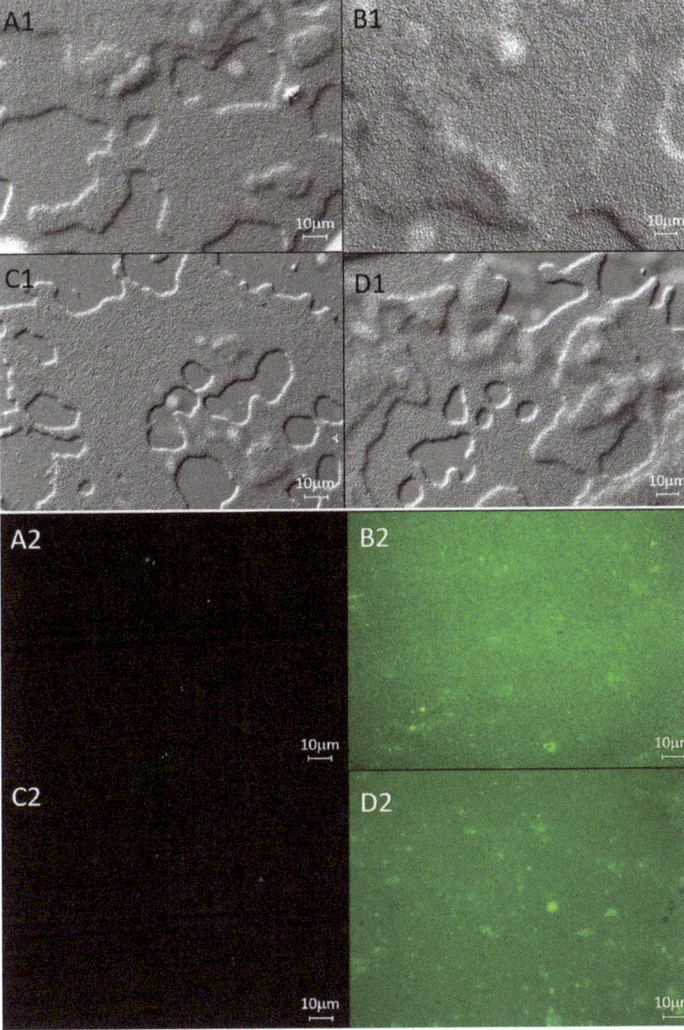

Figure 3. Disruption of *Pseudomonas aeruginosa* biofilm development on glass slides. Images acquired after 24 h in the following conditions: (**A1**): No *SI*PA addition. (**B1**): Heat-inactivated *SI*PA bound to silanized slides. (**C1**): *SI*PA addition in culture media. (**D1**): *SI*PA bound to silanized slides. *SI*PA attachment to slides was determined by using an Alexa Fluor 488 dye goat anti-rabbit whole antibody conjugate. (**A2**): No *SI*PA addition. (**B2**): Heat-inactivated *SI*PA bound to silanized slides. (**C2**): *SI*PA addition in culture media. (**D2**): *SI*PA bound to silanized slides. Microphotographs were obtained using a differential interference contrast (DIC) microscopy (**A1–D1**) at ×1000 total magnification, or an Olympus BX61 epifluorescence microscope (**A2–D2**) at ×400 total magnification.

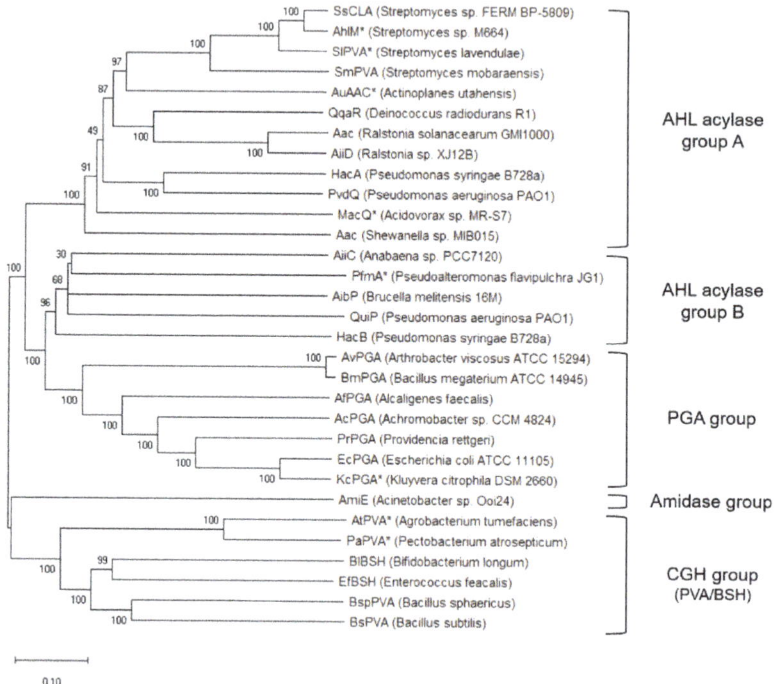

Figure 4. Molecular phylogenetic analysis of *Sl*PA and *Au*AAC in the context of the Ntn-hydrolase superfamily (accession numbers in the NCBI server). Ntn-hydrolases with reported AHL acylase activity: AiiD, from *Ralstonia* sp. XJ12B (AAO41113); PvdQ, from *Pseudomonas aeruginosa* PAO1 (AAG05773); AhlM, from *Streptomyces* sp. M664 (AAT68473); QuiP, from *P. aeruginosa* PAO1 (AAG04421); HacA, from *Pseudomonas syringae* B728a (AAY37014); HacB, from *P. syringae* B728a (AAY39885); AiiC, from *Anabaena* sp. strain PCC7120 (BAB75623); Aac, from *Shewanella* sp. strain MIB015 (BAF94155); QqaR, from *Deinococcus radiodurans* R1 (WP_010889514); MacQ, from *Acidovorax* sp. MR-S7 (BAV56778), PfmA, *Pseudoalteromonas flavipulchra* JG1 (ASS36259); AmiE, *Acinetobacter* sp. Ooi24 (BAP18758); AibP, from *Brucella melitensis* (AAL53453); Aac, from *Ralstonia solanacearum* GMI1000 (WP_011002462); *Sl*PA, from *Streptomyces lavendulae* ATCC 13664 (AAU09670), *Au*AAC, from *Actinoplanes utahensis* NRRL 12052 (WP_043523659). Ntn-hydrolases with reported penicillin G acylase activity: *Ec*PGA, from *Escherichia coli* ATCC 11105 (P06875); *Kc*PGA, from *Kluyvera citrophila* DSM 2660 (P07941); *Ac*PGA, from *Achromobacter* sp. strain CCM 4824 (AAY25991); *Bm*PGA, from *Bacillus megaterium* ATCC 14945 (Q60136); *Af*PGA, from *Alcaligenes faecalis* ATCC 19018 (AAB71221); *Av*PGA, from *Arthrobacter viscosus* ATCC 15294 (P31956); and *Pr*PGA, from *Providencia rettgeri* (AAP86197). Ntn-hydrolases with reported penicillin V acylase activity: *Sm*PVA, from *Streptomyces mobaraensis* (BAF51977); *At*PVA, from *Agrobacterium tumefaciens* (5J9R); *Pa*PVA, from *Pectobacterium atrosepticum* (4WL2); and *Bsp*PVA, from *Bacillus sphaericus* (3PVA). Ntn-hydrolases with reported bile salt hydrolase activity: *Bl*BSH, from *Bifidobacterium longum* (2HF0), and *Ef*BSH, from *Enterococcus feacalis* (4WL3). Ntn-hydrolases with reported cyclic lipopeptide acylase activity were as follows: SsCLA, from *Streptomyces* sp. FERM-BP5809 (BAD07025). Bifunctional acylases (with both penicillin acylase and AHL acylase activities) are indicated by asterisks, but among them only *Sl*PA and *Au*AAC present aculeacin A acylase activity as well. The evolutionary history was inferred using the neighbor-joining (NJ) method within the 3DM and MEGA X packages. The optimal tree with the sum of branch length = 8.78001801 is shown. This analysis involved 31 amino acid sequences. All ambiguous positions were removed for each sequence pair (pairwise deletion option). There was a total of 965 positions in the final dataset.

2.3. Phylogenetic Analysis of SlPA and AuAAC Acylases

To gain further knowledge on the role of the *Sl*PA and *Au*AAC through the analysis of their evolution, the amino acid sequences of *Sl*PA and *Au*AAC were compared in the context of the Ntn-hydrolases superfamily using 3DM a molecular class-specific information system built by multiple structure alignment and automated literature mining [45,46]. The 3DM system contained 12,567 sequences grouped into 13 subfamilies based on 179 available structures. The largest subfamily was "4HSTA" (glutaryl-7-ACA-alpha chain, containing 7155 sequences) (Figure S2). In contrast, both *Sl*PA and *Au*AAC were included in subfamily "5C9IA" (protein related to penicillin acylase, containing 955 sequences). The model protein of subfamily 5C9IA is the aforementioned MacQ, an AHL acylase from *Acidovorax* sp. MR-S7, active towards both AHLs and β-lactam antibiotics [47]. In addition, a phylogenetic study was performed using MEGA X based on the amino acid sequences of 31 Ntn-hydrolases with reported amidase activities (Figure 4). The results showed that *Sl*PA, AHL acylase from *Streptomyces* sp. M664 (AhlM) [31], cyclic lipopeptide acylase from *Streptomyces* sp. FERM BP-5809 (SsCLA) [48] and penicillin V acylase from *Streptomyces mobaraensis* (SmPVA) [49] are very close together in the phylogenetic tree, and these three enzymes are also near *Au*AAC. In this sense, there is moderate sequence identity of *Au*AAC with *Sm*PVA (42.1%), *Sl*PA (42.3%), AhlM (42.7%) and SsCLA (42.9%), whereas *Sl*PA showed high identity with *Sm*PVA (66.3%), SsCLA (85.7%) and AhlM (86.5%), In contrast, both *Sl*PA and *Au*AAC are very distant from those enzymes belonging to the group of penicillin G acylases (PGA group) and penicillin V acylases (PVA group) (Figure 4). In fact, our phylogenetic tree is quite similar to the one described recently by Kusada et al. who reported that Ntn-hydrolase family could be divided into three main groups: a β-lactam acylase group (that would correspond to our PGA group) and two AHL acylase groups (A and B) [34]. Our results are in agreement with this affirmation, and furthermore we have identified new members within each group (Figure 4). In this sense, HacB [38], QuiP [37], AibP [50], PfmA [51] and AiiC [39] would belong to the AHL acylase group B, whereas Aac from *Shewanella* sp. [32], MacQ [34], PvdQ [52], HacA [38], AiiD [35], Aac from *Ralstonia solanacearum* [53], QqaR [40], *Sm*PVA [49], AhlM [31], SsCLA [48], *Sl*PA and *Au*AAC would belong to the AHL acylase group A.

In addition, we have identified two new groups in the phylogenetic tree: an amidase group that includes the AHL acylase from *Acinetobacter* sp. Ooi24 [33], and a cholylglycine hydrolase (CGH) group that includes PVAs from different bacteria such as *A. tumefaciens* (AtPVA), *P. atrosepticum* (PaPVA) [20], *Bacillus sphaericus* (BspPVA) [54] and *Bacillus subtilis* (BsPVA) [55]. Furthermore, members of this PVA group differ in their catalytic N-terminal nucleophile residue (cysteine) and subunit composition (homotetramers) compared to some members of the AHL acylase group A that displayed PVA activity (such as *Sm*PVA, *Sl*PA, and *Au*AAC) which are heterodimers with a catalytic serine at the N-terminal end of their β-subunit. Moreover, genuine PVAs were proposed to be evolutionary related to bile salt hydrolases (BSHs) [56] like BSH from *Bifidobacterium longum* [57] and BSH from *Enterococcus faecalis* [58], forming altogether part of the CGH group of enzymes, and this relationship was confirmed in our phylogenetic tree., Some of the enzymes throughout the phylogenetic tree have been reported to display both penicillin acylase and AHL acylase activities (Figure 4, see enzymes with asterisks), although they belong to different groups. In this sense, Kusada *et al.* [34] had already suggested that those enzymes with bifunctional QQ and antibiotic-acylase activities might be broadly distributed among the phylogeny, and therefore such bifunctionality could be conserved in other acylases of a wide type of microorganisms. Such assumption has been recently confirmed with new reported enzymes such as PfmA (AHL acylase from *Pseudoalteromonas flavipulchra*, able to degrade ampicillin but not penicillin G) [51], and *At*PVA and *Pa*PVA that are able to hydrolyze both AHLs and penicillin V [20]. Our study demonstrates that two other enzymes from Gram-positive bacteria, such as *Sl*PA and *Au*AAC, also present this feature, showing not only penicillin acylase and AHL acylase activities, but also aculeacin A acylase activity [22,24].

Taking into account these results, it should be proved whether other members of the phylogenetic tree (apparently those belonging to the AHL acylase group A) could be able to recognize aliphatic penicillins and/or aculeacin A as substrates. In this sense, some reports might support such hypothesis; for instance, PVA from *S. mobaraensis* is able to catalyze the synthesis of aliphatic penicillins and N-fatty-acylated amino compounds [59], whereas cyclic lipopeptide acylase from *Streptomyces* sp. FERM BP-5809 can deacylate aculeacin A and echinocandin B [48]. On the contrary, Aac from *R. solanacearum* is not able to degrade ampicillin and aculeacin A, although this enzyme was initially identified as a probable aculeacin A acylase transmembrane protein [53].

Furthermore, the combination of structure-guided multiple alignment and literature mining by 3DM enabled confirmation of previously determined residue functions and identification of new potential ones. Using the correlated mutations analysis provided by 3DM, we analyzed the 100 most similar sequences to *Sl*PA and identified clusters of residues which have mutated in a coordinated manner within the subset, pointing to their potential involvement in substrate specificity. The heatmap of mutational correlations generated by 3DM (Figure S3) predicted two of residues at the substrate-binding pocket of *Sl*PA (3DM residue numbers 139 and 148 in Table S1). Interestingly, such residues (identified as Tyrβ24 and Trpβ33, respectively) had been previously suggested to be positioned at the substrate-binding pocket of *Sl*PA [22]. The other seven residues that showed strong mutational correlations (Table S1: 3DM residue numbers 5, 27, 239, 257, 310, 351 and 595 that correspond in *Sl*PA to Tyrα10, Trpα32, Trpβ120, Proβ150, Glnβ203, Proβ266 and Argβ544 respectively) have never been proposed to be located in the substrate-binding pocket of *Sl*PA and may be putatively involved in substrate specificity according to 3DM. Sequence alignment by Clustal Omega allowed to identify the same residues at the equivalent positions in penicillin V acylase from *S. mobaraensis*, AHL acylase from *Streptomyces* sp. M664, and cyclic lipopeptide acylase from *Streptomyces* sp FERM BP-5809 (Figure S4). However, different residues were identified at such positions for other members of the AHL acylase group A, and this result could explain their substrate preference towards AHLs which may differ depending on their acyl-chain length and the presence of the 3-oxo substitution, and their ability to recognize aculeacin A and other echinocandins as substrates. These putative residues might be considered as potential targets for site-directed mutagenesis of this group of enzymes in order to improve their substrate specificity towards different AHLs. Certainly, that would be very useful for some enzymes that show weak activity towards shorter AHLs, such as C_6-HSL and C_4-HSL, QS molecules produced by *Burkholderia* and *Yersinia* [36]. Similarly, mutant variants of these enzymes could be designed with the help of this information and docking simulations in order to more efficient aculeacin A acylases, an approach that has already allowed to prepare mutant cephalosporin acylases that recognize aculeacin A as substrate [60]. Indeed, a similar approach has enabled the obtention of a PvdQ variant with increased C_8-HSL activity that could reduce virulence of the human pathogen *Burkholderia cenocepacia* [61].

In addition, sequence alignment of members of the AHL acylase group A confirmed the presence of a stretch of hydrophobic amino acids encoding a signal peptide, and the conserved glycine, serine, asparagine, histidine, tyrosine and valine residues (Figure S4) that have been demonstrated to be of importance to both autoproteolytic processing and catalysis in Ntn-hydrolases [62–65]. In this sense, Ntn-hydrolases undergo a post-translational processing resulting in a primary pro-peptide that is transformed into an active two-subunit form after the cleavage of signal and spacer peptides [17,63,64,66,67]. Essential to this post-translational modification of the pro-peptide, a conserved glycine–serine pair was clearly located in all members of group A of AHL acylases. In this sense, enzymatic activities of both *Sl*PA and *Au*AAC were abolished when Serβ1 was replaced with cysteine, aspartic acid, histidine or lysine by site-directed mutagenesis [22]. In addition, we have also observed in our enzymes the presence of those reported residues that explained the unusual increased size of the hydrophobic pocket in the crystal structure of PvdQ (Leuα146, Asnβ57, Trpβ186) [52]. These conserved residues would correspond to Valα141, Asnβ61 and Trpβ185 for *Sl*PA, and Leuα136, Asnβ61 and Trpβ181 for *Au*AAC, respectively. All these observations suggest a similar build-up of the substrate-binding site in all members of the AHL acylase group A that would explain their unique substrate preference for long acyl-chains. Nevertheless, 3DM has allowed for the identification of several putative residues (Figure S4) that might explain the reason why some enzymes of this particular group are able to recognize not only long aliphatic acyl chains present in some AHLs and penicillins, but the palmitoyl chain of aculeacin A (such as *Sl*PA and *Au*AAC) or the linoleoyl chain of echinocandin B (such as *Ss*CLA), in comparison to other enzymes that show different amino acid residues in the equivalent positions and cannot deacylate echinocandins (such as Aac from *R. solanacearum*). Additional studies would allow one to determine which residues are actually involved in substrate recognition of different echinocandins, paving the way towards the design of new biocatalysts for the enzymatic production of antifungal compounds.

3. Materials and Methods

3.1. Materials and Bacterial Strains

Unless otherwise specified, all chemical reagents were purchased from Sigma-Aldrich (St. Louis, MO, USA). Cell culture media were from Difco (Detroit, MI, USA). *Streptomyces lividans* CECT 3376 and CECT 3377, overexpressing *Sl*PA and *Au*AAC, respectively, and *Chromobacterium violaceum* CV026 (CECT 5999), were obtained from the Spanish Cell Culture Collection (Valencia, Spain). For sporulation, *S. lividans* CECT 3376 and CECT 3377 were cultured on agar plates containing SFM (Mannitol Soya Flour) medium at 30 °C for 72–96 h. For enzyme production, recombinant *S. lividans* cells were cultured aerobically under submerged conditions in Triptone Soy Broth (TSB) liquid medium at 30 °C and 250 rpm [68]. *C. violaceum* CV026 cells (CECT 5999) were grown in Luria-Bertani (LB) agar plates at 37 °C for bioassays.

3.2. Synthesis of N-Acyl-Homoserine Lactones

Pure L-enantiomers of C_4-HSL, C_6-HSL, C_8-HSL, C_{10}-HSL, C_{12}-HSL and C_{14}-HSL were obtained by chemical synthesis (Thomas *et al.*, 2005), with several modifications: 10 mmol of L-HSL hydrochloride were dissolved in 25 mL of dimethylformamide (DMF), and then 23 mmol of ice-cold triethylamine were added. Later, 14 mmol of corresponding acid chloride was added dropwise with agitation at 4 °C. The mixture was incubated for 2 h at room temperature and the solvent was evaporated. The product was dissolved in dichloromethane and washed first with 1 M Na_2SO_4 and then with a saturated solution of NaCl. Water was eliminated from organic phase by addition of anhydrous $MgSO_4$ and the organic solvent was evaporated. Isolation and purification of the synthesized AHL was carried out by the combination of thin layer chromatography and a silicagel 60 column chromatography (Merck). The elution was performed with an *n*-hexane-EtOAc gradient system. Compounds were visualized in thin layer chromatography by spraying an aqueous solution of potassium permanganate (1% *w/v* $KMnO_4$, 6.67% *w/v* Na_2CO_3 and 0.083% *w/v* NaOH) briefly heating. Structures were elucidated by ^{1}H and ^{13}C NMR analysis.

3.3. Determination of Kinetic Parameters

The substrate specificity of *Au*AAC and *Sl*PA, which were expressed and purified as previously described [22,24], was studied using a wide panel of L-enantiopure AHLs including 3-oxo substituted AHLs: *N*-butyryl-L-homoserine lactone (C_4-HSL), *N*-hexanoyl-L-homoserine lactone (C_6-HSL), *N*-octanoyl-L-homoserine lactone (C_8-HSL), *N*-decanoyl-L-homoserine lactone (C_{10}-HSL), *N*-dodecanoyl-L-homoserine lactone (C_{12}-HSL), *N*-tetradecanoyl-L-homoserine lactone (C_{14}-HSL), *N*-(β-ketocaproyl)-L-homoserine lactone (oxo-C_6-HSL), *N*-(3-oxooctanoyl)-L-homoserine lactone (oxo-C_8-HSL), *N*-(3-oxodecanoyl)-L-homoserine lactone (oxo-C_{10}-HSL), *N*-(3-oxododecanoyl)-L-homoserine lactone (oxo-C_{12}-HSL) and *N*-(3-oxotetradecanoyl)-L-homoserine lactone (oxo-C_{14}-HSL). The methodology employed to detect primary amines released over the course of reactions was similar to that reported in literature [19,41,69]. Kinetic studies were performed at pH 8.0 and 45 °C, and all reactions were carried out in triplicate in 100 μL of final reaction mixture (10 μL of enzymatic solution, 70 μL of 1 M phosphate buffer and 20 μL of substrate dissolved in dimethyl sulfoxide, abbreviated as

DMSO). In order to minimize non-enzymatic conversions, mixtures were kept on ice before initiation and after termination of the reaction. The extent of reactions was then quantified by development with *o*-phthalaldehyde (OPA). In mild alkaline conditions, OPA is stable and reacts readily with primary amines above their isoelectric point in the presence of β-mercaptoethanol to form intensely fluorescent derivatives [70]. Thus, each reaction mixture was mixed directly with 100 μL of OPA solution (Sigma-Aldrich), which stops the reaction and allows quantification of the acylase activity. The resulting mixture was then incubated for 2 min at 25 °C to ensure signal development before reading fluorescence intensity using a FLUOstar Omega (BMG Labtech). Readouts were taken at 25 °C with 355 nm and 460 nm as excitation and emission wavelengths, respectively, with positioning delay of 0.2 s and 5 flashes per well. This signal, which is proportional to the amount of HSL released during the reaction, was interpolated in a calibration curve of pure HSL to enable expression of reaction rates in non-arbitrary units. The range of substrate concentrations was different for each AHL: from 1 to 25 mM for C_4-HSL, from 0.02 to 10.04 mM for C_6-HSL, from 0.02 to 5.02 mM for C_8-HSL, from 0.002 to 2.040 mM for C_{10}-HSL, from 0.002 to 1.000 mM for C_{12}-HSL and from 0.002 to 0.500 mM for C_{14}-HSL; from 0.02 to 1.64 mM for oxo-C_6-HSL, from 0.02 to 8.00 mM for oxo-C_8-HSL, from 0.02 to 1.45 mM for oxo-C_{10}-HSL, from 0.006 to 0.726 mM for oxo-C_{12}-HSL and from 0.002 to 0.080 mM for oxo-C_8-HSL. All reactions were catalyzed by 0.19 and 0.23 μg of *Sl*PA and *Au*AAC, respectively, as these amounts were deemed enough for the hydrolysis of both aliphatic and 3-oxo substituted AHLs. Kinetic parameters were determined by nonlinear regression using Hyper32 program (available on http://homepage.ntlworld.com/john.easterby/hyper32.html). All experiments were performed in triplicate.

3.4. Inhibition of Violacein Production by Chromobacterium Violaceum CV026

C. violaceum, a Gram-negative bacterium commonly found in soil and water, produces a characteristic purple pigment called violacein in response to an AHL-mediated *QS* mechanism [71]. Since *C. violaceum* CV026 is unable to produce C_6-HSL, this mutant strain has been traditionally considered an excellent AHL-biosensor taking into account that violacein synthesis may be induced by AHLs with acyl chains from C_4 to C_8 ("forward bioassay"). Although pigment production is not induced if acyl-chains are longer (from C_{10} to C_{14}), these AHLs can antagonize the common inducers of violacein production ("reverse bioassay") [71]. For the forward assay, *C. violaceum* CV026 was cultured overnight in LB medium, and then 50 μL of culture were used to inoculate the surface of LB agar plates prepared in Petri dishes. Then, 10 μL of reaction mixture containing the enzyme (approximately 1 μg) and the AHLs (20 mM C_6-HSL or 5 mM C_8-HSL) in buffer A (potassium phosphate buffer 0.1 M, pH 8.0 with 40% DMSO) was placed in the center of the plate (forward in vivo assay). Alternatively, said mixture could be previously incubated for 24 h at 40 °C or 45 °C depending on the enzyme (*Sl*PA or *Au*AAC, respectively), and then placed in the center of the plate (forward *in vitro* assay). Positive (C_6-HSL or C_8-HSL in buffer A) and negative (buffer A) controls were included in each assay plate. All plates were incubated in the upright position overnight at 30 °C and then examined for the stimulation of violacein synthesis, as indicated by blue/purple pigmentation of the bacterial lawn. The reverse assays for antagonists were carried out identically, except for the fact that a stimulator (5 μM C_6-HSL or oxo-C_6-HSL) was added to soft agar together with the CV026 strain. In this case, inhibition of violacein synthesis was reported by the presence of white haloes in a purple background. Forward and reverse assays were optimized by checking the detection limits for each AHL. In this sense, serial dilutions of each AHL were tested and detection limits were defined as the lowest quantity of AHL that produced a visible activation or inhibition of violacein synthesis.

The QS interference by AHL degradation of recombinant *S. lividans* strains was tested qualitatively using CV026 strain. Firstly, wild type and recombinant strains of *S. lividans* were grown in agar plates containing TSB broth [68]. Tioestreptone (5 μg/mL) was added to the plates used for recombinant *S. lividans*. Wild type *S. lavendulae* was grown in agar plates containing SYCC broth [68]. Wild type *A. utahensis* was grown in agar plates containing sucrose (3.0 g/L), soy peptone (0.5 g/L), K_2HPO_4

(1.0 g/L), KCl (0.5 g/L), MgSO$_4$·7H$_2$O (0.5 g/L), FeSO$_4$·7H$_2$O (0.002 g/L) pH 6.5 [23]. All actinomycetes culture plates were incubated for 3–4 days at 30 °C. Afterwards, 1 μL of 1 mM C$_8$-HSL solution was added in the growth borderline and then a 15% soft LB agar inoculated with CV026 was layered on top of actinomycetes cultures. After solidification, plates were incubated for 24 h at 30 °C. QS interference activity was observed through violacein inhibition.

3.5. Inhibition of Biofilm Formation by Pseudomonas Aeruginosa on Glass Slides

SIPA immobilization was achieved by using 3-APTS ((3-aminopropil)-triethoxysilane) and glutaraldehyde. In this study, the method was followed as described elsewhere [72]. *Glass slides preparation.* Lab-Tek slides (Lab-Tek II Chamber Slide system, Thermo Fisher) were washed by stirring in methanol (1 mL/well) to remove any organic contaminant, washed 5 times with distilled water and then 2 mL 5 N NaOH was added to the cleaned slides, rinsed 5 times with excess water until NaOH was removed and the water reached pH 7.0. Slides were incubated with 1 mL of freshly prepared 3-APTS 10% (*v/v*) in water during 2 h and then washed thoroughly with abundant water to remove 3-APTS molecules not linked to the surface of support. The following steps were done: Soak derivatized slides in freshly prepared 10% (*v/v*) glutaraldehyde in deionized water during 4 h at 25 °C. Rinse 5 times with distilled water to remove any adsorbed cross-linker. Dry the activated slides in the air. The free terminal aldehyde groups must be subsequently cross-linked to amines groups on the enzyme surface through Schiff's base formation by incubating support in the enzyme solutions. *Enzyme immobilization.* Treated slides were used for immobilization of 0.36 IU/mL *SIPA* (1 mL/well at room temperature/overnight) in phosphate buffer 0.1 M, pH 7.0. One international activity unit (IU) was defined as the amount of enzyme producing 1 μmol/min of 6-APA using penicillin V as substrate under the assay conditions described elsewhere [28]. Control slides were filled with 1 mL phosphate buffer 0.01 M, pH 7.0 overnight at room temperature. After incubation, slides were rinsed 5 times with 3 mL phosphate buffer 0.1 M, pH 7.0 per well and incubated in 2 mL/well 100 mM glycine dissolved in 0.1 M phosphate buffer, pH 7.0 for 30 min to block any unreacted aldehyde groups during 30 min. Slides were soaked 5 times with 3 mL/well 0.1 M phosphate buffer, pH 7.0. The resulting immobilized enzymes can be held at 4 °C prior to use. The amount of immobilized vs. non immobilized enzymes on slides was determined by measuring penicillin V acylase activity [28]. Only a 33% of the initial acylase activity was recovered. Chambered slides were used for *P. aeruginosa* biofilm development by using Luria Bertani broth (LB) as culture media at 28 °C on a rotary bed shaker (100 rpm). Chambered slides were prepared in four different ways as described before: non-treated; treated with *SIPA* bound to silanized slides; treated with heat inactivated *SIPA* (70 °C, 10 min) bound to silanized slides; and *SIPA* incorporated to LB medium and added to untreated slides. *P. aeruginosa* was spotted on LB agar and grown overnight at 37 °C. Then, a cellular suspension of 0.1 O.D. (λ_{500nm}) was prepared and inoculated (100 μL/chamber) containing 2 mL LB. Chambered slides with fitted lids were incubated at 28 °C for 24 h in a closed, humidified container. The slides were then carefully washed to remove planktonic organisms, the chamber was removed and the biofilm development was microscopically followed by DIC microscopy (Nikon Eclipse 80i microscope with a Nikon Digital Sight camera). *Evaluation of enzyme immobilization by epifluorescence microscopy.* This procedure was carried out according to Hormigo et al. [28] with slight variations. Slide-immobilized enzyme was incubated with 5 mL of phosphate-buffered saline (PBS) (8 g of NaCl, 0.2 g of KCl, 0.2 g of KH$_2$PO$_4$ and 1.41 g of Na$_2$HPO$_4$ /2H$_2$O in 1 L of water, pH 7.4) containing 1% (*w/v*) BSA for 30 min. Then, the glass slide was washed three times with 5 mL of PBS containing 0.1% (*w/v*) BSA and incubated for 2 h at 37 °C with 5 mL of an antibody to *SIPA* solution prepared in PBS with 0.1% (*w/v*) BSA. Next, the biocatalysts were washed again with 5 mL of PBS containing 0.1% (*w/v*) BSA, and incubated for 2 h at 25 °C with 5 mL of Alexa Fluor 488 dye goat anti-rabbit whole antibody conjugate prepared in PBS with 0.1% (*w/v*) BSA at a final concentration of 5 mg antibody conjugate/mL. Negative control experiments were carried out to check for non-specific binding of the secondary antibody to the support. Immobilized derivatives were washed three times with 5 mL of PBS containing 0.01% (*w/v*) BSA for 10 min, followed by three washings with 5 mL of PBS for 10 min and finally three washings with 5 mL of deionized water for 5 min. The degree of *Sl*PA immobilized on glass slides was then analyzed using an Olympus BX61 epifluorescence microscope at ×400 total magnification. Activity was also determined in silanized glass chambers.

3.6. Protein Sequence Alignments

The evolutionary history was inferred using the neighbor-joining method [73]. The percentage of replicate trees in which the associated taxa clustered together in the bootstrap test (1000 replicates) was shown next to the branches [74]. The tree was drawn to scale, with branch lengths in the same units as those of the evolutionary distances used to infer the phylogenetic tree. The evolutionary distances were computed using the *p*-distance method [75] and were in the units of the number of amino acid differences per site. All ambiguous positions were removed for each sequence pair (pairwise deletion option). Evolutionary analyses were conducted in 3DM [45,46] and MEGA X [76], whereas multiple sequence alignment of several enzymes was carried out with Clustal Omega. Optimal global alignment of two sequences were performed using the Needleman–Wunsch algorithm (within Clustal Omega) in order to calculate sequence identity.

4. Conclusions

We have demonstrated that reported enzymes, penicillin acylase from *Streptomyces lavendulae* and aculeacin A acylase from *Actinoplanes utahensis*, are an interesting extension of hydrolytic (Ntn) enzymes, with potential for biocatalytic applications. They hydrolyze aliphatic penicillins and the antifungal aculeacin A, and are also able to efficiently hydrolyze the amide bonds of several N-acyl-homoserine lactones (AHLs), quorum sensing molecules from Gram-negative bacteria. Furthermore, both enzymes inhibit the production of violacein by *Chromobacterium violaceum* CV026, and the formation of biofilms by *Pseudomonas aeruginosa*. In addition, the comparative sequence analysis has revealed high identities between both enzymes and AHL quorum quenching acylases.

These results indicate that penicillin acylase from *Streptomyces lavendulae* and aculeacin A acylase from *Actinoplanes utahensis* are involved in QQ processes and both of them could be used for biofouling control in MBR systems and in antimicrobial therapy to prevent colonization of biological surfaces by pathogenic Gram-negative bacteria.

Finally, these enzymes could be considered as versatile biocatalysts, able to hydrolyze the amide bond between an aliphatic acyl side chain and a nucleus containing an amino group, which are present in many substrates (*e.g.*, aliphatic penicillins, aculeacin A and N-acyl-homoserine lactones), in addition to catalyzing the acylation to obtain new semi-synthetic β-lactam antibiotic and echinocandins antifungals. Moreover, the phylogenetic study suggests that *Sl*PA and *Au*AAC could be part of a new family of actinomycete acylases, with a preference towards substrates with long chain aliphatic acyl groups that are involved in QQ processes.

Supplementary Materials: The following are available online at http://www.mdpi.com/2073-4344/10/7/730/s1. Figure S1: Hyperbolic regression of the activity at different concentrations of several AHLs. Figure S2: Structure-guided phylogenetic analysis of the Ntn-hydrolases superfamily as analyzed from 3DM database. Figure S3: Correlated mutations matrix for the 5C9IA subfamily. Figure S4: Sequence alignment of SlPA and AuAAC with other homologues from the AHL acylase group A. Table S1: Amino acids residues at different positions selected in *Sl*PA according to the structure-based alignment from the 3DM database.

Author Contributions: Conceptualization, J.L.G., C.A., A.S., M.A. and I.d.l.M.; methodology, R.V.-B., D.H., J.F.-L., P.T.-A., Y.A.-U., L.S.-A., A.S. and M.A.; software: R.V.-B., D.H., F.R., M.A. and I.d.l.M.; formal analysis R.V.-B., D.H., J.F.-L., P.T.-A., Y.A.-U., A.I.S., A.S., M.A. and I.d.l.M.; investigation, R.V.-B., D.H., J.F.-L., P.T.-A., Y.A.-U., L.S.-A., A.S. and M.A.; data curation R.V.-B., D.H., A.S. and I.d.l.M.; writing—original draft preparation R.V.-B., F.R., A.S., M.A. and I.d.l.M.; writing—review and editing A.I.S., J.L.G., C.A., M.A. and I.d.l.M.; funding acquisition: C.A. and I.d.l.M.; supervision: I.d.l.M. All authors have read and agreed to the published version of the manuscript.

Funding: This research was funded by Ministry of Education and Science, Ministry of Science and Innovation of Spain and Comunidad Autónoma de Madrid, grant numbers BIO2008-03928, DEX-580000-2008-31 and S2009/PPQ-1752, respectively; and CTQ2014-60250-R and CTM2016-76491-P projects from Ministry of Economy, Industry and Competitiveness of Spain as well.

Acknowledgments: This article is dedicated to our long-term mentor and collaborator, Carmen Acebal on the occasion of her retirement. We would also like to express our gratitude to both Maria Pilar Castillón and Carmen Acebal for their unconditional support and friendship over the many years dedicated to the Enzyme Biotechnology Group of Universidad Complutense since its foundation.

Conflicts of Interest: The authors declare no conflict of interest. The funders had no role in the design of the study; in the collection, analyses, or interpretation of data; in the writing of the manuscript, or in the decision to publish the results.

References

1. de Kievit, T.R.; Iglewski, B.H. Bacterial quorum sensing in pathogenic relationships. *Infect. Immun.* **2000**, *68*, 4839–4849. [CrossRef] [PubMed]

2. Miller, M.B.; Bassler, B.L. Quorum sensing in bacteria. *Annu. Rev. Microbiol.* **2001**, *55*, 165–199. [CrossRef] [PubMed]

3. Chen, F.; Gao, Y.X.; Chen, X.Y.; Yu, Z.M.; Li, X.Z. Quorum quenching enzymes and their application in degrading signal molecules to block *quorum sensing* dependent infection. *Int. J. Mol. Sci.* **2013**, *14*, 17477–17500. [CrossRef]

4. Whitehead, N.A.; Welch, M.; Salmond, G.P.C. Silencing the majority. *Nat. Biotechnol.* **2001**, *19*, 735–736. [CrossRef] [PubMed]

5. Soler, A.; Arregui, L.; Arroyo, M.; Mendoza, J.A.; Muras, A.; Alvarez, C.; Garcia-Vera, C.; Marquina, D.; Santos, A.; Serrano, S. Quorum sensing *versus* quenching bacterial isolates obtained from mbr plants treating leachates from municipal solid waste. *Int. J. Environ. Res. Public Health* **2018**, *15*, 1019. [CrossRef]

6. Dickschat, J.S. Quorum sensing and bacterial biofilms. *Nat. Prod. Rep.* **2010**, *27*, 343–369. [CrossRef]

7. Whitehead, N.A.; Barnard, A.M.L.; Slater, H.; Simpson, N.J.L.; Salmond, G.P.C. Quorum sensing in Gram-negative bacteria. *FEMS Microbiol. Rev.* **2001**, *25*, 365–404. [CrossRef] [PubMed]

8. Williams, P. Quorum sensing, communication and cross-kingdom signalling in the bacterial world. *Microbiology* **2007**, *153*, 3923–3938. [CrossRef]

9. Grandclément, C.; Tannières, M.; Moréra, S.; Dessaux, Y.; Faure, D. Quorum quenching: Role in nature and applied developments. *FEMS Microbiol. Rev.* **2016**, *40*, 86–116. [CrossRef]

10. Fetzner, S. Quorum quenching enzymes. *J. Biotechnol.* **2015**, *201*, 2–14. [CrossRef]

11. Czajkowski, R.; Jafra, S. Quenching of acyl-homoserine lactone-dependent quorum sensing by enzymatic disruption of signal molecules. *Acta Biochim. Pol.* **2009**, *56*, 1–16. [CrossRef]

12. Yates, E.A.; Philipp, B.; Buckley, C.; Atkinson, S.; Chhabra, S.R.; Sockett, R.E.; Goldner, M.; Dessaux, Y.; Cámara, M.; Smith, H.; et al. N-acylhomoserine lactones undergo lactonolysis in a pH-,temperature-, and acyl chain length-dependent manner during growth of Yersinia pseudotuberculosis and Pseudomonas aeruginosa. *Infect. Immun.* **2002**, *70*, 5635–5646. [CrossRef]

13. Hibbing, M.E.; Fuqua, C.; Parsek, M.R.; Peterson, S.B. Bacterial competition: Surviving and thriving in the microbial jungle. *Nat. Rev. Microbiol.* **2010**, *8*, 15–25. [CrossRef] [PubMed]

14. Kusada, H.; Zhang, Y.; Tamaki, H.; Kimura, N.; Kamagata, Y. Novel N-acyl Homoserine lactone-degrading bacteria isolated from penicillin-contaminated environments and their quorum-quenching activities. *Front. Microbiol.* **2019**, *10*, 455. [CrossRef] [PubMed]

15. Bzdrenga, J.; Daude, D.; Remy, B.; Jacquet, P.; Plener, L.; Elias, M.; Chabriere, E. Biotechnological applications of quorum quenching enzymes. *Chem. Biol. Interact.* **2017**, *267*, 104–115. [CrossRef] [PubMed]

16. Hong, K.W.; Koh, C.L.; Sam, C.K.; Yin, W.F.; Chan, K.G. Quorum quenching revisited—From signal decays to signalling confusion. *Sensors* **2012**, *12*, 4661–4696. [CrossRef]

17. Oinonen, C.; Rouvinen, J. Structural comparison of Ntn-hydrolases. *Protein. Sci.* **2000**, *9*, 2329–2337. [CrossRef]

18. Utari, P.D.; Vogel, J.; Quax, W.J. Deciphering physiological functions of AHL quorum quenching acylases. *Front. Microbiol.* **2017**, *8*, 1123. [CrossRef]

19. Mukherji, R.; Varshney, N.K.; Panigrahi, P.; Suresh, C.G.; Prabhune, A. A new role for penicillin acylases: Degradation of acyl homoserine lactone quorum sensing signals by *Kluyvera citrophila* penicillin G acylase. *Enzym. Microb. Technol.* **2014**, *56*, 1–7. [CrossRef]

20. Sunder, A.V.; Utari, P.D.; Ramasamy, S.; van Merkerk, R.; Quax, W.; Pundle, A. Penicillin V acylases from gram-negative bacteria degrade N-acylhomoserine lactones and attenuate virulence in Pseudomonas aeruginosa. *Appl. Microbiol. Biotechnol.* **2017**, *101*, 2383–2395. [CrossRef] [PubMed]

21. Torres-Guzmán, R.; de la Mata, I.; Torres-Bacete, J.; Arroyo, M.; Castillón, M.P.; Acebal, C. Substrate specificity of penicillin acylase from Streptomyces lavendulae. *Biochem. Biophys. Res. Commun.* **2002**, *291*, 593–597. [CrossRef] [PubMed]

22. Torres-Bacete, J.; Hormigo, D.; Torres-Gúzman, R.; Arroyo, M.; Castillón, M.P.; García, J.L.; Acebal, C.; de la Mata, I. Overexpression of penicillin V acylase from Streptomyces lavendulae and elucidation of its catalytic residues. *Appl. Environ. Microbiol.* **2015**, *81*, 1225–1233. [CrossRef] [PubMed]

23. Takeshima, H.; Inokoshi, J.; Takada, Y.; Tanaka, H.; Omura, S. A deacylation enzyme for aculeacin A, a neutral lipopeptide antibiotic; from Actinoplanes utahensis: Purification and characterization. *J. Biochem.* **1989**, *105*, 606–610. [CrossRef]

24. Torres-Bacete, J.; Hormigo, D.; Stuart, M.; Arroyo, M.; Torres, P.; Castillón, M.P.; Acebal, C.; García, J.L.; de la Mata, I. Newly discovered penicillin acylase activity of aculeacin A acylase from Actinoplanes utahensis. *Appl. Environ. Microbiol.* **2007**, *73*, 5378–5381. [CrossRef] [PubMed]

25. Arroyo, M.; Torres, R.; de la Mata, I.; Castillón, M.P.; Acebal, C. Interaction of penicillin V acylase with organic solvents: Catalytic activity modulation on the hydrolysis of penicillin V. *Enzym. Microb. Technol.* **1999**, *25*, 378–383. [CrossRef]

26. Arroyo, M.; Torres-Guzman, R.; de la Mata, I.; Castillon, M.P.; Acebal, C. A kinetic examination of penicillin acylase stability in water-organic solvent systems at different temperatures. *Biocatal. Biotransform.* **2002**, *20*, 53–56. [CrossRef]

27. Arroyo, M.; Torres-Guzman, R.; de la Mata, I.; Castillon, M.P.; Acebal, C. Prediction of penicillin V acylase stability in water-organic co-solvent monophasic systems as a function of solvent composition. *Enzym. Microb. Technol.* **2000**, *27*, 122–126. [CrossRef]

28. Hormigo, D.; de la Mata, I.; Castillón, M.P.; Acebal, C.; Arroyo, M. Kinetic and microstructural characterization of immobilized penicillin acylase from Streptomyces lavendulae on Sepabeads EC-EP. *Biocatal Biotransform.* **2009**, *27*, 271–281. [CrossRef]

29. Hormigo, D.; de la Mata, I.; Acebal, C.; Arroyo, M. Immobilized aculeacin A acylase from Actinoplanes utahensis: Characterization of a novel biocatalyst. *Bioresour. Technol.* **2010**, *101*, 4261–4268. [CrossRef]

30. Hormigo, D.; López-Conejo, M.T.; Serrano-Aguirre, L.; García-Martín, A.; Saborido, A.; de la Mata, I.; Arroyo, M. Kinetically controlled acylation of 6-APA catalyzed by penicillin acylase from Streptomyces lavendulae: Effect of reaction conditions in the enzymatic synthesis of penicillin V. *Biocatal. Biotransform.* **2020**, *38*, 253–262. [CrossRef]

31. Park, S.Y.; Kang, H.O.; Jang, H.S.; Lee, J.K.; Koo, B.T.; Yum, D.Y. Identification of extracellular N-acylhomoserine lactone acylase from a *Streptomyces* sp. and its application to quorum quenching. *Appl. Environ. Microbiol.* **2005**, *71*, 2632–2641. [CrossRef]

32. Morohoshi, T.; Nakazawa, S.; Ebata, A.; Kato, N.; Ikeda, T. Identification and characterization of N-acylhomoserine lactone-acylase from the fish intestinal Shewanella sp. strain MIB015. *Biosci. Biotechnol. Biochem.* **2008**, *72*, 1887–1893. [CrossRef] [PubMed]

33. Ochiai, S.; Yasumoto, S.; Morohoshi, T.; Ikeda, T. AmiE, a novel N-acylhomoserine lactone acylase belonging to the amidase family, from the activated-sludge isolate Acinetobacter sp. strain Ooi24. *Appl. Environ. Microbiol.* **2014**, *80*, 6919–6925. [CrossRef]

34. Kusada, H.; Tamaki, H.; Kamagata, Y.; Hanada, S.; Kimura, N. A novel quorum quenching N-acylhomoserine lactone acylase from *Acidovorax* sp Strain MR-S7 mediates antibiotic resistance. *Appl. Environ. Microbiol.* **2017**, *83*, e00080–e00117. [CrossRef]

35. Lin, Y.H.; Xu, J.L.; Hu, J.; Wang, L.H.; Ong, S.L.; Leadbetter, J.R.; Zhang, L.H. Acyl-homoserine lactone acylase from Ralstonia strain XJ12B represents a novel and potent class of quorum quenching enzymes. *Mol. Microbiol.* **2003**, *47*, 849–860. [CrossRef] [PubMed]

36. Sio, C.F.; Otten, L.G.; Cool, R.H.; Diggle, S.P.; Braun, P.G.; Bos, R.; Daykin, M.; Cámara, M.; Williams, P.; Quax, W.J. Quorum quenching by an N-acyl-homoserine lactone acylase from *Pseudomonas aeruginosa* PAO1. *Infect. Immun.* **2006**, *74*, 1673–1682. [CrossRef]

37. Huang, J.J.; Petersen, A.; Whiteley, M.; Leadbetter, J.R. Identification of QuiP, the product of gene PA1032, as the second acyl-homoserine lactone acylase of Pseudomonas aeruginosa PAO1. *Appl. Environ. Microbiol.* **2006**, *72*, 1190–1197. [CrossRef] [PubMed]

38. Shepherd, R.W.; Lindow, S.E. Two dissimilar N-acyl-homoserine lactone acylases of Pseudomonas syringae influence colony and biofilm morphology. *Appl. Environ. Microbiol.* **2009**, *75*, 45–53. [CrossRef] [PubMed]

39. Romero, M.; Diggle, S.P.; Heeb, S.; Cámara, M.; Otero, A. Quorum quenching activity in Anabaena sp. PCC 7120: Identification of AiiC, a novel AHL-acylase. *FEMS Microbiol. Lett.* **2008**, *280*, 73–80. [CrossRef]

40. Koch, G.; Nadal-Jimenez, P.; Cool, R.H.; Quax, W.J. Deinococcus radiodurans can interfere with quorum sensing by producing an AHL-acylase and an AHL-lactonase. *FEMS Microbiol. Lett.* **2014**, *356*, 62–70. [CrossRef]

41. Wahjudi, M.; Papaioannou, E.; Hendrawati, O.; van Assen, A.H.G.; van Merkerk, R.; Cool, R.H.; Poelarends, G.J.; Quax, W.J. PA0305 of Pseudomonas aeruginosa is a quorum quenching acylhomoserine lactone acylase belonging to the Ntn hydrolase superfamily. *Microbiology* **2011**, *157*, 2042–2055. [CrossRef]

42. Davies, D.G.; Parsek, M.R.; Pearson, J.P.; Iglewski, B.H.; Costerton, J.W.; Greenberg, E.P. The involvement of cell-to-cell signals in the development of a bacterial biofilm. *Science* **1998**, *280*, 295–298. [CrossRef] [PubMed]

43. Kohler, T.; Curty, L.K.; Barja, F.; van Delden, C.; Pechere, J.C. Swarming of *Pseudomonas aeruginosa* is dependent on cell-to-cell signaling and requires flagella and pili. *J. Bacteriol.* **2000**, *182*, 5990–5996. [CrossRef] [PubMed]

44. Xia, S.Q.; Zhou, L.J.; Zhang, Z.Q.; Li, J.X. Influence and mechanism of N-(3-oxooctanoyl)-L-homoserine lactone (C_8-oxo-HSL) on biofilm behaviors at early stage. *J. Environ. Sci.* **2012**, *24*, 2035–2040. [CrossRef]

45. Kuipers, R.K.; Joosten, H.J.; van Berkel, W.J.; Leferink, N.G.; Rooijen, E.; Ittmann, E.; van Zimmeren, F.; Jochens, H.; Bornscheuer, U.; Vriend, G.; et al. 3DM: Systematic analysis of heterogeneous superfamily data to discover protein functionalities. *Proteins Struct. Funct. Bioinform.* **2010**, *78*, 2101–2113. [CrossRef] [PubMed]

46. van den Bergh, T.; Tamo, G.; Nobili, A.; Tao, Y.; Tan, T.; Bornscheuer, U.T.; Kuipers, R.K.P.; Vroling, B.; de Jong, R.M.; Subramanian, K.; et al. CorNet: Assigning function to networks of co-evolving residues by automated literature mining. *PLoS ONE* **2017**, *12*, e0176427. [CrossRef]

47. Yasutake, Y.; Kusada, H.; Ebuchi, T.; Hanada, S.; Kamagata, Y.; Tamura, T.; Kimura, N. Bifunctional quorum quenching and antibiotic acylase MacQ forms a 170-kDa capsule-shaped molecule containing spacer polypeptides. *Sci. Rep.* **2017**, *7*, 8946. [CrossRef] [PubMed]

48. Ueda, S.; Shibata, T.; Ito, K.; Oohata, N.; Yamashita, M.; Hino, M.; Yamada, M.; Isogai, Y.; Hashimoto, S. Cloning and expression of the FR901379 acylase gene from Streptomyces sp. no. 6907. *J. Antibiot.* **2011**, *64*, 169–175. [CrossRef]

49. Zhang, D.; Koreishi, M.; Imanaka, H.; Imamura, K.; Nakanishi, K. Cloning and characterization of penicillin V acylase from Streptomyces mobaraensis. *J. Biotechnol.* **2007**, *128*, 788–800. [CrossRef]

50. Terwagne, M.; Mirabella, A.; Lemaire, J.; Deschamps, C.; De Bolle, X.; Letesson, J.J. Quorum sensing and self-quorum quenching in the intracellular pathogen Brucella melitensis. *PLoS ONE* **2013**, *8*, e82514. [CrossRef]

51. Liu, N.; Yu, M.; Zhao, Y.B.; Cheng, J.G.; An, K.; Zhang, X.H. PfmA, a novel quorum quenching N-acylhomoserine lactone acylase from Pseudoalteromonas flavipulchra. *Microbiology* **2017**, *163*, 1389–1398. [CrossRef] [PubMed]

52. Bokhove, M.; Jimenez, P.N.; Quax, W.J.; Dijkstra, B.W. The quorum quenching N-acyl homoserine lactone acylase PvdQ is an Ntn-hydrolase with an unusual substrate-binding pocket. *Proc. Natl. Acad. Sci. USA* **2010**, *107*, 686–691. [CrossRef] [PubMed]

53. Chen, C.N.; Chen, C.J.; Liao, C.T.; Lee, C.Y. A probable aculeacin A acylase from the *Ralstonia solanacearum* GMI1000 is N-acyl-homoserine lactone acylase with quorum quenching activity. *BMC Microbiol.* **2009**, *9*, 89–99. [CrossRef] [PubMed]

54. Pundle, A.; SivaRaman, H. Bacillus sphaericus penicillin V acylase: Purification, substrate specificity, and active-site characterization. *Curr. Microbiol.* **1997**, *34*, 144–148. [CrossRef] [PubMed]

55. Rathinaswamy, P.; Pundle, A.V.; Prabhune, A.A.; Sivaraman, H.; Brannigan, J.A.; Dodson, G.G.; Suresh, C.G. Cloning, purification, crystallization and preliminary structural studies of penicillin V acylase from Bacillus subtilis. *Acta Crystallogr. Sect. F Struct. Biol. Cryst. Commun.* **2005**, *61*, 680–683. [CrossRef] [PubMed]

56. Dong, Z.X.; Lee, B.H. Bile salt hydrolases: Structure and function, substrate preference, and inhibitor development. *Protein Sci.* **2018**, *27*, 1742–1754. [CrossRef]

57. Kumar, R.S.; Brannigan, J.A.; Prabhune, A.A.; Pundle, A.V.; Dodson, G.G.; Dodson, E.J.; Suresh, C.G. Structural and functional analysis of a conjugated bile salt hydrolase from Bifidobacterium longum reveals an evolutionary relationship with penicillin V acylase. *J. Biol. Chem.* **2006**, *281*, 32516–32525. [CrossRef]

58. Chand, D.; Panigrahi, P.; Varshney, N.; Ramasamy, S.; Suresh, C.G. Structure and function of a highly active Bile Salt Hydrolase (BSH) from Enterococcus faecalis and post-translational processing of BSH enzymes. *Biochim. Et Biophys. Acta (BBA)-Proteins Proteom.* **2018**, *1866*, 507–518. [CrossRef]

59. Koreishi, M.; Tani, K.; Ise, Y.; Imanaka, H.; Imamura, K.; Nakanishi, K. Enzymatic synthesis of β-lactam antibiotics and N-fatty-acylated amino compounds by the acyl-transfer reaction catalyzed by penicillin V acylase from Streptomyces mobaraensis. *Biosci. Biotechnol. Biochem.* **2007**, *71*, 1582–1586. [CrossRef]

60. Isogai, Y.; Nakayama, K. Alteration of substrate selection of antibiotic acylase from β-lactam to echinocandin. *Protein Eng. Des. Sel.* **2016**, *29*, 49–56. [CrossRef]

61. Koch, G.; Nadal-Jimenez, P.; Reis, C.R.; Muntendam, R.; Bokhove, M.; Melillo, E.; Dijkstra, B.W.; Cool, R.H.; Quax, W.J. Reducing virulence of the human pathogen Burkholderia by altering the substrate specificity of the quorum quenching acylase PvdQ. *Proc. Natl. Acad. Sci. USA* **2014**, *111*, 1568–1573. [CrossRef] [PubMed]

62. Duggleby, H.J.; Tolley, S.P.; Hill, C.P.; Dodson, E.J.; Dodson, G.; Moody, P.C. Penicillin acylase has a single-amino-acid catalytic centre. *Nature* **1995**, *373*, 264–268. [CrossRef] [PubMed]

63. Brannigan, J.A.; Dodson, G.; Duggleby, H.J.; Moody, P.C.; Smith, J.L.; Tomchick, D.R.; Murzin, A.G. A protein catalytic framework with an N-terminal nucleophile is capable of self-activation. *Nature* **1995**, *378*, 416–419. [CrossRef] [PubMed]

64. Kim, S.; Kim, Y. Active site residues of cephalosporin acylase are critical not only for enzymatic catalysis but also for post-translational modification. *J. Biol. Chem.* **2001**, *276*, 48376–48381. [CrossRef]

65. McVey, C.E.; Walsh, M.A.; Dodson, G.G.; Wilson, K.S.; Brannigan, J.A. Crystal structures of penicillin acylase enzyme-substrate complexes: Structural insights into the catalytic mechanism. *J. Mol. Biol.* **2001**, *313*, 139–150. [CrossRef] [PubMed]

66. Li, Y.; Chen, J.; Jiang, W.; Mao, X.; Zhao, G.; Wang, E. In vivo post-translational processing and subunit reconstitution of cephalosporin acylase from *Pseudomonas* sp. 130. *Eur. J. Biochem.* **1999**, *262*, 713–719. [CrossRef]

67. Hewitt, L.; Kasche, V.; Lummer, K.; Lewis, R.J.; Murshudov, G.N.; Verma, C.S.; Dodson, G.G.; Wilson, K.S. Structure of a slow processing precursor penicillin acylase from *Escherichia coli* reveals the linker peptide blocking the active-site cleft. *J. Mol. Biol.* **2000**, *302*, 887–898. [CrossRef]

68. Kieser, T.; Bibb, M.J.; Buttner, M.J.; Chater, K.F.; Hopwood, D.A. *Practical Streptomyces Genetics*; The John Innes Foundation: Norwich, UK, 2000; ISBN 9780708406236.

69. Clevenger, K.D.; Wu, R.; Er, J.A.V.; Liu, D.; Fast, W. Rational design of a transition state analogue with picomolar affinity for *Pseudomonas aeruginosa* PvdQ, a siderophore biosynthetic enzyme. *ACS Chem. Biol.* **2013**, *8*, 2192–2200. [CrossRef]

70. Švedas, V.J.K.; Galaev, I.J.; Borisov, I.L.; Berezin, I.V. The interaction of amino acids with o-phthaldialdehyde: A kinetic study and spectrophotometric assay of the reaction product. *Anal. Biochem.* **1980**, *101*, 188–195. [CrossRef]

71. McClean, K.H.; Winson, M.K.; Fish, L.; Taylor, A.; Chhabra, S.R.; Camara, M.; Daykin, M.; Lamb, J.H.; Swift, S.; Bycroft, B.W.; et al. Quorum sensing and Chromobacterium violaceum: Exploitation of violacein production and inhibition for the detection of N-acylhomoserine lactones. *Microbiology* **1997**, *143*, 3703–3711. [CrossRef] [PubMed]

72. Torabi, S.F.; Khajeh, K.; Ghasempur, S.; Ghaemi, N.; Siadat, S.O. Covalent attachment of cholesterol oxidase and horseradish peroxidase on perlite through silanization: Activity, stability and co-immobilization. *J. Biotechnol.* **2007**, *131*, 111–120. [CrossRef] [PubMed]

73. Nei, M.; Saitou, N. The neighbor-joining method: A new method for reconstructing phylogenetic trees. *Mol. Biol. Evol.* **1987**, *4*, 406–425. [CrossRef]

74. Felsenstein, J. Confidence limits on phylogenies: An approach using the bootstrap. *Evolution* **1985**, *39*, 783–791. [CrossRef]

75. Nei, M.; Kumar, S. *Molecular Evolution and Phylogenetics*; Oxford University Press: New York, NY, USA, 2000; ISBN 9780199881222.
76. Kumar, S.; Stecher, G.; Li, M.; Knyaz, C.; Tamura, K. MEGA X. Molecular evolutionary genetics analysis across computing platforms. *Mol. Biol. Evol.* **2018**, *35*, 1547–1549. [CrossRef]

 *catalysts*

Article

A Three-Step Process for the Bioconversion of Whey Permeate into a Glucose-Free D-Tagatose Syrup

Fadia V. Cervantes [1], Sawssan Neifar [2], Zoran Merdzo [3], Javier Viña-Gonzalez [1], Lucia Fernandez-Arrojo [1], Antonio O. Ballesteros [1], Maria Fernandez-Lobato [3], Samir Bejar [2] and Francisco J. Plou [1,*]

1   Instituto de Catálisis y Petroleoquímica, CSIC, 28049 Madrid, Spain; fadiacervantes@icp.csic.es (F.V.C.); javier.v@csic.es (J.V.-G.); lucia@icp.csic.es (L.F.-A.); a.ballesteros@icp.csic.es (A.O.B.)
2   Laboratory of Microbial Biotechnology and Engineering Enzymes (LMBEE), Centre of Biotechnology of Sfax (CBS), University of Sfax, Sfax 3018, Tunisia; neifarsawssan@yahoo.fr (S.N.); samir.bejar@cbs.rnrt.tn (S.B.)
3   Centro de Biología Molecular Severo Ochoa, UAM-CSIC, 28049 Madrid, Spain; zoran@cbm.csic.es (Z.M.); mfernandez@cbm.csic.es (M.F.-L.)
*   Correspondence: fplou@icp.csic.es; Tel.: +34-91-5854869

Received: 15 May 2020; Accepted: 8 June 2020; Published: 9 June 2020

Abstract: We have developed a sustainable three-stage process for the revaluation of cheese whey permeate into D-tagatose, a rare sugar with functional properties used as sweetener. The experimental conditions (pH, temperature, cofactors, etc.) for each step were independently optimized. In the first step, concentrated whey containing 180–200 g/L of lactose was fully hydrolyzed by β-galactosidase from *Bifidobacterium bifidum* (Saphera®) in 3 h at 45 °C. Secondly, glucose was selectively removed by treatment with *Pichia pastoris* cells for 3 h at 30 °C. The best results were obtained with 350 mg of cells (previously grown for 16 h) per mL of solution. Finally, L-arabinose isomerase US100 from *Bacillus stearothermophilus* was employed to isomerize D-galactose into D-tagatose at pH 7.5 and 65 °C, in presence of 0.5 mM $MnSO_4$. After 7 h, the concentration of D-tagatose was approximately 30 g/L (33.3% yield, referred to the initial D-galactose present in whey). The proposed integrated process takes place under mild conditions (neutral pH, moderate temperatures) in a short time (13 h), yielding a glucose-free syrup containing D-tagatose and galactose in a ratio 1:2 (*w/w*).

Keywords: biocatalysis; glycosidases; isomerases; *Pichia pastoris*; sweeteners; rare sugars; cheese whey; sustainable chemistry

1. Introduction

D-Tagatose is one of the most promising low-calorie functional sweeteners [1]. It is a ketohexose, namely a C4 epimer of D-fructose. It is a rare sugar only found at small quantities in the gum exudates of the tropical tree *Sterculia setigera* and in several dairy products, e.g., in Ultra-High-Temperature (UHT) sterilized cow's milk [2]. It is heat tolerant, very stable at pH values between 2.0 to 7.0, and highly soluble in water (58% *w/w* at room temperature). It possesses a sucrose-like taste with 92% of its sweetness but contributing only 1.5 kcal/g (38% compared to sucrose) [3], with no cooling effect or aftertaste [4]. D-Tagatose was approved as a novel food in the European Union [3] and has obtained GRAS status by FDA in USA [5]. It is also approved in many other countries and is being widely used in foods, beverages, and dietary supplements [6,7].

Only 20% of the ingested D-tagatose is absorbed in the small intestine [8]. Since the remaining 80% seems to favourably modulate the composition of the gut microbiota [9], this sugar has been proposed as a potential prebiotic. In addition, D-tagatose is able to modulate lipid metabolism, minimize tooth decay, promote blood health, and reduce the symptoms associated with type 2 diabetes, hyperglycemia, anemia, and haemophilia [10–12].

Several chemical [13,14] and chemoenzymatic [15,16] methods have been described for D-tagatose synthesis. In this context, the chemical manufacture of D-tagatose involves the use of metal hydroxides and calcium chloride, the neutralization with mineral acids and the implementation of complex purification steps due to by-product generation [6]. Therefore, enzymatic methods are the most appropriate for D-tagatose production in terms of selectivity, efficiency and environmental impact [17]. Several enzymes have been investigated for D-tagatose synthesis, e.g. phosphoglucose isomerase [18], galactitol 2-dehydrogenase [19] and L-arabinose isomerase (L-AI) [20]. The main reaction catalyzed by L-AI is the bioconversion of L-arabinose into L-ribulose, however, it also promotes efficiently the isomerization of D-galactose to D-tagatose [21]. In fact, most of the publications on the enzymatic synthesis of D-tagatose use L-AI as biocatalyst [22–24].

D-Tagatose production typically begins with D-galactose [25–28], which is quite expensive in its pure form. The valorization of galactose-rich byproducts such as agar [29] or whey [30] is much more attractive from the economic and environmental points of view. Moreover, cheese whey generates significant environmental and health issues due to its large volume production and high organic content [31]. With this aim in mind, several groups have reported the production of D-tagatose from cheese whey in liquid or powdered form [30,32–36], yielding a mixture of D-tagatose, D-galactose and D-glucose. Since the physicochemical properties of the three monosaccharides are quite similar, the isolation of pure D-tagatose is a difficult task [32].

In order to increase the efficiency of the process, several groups have co-expressed β-galactosidase and L-AI in *Pichia pastoris* [37], *Corynebacterium glutamicum* [33] or *Escherichia coli* [30,32,38] for the simultaneous lactose hydrolysis and D-galactose isomerization. The main drawback of this approach is that the optimum pH and temperature of L-AI and β-galactosidase are substantially different, thus limiting the efficiency of the whole process.

The presence of D-glucose in D-tagatose syrups is clearly undesirable, especially because one of the main properties of D-tagatose is its antidiabetic effect [6]. In order to eliminate the glucose released during lactose hydrolysis, Wanarska et al. took advantage of the presence of glucose to cultivate *Pichia pastoris* (which co-expressed L-AI and β-galactosidase) [37]. Another approach addressed by Zhan et al. [39] and Zheng et al. [32] was the implementation of a second step of D-glucose (and eventually D-galactose) fermentation by *Saccharomyces cerevisiae* which may render both D-tagatose and bioethanol.

In this work, we report a three-step process for the bioconversion of concentrated whey permeate into D-tagatose using free enzymes and whole cells. In the first stage, β-galactosidase hydrolyzed lactose. This was followed by treatment with *Pichia pastoris* cells to remove the released D-glucose. Finally, L-AI isomerized D-galactose into D-tagatose. The three stages were independently optimized to obtain the highest efficiency.

2. Results and Discussion

2.1. Chemical Composition of Whey Permeate

It is well reported the effect of metal ions on the activity and stability of many enzymes, including those involved in the transformation of carbohydrates, and in particular in the two enzymes involved in the present work: β-galactosidase [40] and L-arabinose isomerase [41]. In particular, several members of these two families are activated by divalent cations such as Co^{2+}, Mn^{2+} and Mg^{2+} [21,42]. For that reason, we measured by semiquantitative ICP-MS the composition of the concentrated whey permeate employed in this work. Table 1 summarizes the concentration of metal ions present at higher amounts than 0.1 ppm. As shown, the concentration of Mg^{2+} was significant (98 mg/L) but the presence of Co^{2+} and Mn^{2+} was negligible.

Table 1. Metal ions present in whey permeate at concentrations higher than 0.1 ppm, determined by semiquantitative ICP-MS.

Analyte	Concentration (ppm)
Na	390
Mg	98
K	1492
Ca	829
Ti	0.35
Fe	0.18
Zn	1.1
Rb	1.6
Sr	0.45
Mo	0.1

2.2. Hydrolysis of Whey Permeate by β-Galactosidase from Bifidobacterium bifidus

The concentration of lactose in the cheese whey permeates employed in this work varied between 180–200 g/L depending on the batch, as determined by HPAEC-PAD. The pH of whey was adjusted to 6.8. We selected *Bifidobacterium bifidum* β-galactosidase (Saphera®, Novozymes, Bagsværd, Denmark) for the hydrolysis of lactose in whey because we recently observed that at lactose concentrations lower than 200 g/L the main reaction catalyzed by this enzyme is the hydrolysis with negligible transgalactosylation [43]. In order to standardize the dose of enzyme, we measured the activity of Saphera with *o*-nitrophenyl-β-D-galactopyranoside (ONPG). The activity was 1506 ± 0.1 U/mL at 40 °C and pH 6.8, which are the typical preferred conditions for several β-galactosidases from *Bifidobacteria* [44,45].

First, we analyzed the effect of temperature on lactose hydrolysis in whey permeate. We assayed three different temperatures (30, 40 and 45 °C) using 2.5 µL (3.75 ONPG units) per mL of whey. HPAEC-PAD showed that the fastest reaction was the one performed at 45 °C (Table 2). Under these conditions, all the lactose is hydrolyzed in 3 h. We further increased the enzyme concentration up to 7.50 ONPG units per mL but the improvement in the hydrolysis process was not substantial. In order to minimize the cost for the process, we selected the lowest concentration of this enzyme.

Table 2. Effect of temperature on the progress of lactose hydrolysis in whey permeate by Saphera, using 2.5 µL of enzyme (3.75 ONPG units) per mL of whey permeate.

Temperature (°C)	Residual lactose (%) [a]					
	30 min	60 min	90 min	120 min	150 min	180 min
30	89.3	81.5	76.6	65.5	57.1	24.7
40	73.4	45.6	39.7	36.9	21.3	18.7
45	76.3	31.7	29.8	26.1	10.7	0

[a] Measured by HPAEC-PAD.

2.3. Study of the Growth Time and Concentration of Pichia pastoris for Elimination of D-Glucose

Pichia pastoris is capable of using D-glucose, glycerol and methanol as carbon sources but not D-galactose [46]. In this context, Avila-Fernandez et al. successfully employed *P. pastoris* cells to selectively eliminate glucose and fructose in a syrup of fructooligosaccharides obtained by agave fructan hydrolysis [47]. Based on such background, the elimination of D-glucose using *P. pastoris* cells was postulated for the second step of the reaction.

We first developed a simple and fast test to evaluate the effect of *P. pastoris* growing time on the further consumption of glucose. Thus, samples from a *P. pastoris* culture were taken every two hours (after an initial lag phase of 6 h), and the absorbance at 600 nm (as turbidity measurement) of all samples was adjusted to the absorbance at 6 h, to have the same concentration of cells but with different growing times. The *P. pastoris* cells were then mixed with a solution containing 1 g/L of both D-glucose and D-galactose, and the mixture was incubated for 10 min at 30 °C. The cells were separated by centrifugation and the concentration of both sugars was analyzed by HPAEC-PAD (Figure 1). Our results showed that the most voracious cells were those corresponding to 12–16 h growing time. It is well reported that these times correspond the exponential phase of growth of *P. pastoris* [48].

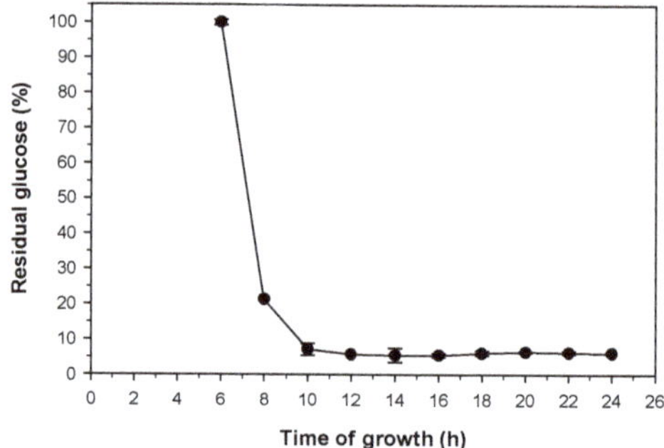

Figure 1. Effect of the growth time of *Pichia pastoris* on the elimination of D-glucose. Experimental conditions: 1 g/L D-glucose, 1 g/L D-galactose, 10 min, 30 °C.

The next step was to determine the amount of cells (grown for 16 h) required to consume the glucose (90–100 g/L) released after lactose hydrolysis in whey permeate. Different amounts of *Pichia pastoris* cells (100–600 mg of wet weight) were added to 1 mL of a solution containing 100 g/L of D-glucose and 100 g/L D-galactose. The residual glucose was analyzed at 1, 2 and 3 h by HPAEC-PAD (Figure 2). As illustrated, the glucose disappeared in one hour using 600 mg of *P. pastoris* cells per mL. However, adding 350 mg of cells per mL, the total disappearance of glucose took place in two hours, and the mass transfer limitations were less significant than with 600 mg cells per mL. On this basis, we selected 350 mg of cells per mL (3 h incubation) to assure the elimination of glucose in the integrated 3-step process for D-tagatose synthesis.

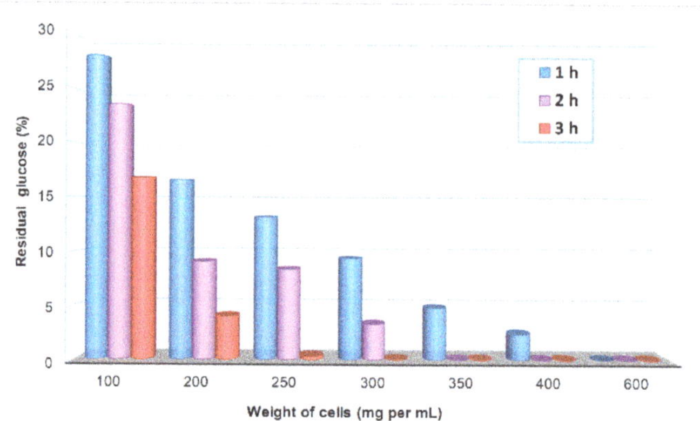

Figure 2. Effect of the amount of cells (wet weight) of *Pichia pastoris* on the elimination at 30 °C of D-glucose in a solution containing 100 g/L of both D-glucose and D-galactose.

2.4. Effect of pH and Temperature on the Stability of L-Arabinose Isomerase

The L-arabinose isomerase (L-AI US100) from *Bacillus stearothermophilus* is a multimeric enzyme formed by four 56 kDa monomers [49]. Compared with other L-arabinose isomerases, L-AI US100 is relatively active at neutral and slightly acidic pH, and at moderate temperatures [50]. These properties are very valuable because at the typical conditions of alkaline pH and high temperature the formation of undesired byproducts in bioconversions of sugars is quite common [51]. For these reasons, L-AI US100 was selected for the transformation of whey permeate into D-tagatose.

In order to establish the optimal conditions for the biotransformation, the stability of L-AI US100 towards pH was evaluated by pre-incubation (at room temperature) of the enzyme for 1 h at pH values between 3.0 and 10.0 employing 0.1 M Britton & Robinson buffer. After that, the L-arabinose isomerase activity was measured following the standard activity assay (at 65 °C and pH 7.5, see Experimental Section). The results shown in Figure 3A represent the relative activity of pre-treated L-AI compared with the non-incubated enzyme. As illustrated, the enzyme is very unstable at pH values below 6.0 and is notably stable at neutral and moderately alkaline pH values.

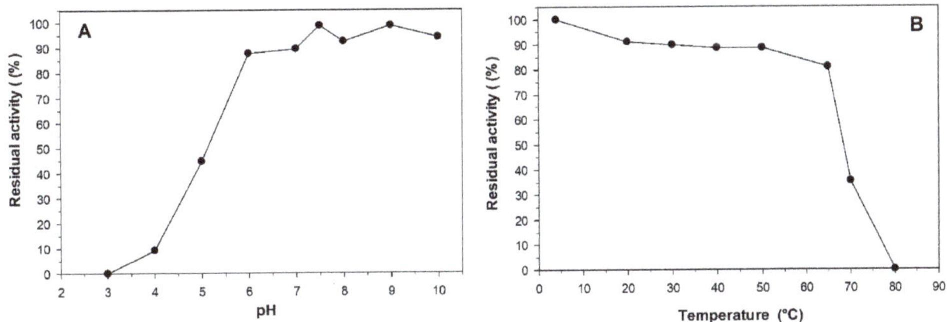

Figure 3. Stability of L-arabinose isomerase US100 from *Bacillus stearothermophilus* towards: (**A**) pH; (**B**) temperature. The pre-incubation time in both cases was 1 h.

The thermal stability of the enzyme was determined by pre-incubating the L-AI US100 for 1 h at temperatures between 4 and 80 °C in 100 mM MOPS buffer (pH 7.5). Following this incubation, the L-AI activity was measured by the standard activity assay (at 65 °C and pH 7.5). As shown in Figure 3B, the enzyme is very stable at temperatures up to 65 °C, and was fast inactivated at temperatures higher than 70 °C.

2.5. Effect of Cofactors, pH and Temperature on the Activity of L-Arabinose Isomerase

The effect of metal cofactors on the activity of L-arabinose isomerase (L-AI US100) from *B. stearothermophilus* at 65 °C and neutral pH was assessed. Four different reactions were performed in absence or presence of 0.5 mM $MnSO_4$ and/or 0.1 mM $CoCl_2$. The reactions were carried out with 20 g/L galactose in 100 mM MOPS buffer (pH 7.5) at 65 °C for 4 h. Results are displayed in Figure 4.

As shown, $CoCl_2$ had a negligible effect on the yield of D-tagatose under these conditions. However, the production of D-tagatose in the presence of 0.5 mM $MnSO_4$ was approximately 40% higher compared with the control.

These results correlate well with previous findings on the effect of metallic ions on the L-AI US100 activity and thermostability. The effect of divalent cations on L-AI US100 must be particularly considered at temperatures higher or equal to 65 °C [49]. In fact, at temperatures above 65 °C both Mn^{2+} and Co^{2+} exert a significant influence on the thermostability of L-AI US100. Since a temperature of 65 °C was selected for the isomerization reaction, and considering that the conversion from D-galactose to D-tagatose was higher in presence of $MnSO_4$ in the reaction medium, we decided to maintain $MnSO_4$ but not $CoCl_2$ in the integrated process from whey to D-tagatose.

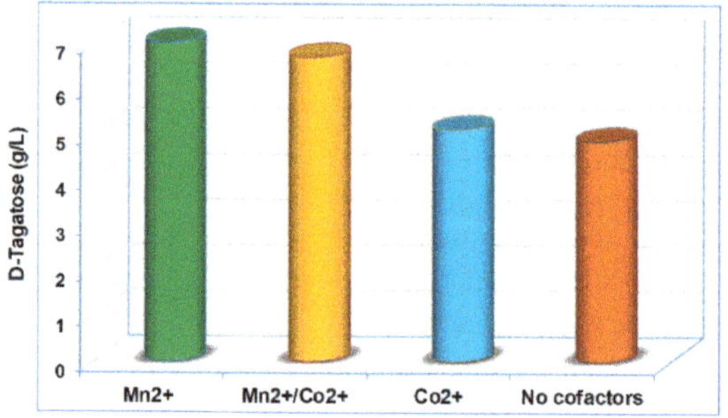

Figure 4. Effect of metal ions on the activity of L-arabinose isomerase US100 from *B. stearothermophilus*, in absence or presence of 0.5 mM $MnSO_4$ and/or 0.1 mM $CoCl_2$. Experimental conditions: 20 g/L D-galactose, 100 mM MOPS (pH 7.5), 0.65 U/mL L-AI, 65 °C. The concentration of D-tagatose was measured after 4 h by HPAEC-PAD.

Furthermore, we analysed the effect of pH and temperature on the production of D-tagatose. We selected the range of pH (7.0–9.5) and temperature (40–65 °C) at which the L-arabinose isomerase from *B. stearothermophilus* was stable (see Figure 3). The reactions were carried out with 20 g/L galactose in 100 mM MOPS buffer for 4 h, in presence of 0.5 mM $MnSO_4$. Results are shown in Figure 5. It remains clear that temperature substantially affects the reaction course. The highest production of D-tagatose in 4 h (7–8 g/L, 35–40% yield) was achieved at 65 °C. The reaction was 6–7 fold faster than at 40 °C. Regarding the pH, its effect was very much lower than that of temperature (Figure 5).

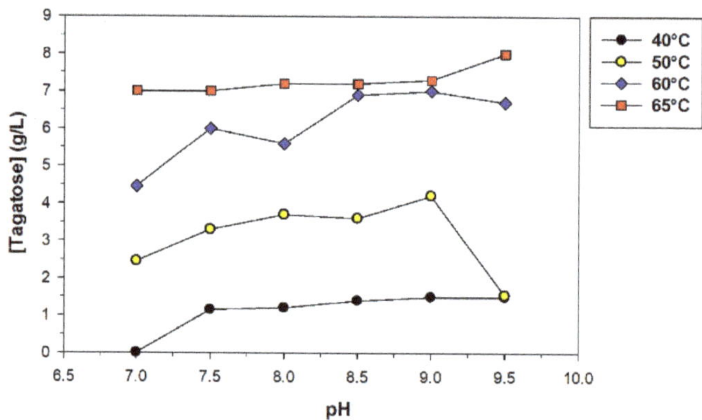

Figure 5. Effect of temperature and pH on the production of D-tagatose by L-arabinose isomerase US100 from *B. stearothermophilus*. Experimental conditions: 20 g/L D-galactose in buffer 100 mM MOPS containing 0.5 mM $MnSO_4$, 0.65 U/mL L-AI. The concentration of D-tagatose was measured after 4 h by HPAEC-PAD.

Considering the results on the effect of pH, temperature and metal cofactors on the stability and activity of L-AI US100, we selected pH 7.5, 65 °C and 0.5 mM $MnSO_4$ as the optimal experimental conditions for the integrated process. These results represent a compromise between activity and stability of the enzyme and correlate well with previous studies reported with this enzyme by using fast spectrophotometric assays [49,50,52] rather than chromatographic analysis by HPAEC-PAD performed in this work.

2.6. Sequential Biotransformation of Whey Permeate into D-Tagatose

We propose a three-step process for D-tagatose synthesis from whey permeate based on the experimental conditions that were previously optimized. In the first stage, the lactose (180–200 g/L) contained in whey is hydrolyzed by β-galactosidase from *Bifidobacterium bifidum* (7.5 U/mL). Total hydrolysis is achieved in 3 h at 45 °C. Figure 6 shows a HPAEC-PAD chromatogram illustrating the complete disappearance of lactose in the reaction medium.

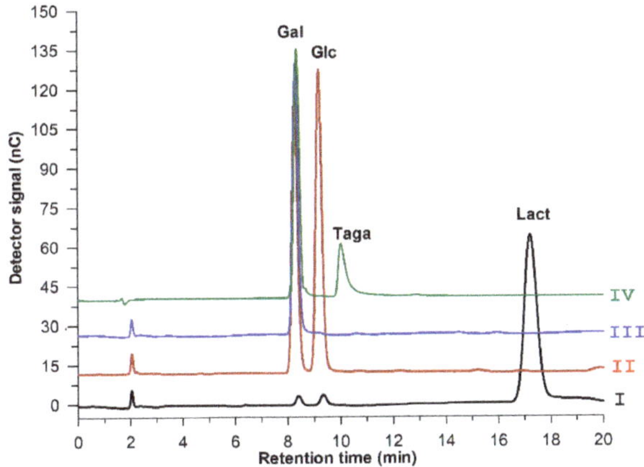

Figure 6. HPAEC-PAD chromatograms showing the bioconversion of whey permeate into D-tagatose syrup. (I) Original whey permeate; (II) After treatment with β-galactosidase from *B. bifidum*; (III) After removal of glucose with *P. pastoris* cells; (IV) After isomerization with of L-arabinose isomerase US100. Gal: D-Galactose; Glc: D-glucose; Taga: D-tagatose; Lact: D-Lactose.

In the second step, the hydrolyzate is treated with *Pichia pastoris* cells (350 mg of cells per mL, obtained as described elsewhere) during 3 h at 30 °C. As shown in Figure 6, all the glucose is metabolized by the yeast and the amount of D-galactose remains intact. After centrifugation to remove the cells, the pH of supernatant is adjusted to 7.5 (because the medium is acidified during the treatment with *P. pastoris*) and $MnSO_4$ is added to reach a final concentration of 0.5 mM. These adjustments prepare the mixture for the third step.

Finally, L-arabinose isomerase L-AI US100 is added (0.65 U per mL in the mixture, measured by the standard assay). The reaction is incubated at 65 °C and different aliquots are taken to follow the progress of the reaction. As depicted in Figure 6, D-tagatose is formed and the final product contains a mixture of D-tagatose and D-galactose.

Figure 7 shows the progress of D-tagatose formation over time. The concentration of D-tagatose increases rapidly during the first four hours and then tends to stabilize in approximately 30 g/L. Starting of a whey permeate with 180 g/L lactose, the yield of D-tagatose referred to lactose was 16.7%, and 33.3% referred to the D-galactose present in whey. We believe that this yield could be improved by using a higher concentration of L-AI thus minimizing the enzyme inactivation effects (see Figure 3).

Wanarska et al. reported a 30% yield of D-tagatose (referred to D-galactose) using a strain of *Pichia pastoris* that co-expressed L-AI and β-galactosidase, starting of a whey permeate containing 110 g/L lactose [37]. Zheng et al. employed a whey with 110 g/L lactose to get 43.6% yield of D-tagatose (referred to D-galactose) using a genetically engineered strain of *Escherichia coli* that expressed L-AI [32]. Similar results (40.4% yield) were reported by Xu et al. employing *E. coli* that co-expressed a β-galactosidase from *Thermus thermophilus* and L-arabinose isomerase from *Lactobacillus fermentum* [38,39]. Jayamuthunagai et al. reported 38% yield of D-tagatose (referred to D-galactose) from a hydrolyzed whey permeate containing 300 g/L D-galactose, using permeabilized and alginate-entrapped cells of *Lactobacillus plantarum* [36]. Shen et al. expressed xylose isomerase and L-AI in a strain of *Corynebacterium glutamicum* able to metabolize lactose. Starting from whey containing 98 g/L lactose, they obtained 20.4 g/L D-tagatose (44% yield referred to D-galactose) [33].

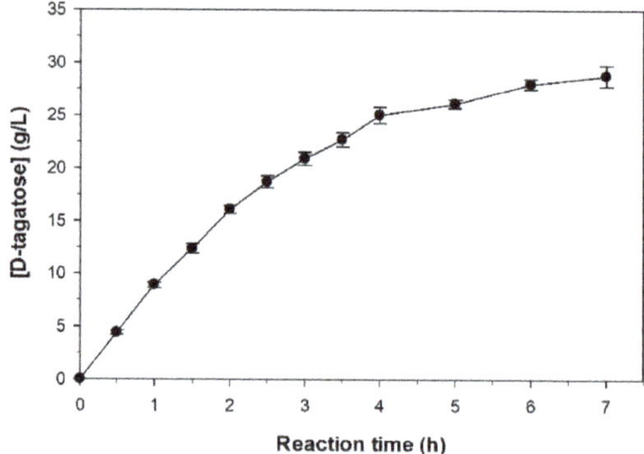

Figure 7. Progress of formation of D-tagatose from D-galactose syrup obtained by hydrolysis of whey permeate. Experimental conditions: 90 g/L D-galactose (coming from 180 g/L lactose in whey permeate), 0.65 U/mL L-arabinose isomerase US100, 0.5 mM $MnSO_4$, pH 7.5, 65 °C.

A scheme summarizing the different steps of the integrated process is presented in Figure 8.

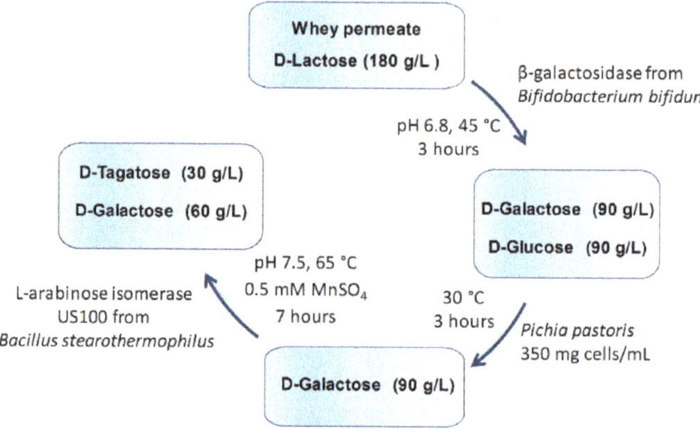

Figure 8. Scheme of the integrated process for bioconversion of whey permeate into D-tagatose syrup.

3. Materials and Methods

3.1. Enzymes and Reagents

L-Arabinose isomerase (L-AI) from *Bacillus staerothermophilus* US100 (L-AI US 100) was recombinantly produced in *E. coli* as described in previous publications [50,52,53]. The β-galactosidase from *Bifidobacterium bifidus* (Saphera®, Novozym 46091) was gently donated by Novozymes A/S (Bagsværd, Denmark). *Pichia pastoris* GS115 (*his4*) was obtained from Invitrogen (Carlsbad, CA, USA). Concentrated whey permeate (180–200 g/L) was kindly donated by Innolact (Castro de Rei, Lugo, Spain). D-Galactose and D-glucose were purchased from Sigma-Aldrich (Madrid, Spain). D-Tagatose was acquired from Tokyo Chemical Industry Co. (Tokyo, Japan). Lactose and sodium acetate trihydrate were from Fisher Chemical (Madrid, Spain). NaOH 50% (*v/v*) was from Acros Organics (Geel, Belgium). All other reagents and solvents were of the highest purity grade available.

3.2. Elemental Analysis of Whey Permeate using Inductive Coupled Plasma Mass Spectrometry

Semiquantitative analysis of metals in whey permeate was performed on a NexION 300X Inductively Coupled Plasma-Mass Spectrometer (ICP-MS, PerkinElmer, Waltham, MA, USA) equipped with Universal Cell Technology, as previously described [54].

3.3. β-Galactosidase Activity Assay

The assay of β-galactosidase activity was performed using *o*-nitrophenyl-β-D-galactopyranoside (ONPG) as substrate. The activity was measured at 40 °C following *o*-nitrophenol (ONP) release at 405 nm using a microplate reader (Versamax, Molecular Devices, San Jose, CA, USA). The reaction was started by adding 10 μL of the enzyme solution (properly diluted) to 190 μL of 15 mM ONPG in 0.1 M sodium phosphate buffer (pH 6.8). The increase of absorbance at 405 nm was followed continuously at 40 °C during 5 min. The molar extinction coefficient of *o*-nitrophenol at pH 6.8 was 1627 M^{-1} cm^{-1}. One unit of activity (*U*) was defined as the corresponding to the hydrolysis of 1 μmol of ONPG per min under the above specified conditions.

3.4. Lactose Hydrolysis in Whey by Bifidobacterium Bifidum β-Galactosidase

Concentrated whey permeate (2 mL, containing 180–200 g/L lactose) was mixed with β-galactosidase from *Bifidobacterium bifidum* (Saphera®, 5–10 μL, 3.75–7.5 ONPG units per mL). The mixture was incubated at 30, 40 or 45 °C in an orbital shaker (Vortemp 1550, Labnet International, Big Flats, NY, USA) at 200 rpm. 200 μL aliquots were taken from the reaction vial each 30 minutes until 3 hours. The enzyme was then inactivated by incubating the samples in a Thermomixer (Eppendorf, Hamburg, Germany) for 10 min at 95 °C. Samples were then filtered using micro-centrifuge filter tubes, with 0.45 μm cellulose acetate filters (National Scientific, Claremont, CA, USA) at 6000 rpm for 5 min. The samples were diluted with water (1:400 and 1:4000) and then analysed using HPAEC-PAD.

3.5. Effect of Growth Time and Concentration of Pichia pastoris on the Elimination of D-Glucose

Pichia pastoris preculture was grown in 5 mL of 1% yeast extract, 1% peptone and 2% dextrose (all *w/v*) liquid medium (YEPD) at 30 °C and 1200 rpm for 24 h. After this time, the absorbance was measured, and 1 mL of preculture was added to a 250 mL flask containing 25 mL of YEPD, which was left growing at 30 °C and 1200 rpm. The first sample was taken at 6 h. The optical density (OD) was measured at 600 nm in a UV-1800 spectrophotometer (Shimadzu, Kioto, Japan) taking this OD as the reference value. The cells were centrifuged and washed three times with distilled water to remove the remaining medium from the cells. Samples were taken every 2 h until 24 h. Before centrifugation, all the samples were diluted to reach the reference value of absorbance, with the aim of having the same amount of cells in the experiments but with different growing times. Reactions with *P. pastoris* cells were performed adding 500 μL of a solution containing 1 g/L D-glucose and 1 g/L D-galactose to the cells. Reactions were incubated in a tube rotator (Argos Technologies Inc., Vernon Hills, IL, USA) for 10 min at 30 °C, then samples were centrifuged, the supernatant was inactivated at 95 °C for 10 min and analyzed by HPAEC-PAD. To determine the amount of cells required to consume the glucose expected in the sequential process for production of D-tagatose, different amounts of *P. pastoris* cells (300, 400, 500 and 600 mg of wet weight) were added to 1 mL of a solution containing 100 g/L D-glucose and 100 g/L D-galactose. Aliquots were withdrawn at 1, 2 and 3 h. After each extraction, the samples were centrifuged at 10,000 rpm for 3 min and the concentration of glucose and galactose in the supernatant was analyzed by HPAEC-PAD.

3.6. L-Arabinose Isomerase Activity Assay

The L-arabinose isomerase (L-AI) activity was determined adding the enzyme to a solution containing 20 g/L of D-galactose in 100 mM MOPS buffer (pH 7.5) containing 0.5 mM $MnSO_4$ and 0.1 mM $CoCl_2$. The mixture was incubated at 65 °C for 4 h, and the reaction stopped in a water bath at

90 °C for 10 min. The concentration of D-tagatose was measured by HPAEC-PAD. One unit of L-AI activity (U) was defined as the corresponding to the formation of 1 µmol of D-tagatose per minute, under the conditions specified above.

3.7. Stability of L-Arabinose Isomerase

The stability of L-arabinose isomerase (L-AI US 100) from *B. stearothermophilus* towards pH was assessed pre-incubating the enzyme at different pH values (3.0 to 10.0) in 0.1 M Britton & Robinson buffer [55] at room temperature for 1 h. The thermal stability of the enzyme was determined by pre-incubating the L-AI at 4, 20, 30, 40, 50, 65, 70 and 80 °C in 100 mM MOPS buffer (pH 7.5) for 1 h. After incubation, the remaining L-arabinose isomerase activity was measured following the standard activity assay (Section 3.6). The activity of pre-treated L-AI US100 was compared with the activity of non-pre-incubated enzyme that was taken as control (100%).

3.8. Effect of Metals, pH and Temperature on the Activity of L-Arabinose Isomerase

To analyze the effect of metal cofactors on the activity of L-arabinose isomerase (L-AI US 100), different reactions were carried out in presence of 0.5 mM $MnSO_4$ and/or 0.1 mM $CoCl_2$. The reactions were carried out with 20 g/L galactose in 100 mM MOPS buffer (pH 7.5) at 65 °C for 4 h, using 0.65 U/mL L-AI, and the concentration of D-tagatose was determined by HPAEC-PAD. A control reaction without cofactors was also performed. To determine the best reaction conditions for the enzyme, different values of pH (7, 7.5, 8, 8.5, 9, 9.5) and temperature (40, 50, 60 and 65 °C) were tested, in a reaction containing 20 g/L of D-galactose, 0.5 mM of $MnSO_4$ in 100 mM MOPS buffer (pH 7.5) and 0.65 U/mL L-AI. After 4 h, the reaction mixtures were analyzed by HPAEC-PAD.

3.9. Sequential Biotransformation of Whey Permeate into D-Tagatose

The biosynthesis of D-tagatose syrup from whey permeate was performed in three steps. Initially, the pH of whey was adjusted to 6.8. In the first stage, β-galactosidase from *Bifidobacterium bifidum* (5 µL, 7.5 U/mL measured with ONPG) was added to 2 mL of concentrated whey permeate (containing 180–200 g/L lactose, depending on the batch). The mixture was incubated for 3 h in an Envirogenie orbital stirrer (Scientific Industries Inc., Bohemia, NY, USA) at 45 °C. The second step involved the treatment of the hydrolyzate with 700 mg (wet weight) of *Pichia pastoris* cells (previously grown for 16 h) and the mixture was incubated in an orbital shaker (Orbitron, Infors HT, Surrey, UK) for 3 h at 30 °C. Then, the cells were removed by centrifugation (model 5810, Eppendorf) at 5000 rpm for 20 min at 4 °C. The supernatant was then separated from the cells and inactivated in a water bath at 90 °C for 10 min. In the third stage, the pH of supernatant was adjusted to 7.5 and $MnSO_4$ was added up to a final concentration of 0.5 mM. Then, 1.3 units of L-arabinose isomerase L-AI US100 were added (0.65 U/mL in the mixture). The reaction was incubated in a thermoshaker (model TS-100, Biosan, Riga, Latvia) at 65 °C and 1500 rpm. Aliquots were taken every 30 min, filtered on UltraFree centrifugal filters (0.45 µm, Millipore, Burlington, MA, USA) and analysed by HPAEC-PAD.

3.10. HPAEC-PAD Analysis

Sugar (D-galactose, D-glucose, D-tagatose, D-fructose and D-lactose) analysis was carried out by high performance anion-exchange chromatography coupled with pulsed amperometric detection (HPAEC-PAD) on an ICS3000 system (Dionex, Thermo Fischer Scientific Inc., Waltham, MA, USA) consisting of a SP gradient pump, an electrochemical detector with a gold working electrode and Ag/AgCl as reference electrode, and an autosampler (model AS-HV). All eluents were degassed by flushing with helium. A pellicular anion-exchange 4 × 250 mm Carbo-Pack PA-1 column (Dionex) connected to a 4 × 50 mm CarboPac PA-1 guard column was used at 30 °C. Eluent preparation was performed with MilliQ water, 50% (*w/v*) NaOH and sodium acetate trihydrate. The compounds were eluted by an isocratic method in which the mobile phase contained 10 mM NaOH and 2 mM sodium acetate, using a flow rate of 1 mL/min for 25 min. The peaks were analyzed using the Chromeleon software. Identification of the different carbohydrates was carried out employing commercially available standards.

4. Conclusions

The present work describes an environmentally friendly process for the bioconversion of cheese whey, a lactose-rich byproduct of the food industry, into D-tagatose, a rare sugar that has become one of the most promising low-calorie sweeteners in the market due to its functional properties. The methodology is based in the complete hydrolysis of lactose by a bifidobacterial β-galactosidase, followed by the selective removal of glucose with *Pichia pastoris* cells, and finally the isomerization of the remaining D-galactose into D-tagatose by L-arabinose isomerase from *Bacillus stearothermophilus*. The three steps were optimized independently in such a way that the integrated process is carried out in a short time (13 h) yielding 33.3% of D-tagatose (referred to the initial D-galactose). The total time could be even reduced by using a higher concentration of L-AI in the third step. One of the main advantages of the proposed project is its sustainability, as the three steps take place under mild conditions of pH (6.8–7.5), moderate temperatures (30–65 °C) and with completely biodegradable catalysts. The final syrup is free of glucose, and contains D-tagatose and D-galactose in a ratio 1:2 (*w/w*). This work opens new possibilities for the synthesis of D-tagatose and for the development of different reaction engineering strategies including enzyme immobilization.

Author Contributions: F.J.P., S.B. and A.O.B. conceived and designed the experiments; F.V.C. carried out most of the experiments; S.N. and S.B. produced and characterized the L-AI US100; Z.M., J.V.-G. and M.F.-L. were in charge of the microbiological issues in this work; L.F.-A. contributed to the chromatographic analysis of carbohydrates; F.J.P. wrote the paper, which was improved by the rest of authors. All authors have read and agreed to the published version of the manuscript.

Funding: This work was supported by grants from the Spanish Ministry of Economy and Competitiveness (Grants BIO2016-76601-C3-1-R and C3-2-R) and Fundación Ramón Areces (XIX Call of Research Grants in Life and Material Sciences). We thank Fundación Ramón Areces for an institutional grant to the Center of Molecular Biology Severo Ochoa. F.V. Cervantes thanks CONACYT (Mexico) for her Ph.D. fellowship (Ref. 440242).

Acknowledgments: We thank Ramiro Martinez (Novozymes A/S, Spain) for supplying Saphera and for useful suggestions. We are grateful to Innolact (Castro de rei, Lugo, Spain) for donating concentrated whey permeate. We acknowledge support of the publication fee by the CSIC Open Access Publication Support Initiative through its Unit of Information Resources for Research (URICI).

Conflicts of Interest: The authors declare no conflict of interest.

References

1. Torrico, D.D.; Tam, J.; Fuentes, S.; Viejo, C.G.; Dunshea, F.R. D-tagatose as a sucrose substitute and its effect on the physico-chemical properties and acceptability of strawberry-flavored yogurt. *Foods* **2019**, *8*, 256. [CrossRef] [PubMed]

2. Oh, D.K. Tagatose: Properties, applications, and biotechnological processes. *Appl. Microbiol. Biotechnol.* **2007**, *76*, 1–8. [CrossRef] [PubMed]

3. Von Rymon Lipinski, G.-W. Sweeteners. In *Biotechnology of Food and Feed Additives*; Zorn, H., Czermak, P., Eds.; Springer: Berlin, Germany, 2014; pp. 1–28. [CrossRef]

4. Acevedo, W.; Capitaine, C.; Rodríguez, R.; Araya-Durán, I.; González-Nilo, F.; Pérez-Correa, J.R.; Agosin, E. Selecting optimal mixtures of natural sweeteners for carbonated soft drinks through multi-objective decision modeling and sensory validation. *J. Sens. Stud.* **2018**, *33*, 6. [CrossRef]

5. Levin, G.V. Tagatose, the new GRAS sweetener and health product. *J. Med. Food* **2002**, *5*, 23–36. [CrossRef] [PubMed]

6. Roy, S.; Chikkerur, J.; Roy, S.C.; Dhali, A.; Kolte, A.P.; Sridhar, M.; Samanta, A.K. Tagatose as a potential nutraceutical: Production, properties, biological roles, and applications. *J. Food Sci.* **2018**, *83*, 2699–2709. [CrossRef] [PubMed]

7. Rubio-Arraez, S.; Ferrer, C.; Capella, J.V.; Ortolá, M.D.; Castelló, M.L. Development of lemon marmalade formulated with new sweeteners (isomaltulose and tagatose): Effect on antioxidant, rheological and optical properties. *J. Food Process Eng.* **2017**, *40*, e12371. [CrossRef]

8. Manzo, R.M.; Antunes, A.S.L.M.; de Sousa Mendes, J.; Hissa, D.C.; Gonçalves, L.R.B.; Mammarella, E.J. Biochemical characterization of heat-tolerant recombinant L-arabinose isomerase from *Enterococcus faecium* DBFIQ E36 strain with feasible applications in D-tagatose production. *Mol. Biotechnol.* **2019**, *61*, 385–399. [CrossRef] [PubMed]

9. Liang, Y.X.; Wen, P.; Wang, Y.; Ouyang, D.M.; Wang, D.; Chen, Y.Z.; Song, Y.; Deng, J.; Sun, Y.M.; Wang, H. The constipation-relieving property of D-tagatose by modulating the composition of gut microbiota. *Int. J. Mol. Sci.* **2019**, *20*, 5721. [CrossRef] [PubMed]

10. Espinosa, I.; Fogelfeld, L. Tagatose: From a sweetener to a new diabetic medication? *Expert Opin. Investig. Drugs* **2010**, *19*, 285–294. [CrossRef] [PubMed]

11. Lu, Y.; Levin, G.V.; Donner, T.W. Tagatose, a new antidiabetic and obesity control drug. *Diabetes Obes. Metab.* **2008**, *10*, 109–134. [CrossRef] [PubMed]

12. Nagata, Y.; Mizuta, N.; Kanasaki, A.; Tanaka, K. Rare sugars, D-allulose, D-tagatose and D-sorbose, differently modulate lipid metabolism in rats. *J. Sci. Food Agric.* **2018**, *98*, 2020–2026. [CrossRef] [PubMed]

13. Cheng, S.; Metzger, L.E.; Martínez-Monteagudo, S.I. One-pot synthesis of sweetening syrup from lactose. *Sci. Rep.* **2020**, *10*, 1–9. [CrossRef] [PubMed]

14. Drabo, P.; Delidovich, I. Catalytic isomerization of galactose into tagatose in the presence of bases and Lewis acids. *Catal. Commun.* **2018**, *107*, 24–28. [CrossRef]

15. Freimund, S.; Huwig, A.; Giffhorn, F.; Köpper, S. Convenient chemo-enzymatic synthesis of D-tagatose. *J. Carbohydr. Chem.* **1996**, *15*, 115–120. [CrossRef]

16. Sha, F.; Zheng, Y.; Chen, J.; Chen, K.; Cao, F.; Yan, M.; Ouyang, P. D-Tagatose manufacture through bio-oxidation of galactitol derived from waste xylose mother liquor. *Green Chem.* **2018**, *20*, 2382–2391. [CrossRef]

17. Bober, J.R.; Nair, N.U. Galactose to tagatose isomerization at moderate temperatures with high conversion and productivity. *Nat. Commun.* **2019**, *10*, 4548. [CrossRef] [PubMed]

18. Patel, M.J.; Patel, A.T.; Akhani, R.; Dedania, S.; Patel, D.H. Bioproduction of D-tagatose from D-galactose using phosphoglucose isomerase from *Pseudomonas aeruginosa* PAO1. *Appl. Biochem. Biotechnol.* **2016**, *179*, 715–727. [CrossRef] [PubMed]

19. Jagtap, S.S.; Singh, R.; Kang, Y.C.; Zhao, H.; Lee, J.K. Cloning and characterization of a galactitol 2-dehydrogenase from *Rhizobium legumenosarum* and its application in D-tagatose production. *Enzyme Microb. Technol.* **2014**, *58–59*, 44–51. [CrossRef] [PubMed]

20. Boudebbouze, S.; Maguin, E.; Rhimi, M. Bacterial L-arabinose isomerases: Industrial application for D-tagatose production. *Recent Pat. DNA Gene Seq.* **2011**, *5*, 194–201. [CrossRef] [PubMed]

21. Chouayekh, H.; Bejar, W.; Rhimi, M.; Jelleli, K.; Mseddi, M.; Bejar, S. Characterization of an L-arabinose isomerase from the Lactobacillus plantarum NC8 strain showing pronounced stability at acidic pH. *FEMS Microbiol. Lett.* **2007**, *277*, 260–267. [CrossRef] [PubMed]

22. De Sousa, M.; Melo, V.M.M.; Hissa, D.C.; Manzo, R.M.; Mammarella, E.J.; Antunes, A.S.L.M.; García, J.L.; Pessela, B.C.; Gonçalves, L.R.B. One-step immobilization and stabilization of a recombinant *Enterococcus faecium* DBFIQ E36 L-arabinose isomerase for D-tagatose synthesis. *Appl. Biochem. Biotechnol.* **2019**, *188*, 310–325. [CrossRef] [PubMed]

23. Du, M.; Zhao, D.; Cheng, S.; Sun, D.; Chen, M.; Gao, Z.; Zhang, C. Towards efficient enzymatic conversion of D-galactose to D-tagatose: Purification and characterization of L-arabinose isomerase from Lactobacillus brevis. *Bioprocess Biosyst. Eng.* **2019**, *42*, 107–116. [CrossRef] [PubMed]

24. Mei, W.; Wang, L.; Zang, Y.; Zheng, Z.; Ouyang, J. Characterization of an L-arabinose isomerase from Bacillus coagulans NL01 and its application for D-tagatose production. *BMC Biotechnol.* **2016**, *16*, 55. [CrossRef] [PubMed]

25. Nguyen, T.K.; Hong, M.G.; Chang, P.S.; Lee, B.H.; Yoo, S.H. Biochemical properties of L-arabinose isomerase from Clostridium hylemonae to produce D-tagatose as a functional sweetener. *PLoS ONE* **2018**, *13*, e0196099. [CrossRef] [PubMed]

26. Zheng, Z.; Mei, W.; Xia, M.; He, Q.; Ouyang, J. Rational DESIGN of *Bacillus coagulans* NL01 L-arabinose isomerase and use of its F279I variant in D-tagatose production. *J. Agric. Food Chem.* **2017**, *65*, 4715–4721. [CrossRef] [PubMed]

27. Liu, Y.; Li, S.; Xu, H.; Wu, L.; Xu, Z.; Liu, J.; Feng, X. Efficient production of D-tagatose using a food-grade surface display system. *J. Agric. Food Chem.* **2014**, *62*, 6756–6762. [CrossRef] [PubMed]

28. Guo, Q.; An, Y.; Yun, J.; Yang, M.; Magocha, T.A.; Zhu, J.; Xue, Y.; Qi, Y.; Hossain, Z.; Sun, W.; et al. Enhanced D-tagatose production by spore surface-displayed L-arabinose isomerase from isolated Lactobacillus brevis PC16 and biotransformation. *Bioresour. Technol.* **2018**, *247*, 940–946. [CrossRef] [PubMed]

29. Jeong, D.W.; Hyeon, J.E.; Shin, S.K.; Han, S.O. Trienzymatic complex system for isomerization of agar-derived D-galactose into D-tagatose as a low-calorie sweetener. *J. Agric. Food Chem.* **2020**, *68*, 3195–3202. [CrossRef] [PubMed]

30. Zhang, G.; Zabed, H.M.; Yun, J.; Yuan, J.; Zhang, Y.; Wang, Y.; Qi, X. Two-stage biosynthesis of D-tagatose from milk whey powder by an engineered Escherichia coli strain expressing L-arabinose isomerase from Lactobacillus plantarum. *Bioresour. Technol.* **2020**, *305*, 123010. [CrossRef] [PubMed]

31. Illanés, A. Whey upgrading by enzyme biocatalysis. *Electron. J. Biotechnol.* **2011**, *14*, 9. [CrossRef]

32. Zheng, Z.; Xie, J.; Liu, P.; Li, X.; Ouyang, J. Elegant and efficient biotransformation for dual production of D-tagatose and bioethanol from cheese whey powder. *J. Agric. Food Chem.* **2019**, *67*, 829–835. [CrossRef] [PubMed]

33. Shen, J.; Chen, J.; Jensen, P.R.; Solem, C. Sweet as sugar—Efficient conversion of lactose into sweet sugars using a novel whole-cell catalyst. *J. Agric. Food Chem.* **2019**, *67*, 6257–6262. [CrossRef] [PubMed]

34. Torres, P.; Batista-Viera, F. Production of D-tagatose and D-fructose from whey by co-immobilized enzymatic system. *Mol. Catal.* **2019**, *463*, 99–109. [CrossRef]

35. Nath, A.; Verasztó, B.; Basak, S.; Koris, A.; Kovács, Z.; Vatai, G. Synthesis of lactose-derived nutraceuticals from dairy waste whey—A review. *Food Bioprocess Technol.* **2016**, *9*, 16–48. [CrossRef]

36. Jayamuthunagai, J.; Srisowmeya, G.; Chakravarthy, M.; Gautam, P. D-Tagatose production by permeabilized and immobilized *Lactobacillus plantarum* using whey permeate. *Bioresour. Technol.* **2017**, *235*, 250–255. [CrossRef] [PubMed]

37. Wanarska, M.; Kur, J. A method for the production of D-tagatose using a recombinant *Pichia pastoris* strain secreting β-D-galactosidase from *Arthrobacter chlorophenolicus* and a recombinant L-arabinose isomerase from Arthrobacter sp. 22c. *Microb. Cell Fact.* **2012**, *11*, 113. [CrossRef] [PubMed]

38. Xu, Z.; Xu, Z.; Tang, B.; Li, S.; Gao, J.; Chi, B.; Xu, H. Construction and co-expression of polycistronic plasmids encoding thermophilic L-arabinose isomerase and hyperthermophilic β-galactosidase for single-step production of D-tagatose. *Biochem. Eng. J.* **2016**, *109*, 28–34. [CrossRef]

39. Zhan, Y.; Xu, Z.; Li, S.; Liu, X.; Xu, L.; Feng, X.; Xu, H. Coexpression of β-D-galactosidase and L-arabinose isomerase in the production of D-tagatose: A functional sweetener. *J. Agric. Food Chem.* **2014**, *62*, 2412–2417. [CrossRef] [PubMed]

40. Plou, F.J.; Polaina, J.; Sanz-Aparicio, J.; Fernandez-Lobato, M. β-Galactosidases for lactose hydrolysis and galactooligosaccharide synthesis. In *Microbial Enzyme Technology in Food Applications*; Ray, R.C., Rosell, C.M., Eds.; CRC Press: Boca Raton, FL, USA, 2016; pp. 123–146.

41. Jørgensen, F.; Hansen, O.C.; Stougaard, P. Enzymatic conversion of D-galactose to D-tagatose: Heterologous expression and characterisation of a thermostable L-arabinose isomerase from Thermoanaerobacter mathranii. *Appl. Microbiol. Biotechnol.* **2004**, *64*, 816–822. [CrossRef] [PubMed]

42. Santibáñez, L.; Fernández-Arrojo, L.; Guerrero, C.; Plou, F.J.; Illanes, A. Removal of lactose in crude galacto-oligosaccharides by β-galactosidase from *Kluyveromyces lactis*. *J. Mol. Catal. B Enzym.* **2016**, *133*, 85–91. [CrossRef]

43. Füreder, V.; Rodriguez-Colinas, B.; Cervantes, F.; Fernandez-Arrojo, L.; Poveda, A.; Jimenez-Barbero, J.; Ballesteros, A.O.; Plou, F.J. Selective synthesis of galactooligosaccharides containing β(1→3) linkages with β-galactosidase from Bifidobacterium bifidum (Saphera). *J. Agric. Food Chem.* **2020**, *68*, 4930–4938. [CrossRef] [PubMed]

44. Hsu, C.A.; Lee, S.L.; Chou, C.C. Enzymatic production of galactooligosaccharides by beta-galactosidase from Bifidobacterium longum BCRC 15708. *J. Agric. Food Chem.* **2007**, *55*, 2225–2230. [CrossRef] [PubMed]

45. Goulas, A.; Tzortzis, G.; Gibson, G.R. Development of a process for the production and purification of alpha- and beta-galactooligosaccharides from Bifobacterium bifidum NCIMB 41171. *Int. Dairy J.* **2007**, *17*, 648–656. [CrossRef]

46. Mattanovich, D.; Graf, A.; Stadlmann, J.; Dragosits, M.; Redl, A.; Maurer, M.; Kleinheinz, M.; Sauer, M.; Altmann, F.; Gasser, B. Genome, secretome and glucose transport highlight unique features of the protein production host Pichia pastoris. *Microb. Cell Fact.* **2009**, *8*, 29. [CrossRef] [PubMed]

47. Ávila-Fernández, Á.; Galicia-Lagunas, N.; Rodríguez-Alegría, M.E.; Olvera, C.; López-Munguía, A. Production of functional oligosaccharides through limited acid hydrolysis of agave fructans. *Food Chem.* **2011**, *129*, 380–386. [CrossRef] [PubMed]

48. Potvin, G.; Zhang, Z.; Defela, A.; Lam, H. Screening of alternative carbon sources for recombinant protein production in Pichia pastoris. *Int. J. Chem. React. Eng.* **2016**, *14*, 251–257. [CrossRef]

49. Rhimi, M.; Bejar, S. Cloning, purification and biochemical characterization of metallic-ions independent and thermoactive L-arabinose isomerase from the Bacillus stearothermophilus US100 strain. *Biochim. Biophys. Acta Gen. Subj.* **2006**, *1760*, 191–199. [CrossRef] [PubMed]

50. Rhimi, M.; Aghajari, N.; Juy, M.; Chouayekh, H.; Maguin, E.; Haser, R.; Bejar, S. Rational design of Bacillus stearothermophilus US100 L-arabinose isomerase: Potential applications for D-tagatose production. *Biochimie* **2009**, *91*, 650–653. [CrossRef] [PubMed]

51. Lee, S.-J.; Lee, D.-W.; Choe, E.-A.; Hong, Y.-H.; Kim, S.-B.; Kim, B.-C.; Pyun, Y.-R. Characterization of a Characterization of a thermoacidophilic L-arabinose isomerase from *Alicyclobacillus acidocaldarius*: Role of Lys-269 in pH optimum. *Appl. Environ. Microbiol.* **2005**, *71*, 7888. [CrossRef] [PubMed]

52. Rhimi, M.; Juy, M.; Aghajari, N.; Haser, R.; Bejar, S. Probing the essential catalytic residues and substrate affinity in the thermoactive Bacillus stearothermophilus US100 L-arabinose isomerase by site-directed mutagenesis. *J. Bacteriol.* **2007**, *189*, 3556–3563. [CrossRef] [PubMed]

53. Rhimi, M.; Messaoud, E.B.; Borgi, M.A.; khadra, K.B.; Bejar, S. Co-expression of L-arabinose isomerase and D-glucose isomerase in E. coli and development of an efficient process producing simultaneously D-tagatose and D-fructose. *Enzyme Microb. Technol.* **2007**, *40*, 1531–1537. [CrossRef]

54. Zuluaga, J.; Rodríguez, N.; Rivas-Ramirez, I.; De La Fuente, V.; Rufo, L.; Amils, R. An improved semiquantitative method for elemental analysis of plants using inductive coupled plasma mass spectrometry. *Biol. Trace Elem. Res.* **2011**, *144*, 1302–1317. [CrossRef] [PubMed]

55. Britton, H.T.S.; Robinson, R.A. Universal buffer solutions and the dissociation constant of veronal. *J. Chem. Soc.* **1931**, 1456–1462. [CrossRef]

Review

Aspergillus: A Powerful Protein Production Platform

Fani Ntana [1], Uffe Hasbro Mortensen [2], Catherine Sarazin [1,*] and Rainer Figge [1]

[1] Unité de Génie Enzymatique et Cellulaire, UMR 7025 CNRS/UPJV/UTC, Université de Picardie Jules Verne, 80039 Amiens, France; fani.ntana@u-picardie.fr (F.N.); rainer.figge@u-picardie.fr (R.F.)

[2] Department of Biotechnology and Biomedicine, Technical University of Denmark, Søltofts Plads, Building 223, 2800 Kongens Lyngby, Denmark; um@bio.dtu.dk

* Correspondence: catherine.sarazin@u-picardie.fr; Tel.: +33-3-22-82-75-95

Received: 29 August 2020; Accepted: 11 September 2020; Published: 16 September 2020

Abstract: Aspergilli have been widely used in the production of organic acids, enzymes, and secondary metabolites for almost a century. Today, several GRAS (generally recognized as safe) Aspergillus species hold a central role in the field of industrial biotechnology with multiple profitable applications. Since the 1990s, research has focused on the use of Aspergillus species in the development of cell factories for the production of recombinant proteins mainly due to their natively high secretion capacity. Advances in the Aspergillus-specific molecular toolkit and combination of several engineering strategies (e.g., protease-deficient strains and fusions to carrier proteins) resulted in strains able to generate high titers of recombinant fungal proteins. However, the production of non-fungal proteins appears to still be inefficient due to bottlenecks in fungal expression and secretion machinery. After a brief overview of the different heterologous expression systems currently available, this review focuses on the filamentous fungi belonging to the genus Aspergillus and their use in recombinant protein production. We describe key steps in protein synthesis and secretion that may limit production efficiency in Aspergillus systems and present genetic engineering approaches and bioprocessing strategies that have been adopted in order to improve recombinant protein titers and expand the potential of Aspergilli as competitive production platforms.

Keywords: Aspergillus; fermentation; filamentous fungi; genetic engineering; heterologous expression; recombinant protein; secretion; transcriptional regulation

---

1. Introduction

Proteins are functionally versatile biomolecules (e.g., enzymes, structural proteins, and hormones) involved in multiple biological processes in the cell. Despite their role in supporting biological systems, proteins have been extensively studied for their potential in the formulation of commercial products. They often find applicability in the production of pharmaceuticals, food, beverages, biofuels, cosmetics, detergents, etc. [1,2].

Market demand for industrially relevant proteins has guided research into exploring practices that can lead to large-scale production levels [3]. The development of recombinant DNA technology has opened up the possibility of producing recombinant proteins in heterologous expression systems that can support high production yields. In that respect, any gene can now be transferred into a production host able to generate large quantities of the corresponding protein of interest, avoiding limitations related to the conventional extraction of the protein from its native host [4]. Human insulin produced in *E. coli* cells was the first recombinant protein that was actually approved by the FDA for clinical use. The recombinant insulin Humulin®, originally developed by Genentech, was eventually commercialized in 1982 [3]. Since then, a plethora of other proteins with pharmaceutical and industrial applications have successfully been synthesized in heterologous expression systems and have made their way into the market [1].

Today, recombinant proteins can be synthesized using a wide range of production platforms, including bacteria, yeasts and filamentous fungi, mammalian or insect cells, and transgenic plants, to name a few. Every heterologous production system though comes with certain advantages and drawbacks (Table 1). In most cases, the structure and function of the protein of interest determines which production system is the most appropriate to be used. For example, when it comes to manufacturing therapeutic proteins of high quality, mammalian cell lines are predominantly used, as they can produce complex, human-like glycosylated proteins that are safe for patients. In fact, almost 84% of the biopharmaceutical proteins are currently produced by Chinese Hamster Ovarian (CHO) cell lines [5].

For the production of non-medicinal proteins, a more economical approach is usually followed, using either bacterial or fungal production hosts [1,6,7]. While bacteria are often suitable for smaller proteins that do not require complex post-translational modifications, production of larger and more complex proteins is usually performed in yeast, e.g., *Pichia pastoris* [8]. However, yeasts have the tendency to hyperglycosylate secreted proteins, and thus reduce their in vivo half-life and affect their efficacy [9]. Additional limitations including low expression levels and plasmid instability have restricted the use of some yeasts (e.g., *S. cerevisiae*) in the production of industrial enzymes [10]. An alternative production platform that can support low-cost synthesis of large proteins with complex modifications, but with a lesser degree of hypermannosylation during glycosylation compared to yeast is filamentous fungi. In addition, due to their saprophytic lifestyle, most filamentous fungi have already developed the ability to produce and secrete a vast amount of enzymes in order to break down and feed on organic matter [11]. Strains belonging to the genera Aspergillus, Trichoderma, and Neurospora are in fact widely used for production of recombinant proteins with industrial applications [12–15]. Several reviews have described the potentials of filamentous fungi in the production of pharmaceutical and other industrial proteins, as well as the genetic engineering approaches followed to maximize production levels [7,16,17]. In this review, we specifically focus on the use of Aspergillus species in the manufacturing of recombinant proteins. Bottlenecks in protein synthesis and secretion are discussed, while our comprehensive literature search provides a general overview of the most important genetic engineering projects and bioprocessing strategies applied over the past 30 years to improve recombinant protein yields in Aspergillus.

Table 1. Comparison of the most commonly used heterologous expression systems in the field of recombinant protein production.

Expression Platform	Genetic Manipulation	Growth Rate	Product Titers	Product Quality	Product Purification	Contamination Risk	Production Cost	Relevant Literature
Bacteria (*Escherichia coli*)	Simple	Fast	High	Products can be non-functional (codon bias, no adequate post-translation modifications)	Can be problematic (e.g., inclusion bodies)	Medium (endotoxins)	Low	[18,19]
Yeasts (*Saccharomyces cerevisiae, Pichia pastoris*, etc.)	Simple	Fast	*S. cerevisiae* limited *P. pastoris* higher	Hypermannosylation of glycoproteins often occurs (shortens half-life of the protein in vivo, leads to immunogenic reactions)	Feasible	Low	Low	[9,10]
Filamentous fungi (*Aspergillus niger, Trichoderma reesei, Neurospora crassa*)	Feasible	Medium	High	Less hypermannosylation compared to yeasts, but still differences from mammalian glycosylation patterns	Simple	Medium (mycotoxins)	Low	[7]
Insect cells (*Spodoptera frugiperda, Drosophila melanogaster*)	Laborious	Fast	High	Not able to carry out N-glycosylation	Feasible	Very low	High	[20]
Mammalian cells (CHO cells, Human cell lines)	Laborious	Slow	Low	High quality therapeutic proteins, human-like glycosylation pattern	Simple	High (viruses and prions)	High	[21]
Transgenic animals (goats, chickens)	Laborious	Very slow	High	High quality therapeutic proteins	Simple	High (viruses and prions)	High, ethically questionable	[22]
Transgenic plants (rice, bananas, carrots, potatoes)	Feasible	Slow	High	Some differences in glycan structures from human-like pattern	Complex and expensive downstream processing	Very low	Medium	[23,24]

2. Industrial Application of Aspergilli

2.1. Traditional Uses of Aspergillus Species

The use of Aspergillus species in biotechnology begun approximately a century ago, when James Currie, a food chemist, discovered that the filamentous mold *A. niger* was able to produce citric acid, a food and beverage additive that was conventionally extracted from citrus fruits [25]. Since then, production of citric acid, now performed in *A. niger* cultures that grow on inexpensive sugar-based minimal media, has turned into a multibillion dollar business [26].

Nonetheless, industrial applications of Aspergilli are not limited to the production of citric acid. Several species have been used as prolific producers of other organic acids (e.g., itaconic), secondary metabolites, and enzymes of biotechnological significance [11]. For example, *A. niger* produces several enzymes used in food and feed production such as glucoamylases, proteases, and phytases [26]. *A. oryzae*, traditionally used in Asian cuisine, has been exploited as a cell factory for producing malate, which is used in the development of food and pharmaceutical products [27]. *A. terreus* has attracted interest due to its ability to produce a group of secondary metabolites called statins that are used in the production of cholesterol-lowering drugs [28]. In fact, AB Enzymes, BASF, Chr. Hansen, DuPont, and Novozymes are only a few examples of companies that have been or are still using Aspergillus species in large-scale manufacturing of commercial products such as organic acids, enzymes, proteins, and secondary metabolites [29].

2.2. The Use of Aspergillus Species in Heterologous Protein Production

Filamentous fungi are generally considered promising hosts for production of recombinant proteins, mainly due to their secretory capacity and metabolic versatility. However, only a few species appear to be able to produce competitive recombinant protein levels and even fewer have been developed into industrial production platforms. This can be attributed mainly to our incomplete knowledge of fungal physiology. For example, the mechanisms behind protein production and secretion in fungal cells are not yet fully understood for most of the species. In addition, the presence of unwanted metabolites (e.g., mycotoxins) has excluded several fungi from industrial production [29].

Aspergillus is a genus that has been studied extensively due to its value as a model organism in fungal research (*A. nidulans*) and its industrial importance in citric acid and enzyme production (*A. niger*, *A. oryzae*) [26]. Several molecular tools (e.g., synthetic promoters and terminators, selection markers, RNA interference-RNAi, and CRISPR-Clusters of Regularly Interspaced Short Palindromic Repeats-associated technologies), suitable for Aspergillus species, have also been developed, facilitating efficient and targeted manipulation of their genomes [30,31]. CRISPR/Cas, for example, a system developed to create site-specific double strand DNA breaks, has been successfully applied in editing the genome of *A. niger* [32–35], *A. nidulans* [35], *A. oryzae* [36], *A. fumigatus* [37], and other aspergilli [35]. With a relatively well-understood physiology (growth and development, gene expression, and secretion machinery) and several molecular tools available, the GRAS *A. niger* has already been used in industrial production of recombinant proteins, such as calf chymosin [38], human lactoferrin [39], and the plant-derived sweetener neoculin [40]. Nevertheless, heterologous protein production in Aspergillus species is not always efficient, leading to low production titers. In such cases, strategies that are usually applied to improve titers involve genetic engineering of the production strains and establishing the appropriate fermentation conditions.

3. Genetic Engineering Approaches for Aspergillus Strain Improvement

Due to their capacity to secrete large quantities of proteins into the culture medium, Aspergillus species, and especially *A. niger*, are considered promising candidates for the development of large-scale heterologous protein production platforms. However, production yields for heterologous proteins are usually much lower compared to the ones detected for the native proteins. Failure to achieve the desired protein amounts in Aspergillus cultures can be attributed to limitations related to transcription, translation, and the post-translation processing and modifications during protein production. Additionally, bottlenecks in the fungal secretion machinery and the problem of extracellular degradation by fungal proteases further hinder the efficient production of foreign proteins in Aspergillus species [41]. These limitations during protein production in aspergilli will be discussed in detail in the following paragraphs.

3.1. Transcriptional Regulation

3.1.1. Promoters

Regulation of protein synthesis begins on the level of transcription. The first step for achieving high protein yields in heterologous production systems is the use of strong promoters that can drive high gene expression. A variety of constitutively active (e.g., PgpdA, glyceraldehyde-3-phosphate dehydrogenase promoter; PadhA, aldehyde dehydrogenase promoter; Ptef1, translation elongation factor 1 promoter; and Ph4h3, histones H4.1 and H3 bidirectional promoter) and inducible promoters (e.g., PglaA, glucoamylase promoter; PalcC, alcohol dehydrogenase promoter; and PamyA, amylase promoter) are currently available for Aspergillus species [42,43].

Native inducible promoters are commonly used as being more efficient in achieving high protein titers, as they allow separation of protein synthesis from biomass formation. This separation can also be extremely useful when the protein to be produced is toxic for the fungus [42]. The inducible promoter of the *A. niger* glucoamylase gene (PglaA) is frequently used in many Aspergillus expression systems. Expression of *gla*A is highly induced when maltose or starch are used as carbon sources, but repressed in the presence of xylose. High *gla*A expression levels have been correlated with a 5′ cis-regulatory element, and specifically the region within 500 bp upstream of the translational start codon. This region (−464 to −426) contains a protein-binding CCAAT motif, crucial for the high activity of PglaA [44]. Insertion of eight copies of this region into the PglaA sequence significantly increased expression levels of a heterologous gene (*Vitreoscilla* haemoglobin), multiplying protein production by almost 20-fold [45] (Table 2).

Table 2. Approaches for improving recombinant protein production through promoter engineering.

Process	Modification	Performance	Improvement Factor	Reference
Promoters	Use of several promoters (P) in *A. awamori*	PB2 from *Acremonium chrysogenum*: 0.25–2 mg/L thaumatin PpcbC from *Penicillium chrysogenum*: 0.25–2 mg/L thaumatin PgdhA from *A. awamori*: 1–9 mg/L thaumatin PgpdA from *A. nidulans*: 0.75–11 mg/L thaumatin	-	[46]
	Insertion of multiple copies of an activator protein-binding site from the cis-regulatory region of *A. niger gla*A to the new promoter in *A. niger*	396.0 ± 51.5 mg/L of *Vitreoscilla* hemoglobin compared to 19.7 ± 4.8 mg/L from the strain with 1 copy	20	[45]
	Use of hybrid promoters (combination of a human hERa-activated promoter (pERE), *S. cerevisiae URA3* promoter and *A. nidulans nirA* promoter) in *A. nidulans*	pERE-RS-*nirA* + *lacZ*: 25 U of β-galactosidase activity/mg of protein	-	[47]
		pERE-URA-*nirA* + *lacZ*: 100 U of β-galactosidase activity/mg of protein	4	
		pERE-URA-RS + *lacZ*: 1400 U of β-galactosidase activity/mg of protein [1 pM inducer (DES)]	56	

Table 2. *Cont.*

Process	Modification	Performance	Improvement Factor	Reference
	Use of a hemolysin-like protein promoter (Phyl) for heterologous production in *A. oryzae*	Reporter gene: Endoglucanase Cel B Pamy: 24.1 ± 5.5 U/mL, Phyl: 57.9 ± 17.4 U/mL	2.4	
		Reporter gene: *Trichoderma* endoglucanase I Pamy: 7.7 ± 3.9 U/mL, Phyl: 27.8 ± 1.3 U/mL	3.6	[48]
		Reporter gene: *Trichoderma* endoglucanase III Pamy:4.0 ± 0.6 U/mL, hyl:31.7 ± 3.3 U/mL	7.9	
	Regulatory elements (TerR and PterA) from *A. terreus* terrain gene cluster *for* *E. coli lacZ* expression in *A. niger*	Promoter activity ~5000 mU/mg when TerR under PgpdA (No activity when TerR under the native promoter)	-	
		Promoter activity ~10,000 mU/mg (when TerR under PgpdA in 2 copies)	2	[49]
		Promoter activity ~15,000 mU/mg (when TerR under PamyB)	3	
	A. niger α-glucosyltransferase produced under *the A. niger* pyruvate kinase promoter	2000 U/mL total activity of α-glucosyltransferase compared to 600 U/mL in the wild type	3.3	[50]
	Overexpression of the transcription factor RsmA, while the aflR promoter was inserted in front of the *pslcc* in *A. nidulans*	0.06 U/mL of *Pycnoporus sanguineus* laccase compared to 0.004 U/mL in the control strain	15	[51,52]
	A novel promoter from *Talaromyces emersonii* (Pglucan1200) for expressing *glaA* in *A. niger*	6000 U/mL of GlaA, enzyme activity increased by about 25% compared to 5000 U/mL in the strain with the PglaA	1.2	[53]
	The constitutive promoter of *ecm33* (Pecm33) from *A. niger* in *A. niger* — Maltose:	Pecm33 activity induced by 1.7 compared to PglaA activity that induced by 2.7		
	Glucose:	Pecm33 activity induced by 1.1 compared to PglaA activity that induced by 1.8	-	[54]
	Xylose:	Pecm33 activity induced by 2 compared to PglaA activity that induced by 1.3 Increased Pecm33 activity at 37 °C		

Although the majority of the promoters used are derived from the primary metabolism, there have also been attempts to develop expression cassettes using regulatory elements from fungal secondary metabolite pathways (Table 2) [49,51,52]. In such cases, the strain carries the heterologous gene under the control of an inducible promoter and it is engineered to overexpress the gene that encodes the transcription factor, which activates the specific promoter, thus achieving high expression levels.

Apart from the endogenous promoters discussed above, non-endogenous, or synthetic, tunable promoter systems have also been developed for Aspergillus species [47,55–57]. One of them, the Tet On/Off system, was adapted from the mechanism regulating the tetracycline resistance operon in *E. coli*. It is also based on a dual player system, where the production strain is co-transformed with a plasmid carrying the heterologous gene regulated by the activity of a tetracycline-responsive promoter and a plasmid encoding a tetracycline transactivator (tTA). In the absence of the tetracycline or its derivative, doxycycline (DOX), the tTA binds to the promoter, inducing heterologous gene expression. However, where DOX is added, tTA disassociates from the binding sites and expression shuts down [56].

3.1.2. Gene Copy Number and Integration Site

In general, it has been suggested that integration of multiple gene copies can result in increased protein production levels. This has been observed frequently in Aspergillus systems expressing native proteins, such as glucoamylase or amylase (Table 3) [58–60]. In fact, *A. niger* strains containing multiple *glaA* copies (20 and 80) were able to secrete five to eight times more glucoamylase [59]. Similarly, introducing additional copies of heterologous genes also leads to higher amounts of protein produced (Table 3) [47,61]. However, this is not always the case. While studying the effect of copy number on heterologous expression in *A. nidulans*, Lubertozzi and Keasling (2006) observed that β-galactosidase activity of the transformants did not consistently correlate with the *lacZ* dosage. They suggested that this could be due to a gene silencing mechanism, previously observed in filamentous fungi, or due to pleiotropic effects of random integration [62]. In addition, Verdoes et al. (1995) suggests that the reasons for this limitation in protein production by strains harboring multiple copies is both the site of integration and the availability of trans-regulatory factors involved in transcription [63].

Table 3. Approaches for improving recombinant protein production through integration of multiple gene copies.

Process	Modification	Performance	Improvement Factor	Reference
Copy number	Integration of up to 200 additional copies of the *glaA* in *A. niger*	355 mg/L of glucoamylase compared to parental strain (50 mg/L)	7.1	[58]
	Integration of 80 additional copies of the *glaA* in *A. niger*	1268 mU of glucoamylase/mL culture filtrate compared to 280 mU/mL in the parental strain	4.5	[59]
	Integration of multiple copies of cassettes with thaumatin gene in *A. awamori* (two types of cassettes, 1 with PB2 and 1 with PgdhA)	8 copies: 10 ± 0.4 mg/L thaumatin	-	[46]
		11 copies: 14 ± 1.1 mg/L thaumatin	1.4	
		14 copies: 11 ± 0.8 mg/L thaumatin	1.1	
		10 copies: 14 ± 1.3 mg/L thaumatin	1.4	
	Insertion of multiple copies of the cassette pERE-URA-RS + *lacZ* in *A. nidulans*	1 copy of pERE-URA-RS + *lacZ*: 9500 U of β-galactosidase/mg of protein	5.4	[47]
		Multiple copies of pERE-URA-RS + *lacZ*: 51,000U of β-galactosidase/mg of protein [1 nM inducer (DES)]		
	Integration of an additional copy of the *glaA*- RFP (2 in total) in *A. nidulans*	A 70% increase in maximum fluorescence level (quantification data not available)	1.7	[61]

The genomic site, where the expression cassette is integrated during transformation, is indeed an important factor that influences transcription of the heterologous gene and consequently protein production [62,63]. In order to tackle unpredicted limitations related to random insertion, site-specific integration systems are usually applied [32,64]. Moreover, identification of genomic loci with high transcriptional activity, followed by targeted integration of expression cassettes in the specific sites, provides an additional approach for boosting transgene expression and recombinant protein synthesis [65]. A promising integration site for expression and characterization of heterologous genes was identified in the genome of *A. nidulans*. Insertion of heterologous genes in the specific locus (Integration Site 1-IS1) did not interfere with the fitness of the strain, while it allowed for high and stable expression levels in a tissue-independent manner [66].

3.2. Translational Regulation

Codon Usage and mRNA Stability

Most of the amino acids found in nature are encoded by more than one codon (synonymous codons). The preference for one codon over another, known as codon usage, varies among organisms and can be a limitation when heterologous genes are expressed in a host with different codon usage compared to the codon usage of the organism from which the genes has been isolated [67]. Codon optimization is commonly used in such cases, where rare codons found in heterologous genes are replaced by synonymous codons that encode the same amino acid but are more frequently used in the expression host. Codon optimization has been proposed as a successful practice for improving mRNA stability and translational efficiency, leading to higher levels of heterologous protein production [68,69].

Several studies in recombinant protein production in Aspergillus species have reported that codon optimized genes are expressed more efficiently, resulting in improved heterologous protein yields (Table 4) [69–72]. For example, *A. oryzae* strains carrying a codon optimized *der f7*, a gene that encodes *Dermatophagoides farina* mite allergen 7, showed increased gene expression levels and it produced almost three to five more protein than the strains with the non-optimized gene [70]. Tanaka et al. (2012) showed later that the increase in transcriptional and translational efficiency was clearly assigned to improved mRNA stability due to codon optimization of *der f7* [73].

Table 4. Approaches for improving recombinant protein production through codon optimization of the expressed sequences.

Process	Modification	Performance	Improvement Factor	Reference
Codon usage	A *Cyamopsis tetragonoloba* α-galactosidase gene optimized based on *Saccharomyces cerevisiae* codon usage and expressed in *A. awamori*	Synthetic gene: 0.4 mg/L α-galactosidase Wild type gene: Undetectable levels	-	[72]
	A *Solanum tuberosum* α-glucan phosphorylase synthetic gene optimized based on *A. niger*-preferred codon usage for production in *A. niger*	Synthetic gene: 39.6–94.6 mg/L α-glucan phosphorylase Wild type gene: <0.1 mg/L	-	[71]
	A codon optimized *Dermatophagoides farina der f7* gene based on *A. oryzae* codon usage and expressed in *A. oryzae*	Non-fused: undetectable level to a detectable level Fused to GlaA: 3 to 5 fold increase (Signal intensity quantification of the bands from the SDS-PAGE)	3–5	[70]

However, the codon usage of the native gene "donor" hosts is not trivial. It has been shown that codon usage influences local rates of translation elongation (preferred codons speed up elongation and rare codons slow it down), assisting in proper co-translational protein folding. This means that codon optimization of a genetic sequence that results in increased translation velocity, can also disrupt protein folding, secretion and activity [74].

In general, post-transcriptional events, such as mRNA processing (addition of the 3′polyA-tail or the 5′cap) and base modifications, the presence of rare codons or destabilizing sequences can have an impact on the length and stability of mRNA, thus limiting efficient translation and protein synthesis [75]. Limitations related to mRNA stability can also be addressed by using gene fusions, which contain heterologous proteins fused to a well-secreted carrier protein [75], a strategy that will be discussed later in the review.

3.3. Glycosylation

Glycosylation is a post-translational modification of eukaryotic proteins, during which glycans are added on specific amino acid residues. Two main types of glycosylation have been described: the *O*-linked glycosylation, where glycans are added on the side-chain hydroxyl groups of serine or threonine residues [76], and the *N*-linked glycosylation, which takes place on the side-chain amino groups of asparagine residues (at a N-X-S/T motif, where X is any amino acid except proline) [77].

Glycosylation is crucial for glycoproteins, as it influences their stability, activity, but also their passage through the secretory pathway [78,79] (see also below Section 3.4.1. The fungal Secretory Pathway—Glycosylation).

Filamentous fungi, and thus aspergilli, have the ability to perform post-translational modifications, including glycosylation, making them appropriate hosts for production of eukaryotic proteins. However, when it comes to production of mammalian proteins, fungal glycosylation is a bottleneck [78]. Filamentous fungi typically produce *N*-glycans that differ in composition and structure from mammalian ones. This can create problems during heterologous production of mammalian proteins in fungal expression systems, such as incorrect folding and subsequent elimination of the aberrant proteins by the cell quality control mechanisms. Moreover, addition of unusual fungal glycan structures on the heterologous protein may affect its stability and activity, or in the case of therapeutic proteins, increase their immunogenicity [80].

During *N*-linked glycosylation, the glycan (Glc3Man9GlcNAc2), consisting of three glucose (Glc3), nine mannose (Man9), and two N-acetylglucosamine (GlcNAc2) residues, is transferred to the conserved consensus sequence Asn-X-Ser/Thr. Shortly after, the removal of three Glc and one Man sugar results in the Man8GlcNAc2 glucan, a process conserved among all eukaryotes. However, further modifications on Man8GlcNAc2 differ between mammalian and fungal cells. Filamentous fungi and yeasts usually produce small, high-mannose *N*-glycans, while the mammalian glycans are more complex, containing *N*-acetylglucosamine, galactose, fucose, and sialic acid.

Several attempts have been made to engineer the glycosylation pathway in aspergilli towards the synthesis of complex mammalian-like glycans [81–84]. Kainz et al. (2008) focused on two crucial steps in the pathway for obtaining glycoproteins of mammalian type: First, the trimming of terminal mannose residues from Man8GlcNAc2 to Man5GlcNAc2 structures, a process catalyzed by a mannosidase, and second, the subsequent transfer of *N*-acetylglucosamine to yield GlcNAcMan5GlcNAc2, catalyzed by a glycosyltransferase. Insertion of genes encoding the α-1,2-mannosidase and β-1,2-*N*-acetylglucosaminyltransferase I in *A. niger* and *A. nidulans* resulted in the synthesis of glycans with the desirable structure described before. In addition, deleting a gene (*algC*) involved in the early steps of fungal glycosylation, further contributed to the synthesis of glycan structures resembling those of humanized glycoproteins [81].

3.4. Secretion

Another advantage of using Aspergillus species as protein production systems is their natural capacity to secrete high amounts of protein in the extracellular environment. Nevertheless, heterologous proteins often lack specific features of the native secreted proteins, leading to low secretion efficiency and low yields. Therefore, several studies have looked into unraveling the secretion processes in filamentous fungi [85,86]. In fact, multiple bottlenecks of the fungal secretory pathway have been identified and key factors of secretion have been engineered, improving production of heterologous proteins. Alternatively, fusion of these proteins to native, secreted proteins (carrier proteins) has also been able to enhance secretion, bypassing the complexity of engineering steps of fungal secretion [42].

3.4.1. The fungal Secretory Pathway

Proteins moving through the fungal secretory pathway are subjected to several quality control tests until they are finally secreted. Attempts to engineer fungal secretion have been made in all the possible rate-limiting steps, starting from the processing that occurs in the endoplasmic reticulum (ER), to protein transport and degradation pathways (Figure 1).

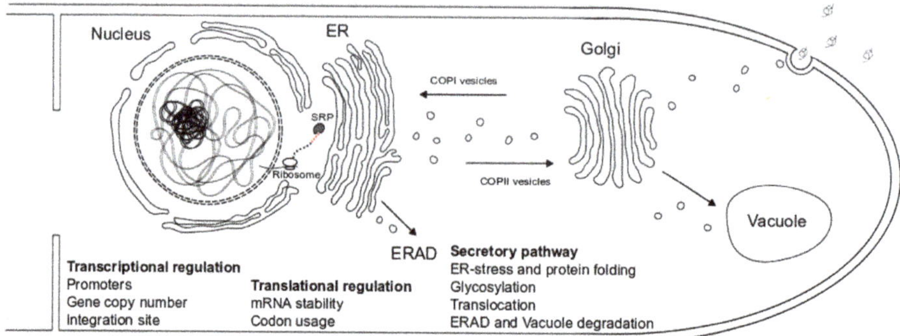

Figure 1. Protein synthesis and secretion in a fungal (e.g., *A. niger*) cell are schematically described. The figure also presents key steps that can be bottlenecks in the production of recombinant proteins in filamentous fungi. At transcription level, high expression of the gene of interest can be achieved by using strong promoters, integrating multiple gene copies and choosing integration sites that positively influences recombinant gene expression. Translation can also be a limiting step in recombinant protein production and can be improved by increasing mRNA stability and adjusting codon usage of the heterologous coding sequence to the fungal host. Following translation, proteins guided by the signal recognition particle (SRP) enter the ER lumen, where they receive post-translation modifications. At this stage, accumulation of unfolded proteins due to high gene expression can lead to ER-stress. Activation of several ER-resident chaperones and foldases (see text for details) can assist with proper folding of most proteins, relieving the overloaded ER. Proteins that fail to be properly folded are guided to the cytoplasm where they are degraded through the ERAD (ER-associated protein degradation) pathway by the proteasome. Proper folding, and consequently efficient secretion, is also dependent on the glycosylation process in the ER. Properly folded and glycosylated proteins destined to be secreted are packed into Coat Protein complex II (COPII) vesicles and transported to the Golgi complex, where further glycosylation takes place. Through a vesicle-mediated process, proteins finally reach the cell surface, where they are released into the periplasmic region. Additional protein translocation pathways, such as retrograde trafficking and vacuole degradation, can indirectly affect protein secretion efficiency. In retrograde trafficking, ER-resident proteins that have escaped or misfolded Golgi-resident proteins are transported from the Golgi back to the ER packed in COPI vesicles. Finally, misfolded and dysfunctional proteins that fail to pass the quality control of the secretory pathway can be transported to the vacuole by Vacuolar protein sorting (Vps) receptors and are degraded (autophagy-related degradation).

- ER-stress and protein folding

Secretion starts with the protein being transported to the ER, where post-translational modifications, like glycosylation, disulfide bond formation, and folding, take place. Accumulation of misfolded or unfolded proteins in the ER can activate stress response pathways in the cell, including the unfolded-protein response (UPR) and the ER-associated degradation (ERAD) pathway (Figure 1). Activation of the UPR usually increases expression of genes related to post-translation protein processing (e.g., lectins, chaperones, foldases, and protein disulfide isomerases), resulting in proper protein folding, and eventually alleviates ER-stress [87].

ER-overload is often observed in heterologous expression systems, where high transcriptional activity leads to buildup of incompletely folded proteins. As a result, efficient protein secretion is hindered, resulting in low production yields of heterologous proteins. In such cases, engineering the UPR is considered to be a promising practice to improve efficiency of heterologous expression systems [85]. Induction of the UPR by overexpressing the gene encoding the UPR transcription factor HacA in *A. awamori* increased production yields of the *Trametes versicolor* laccase by sevenfold and these of bovine preprochymosin by almost threefold [88]. In addition, constitutive expression of *hacA* resulted in 1.5 times higher production levels of neoculin, a plant nonglycemic sweetener, in *A. oryzae* cultures [40]. Many genetic engineering projects have also targeted single UPR components (e.g., ER-resident chaperones and foldases) for enhancing protein secretion in several Aspergillus species (Table 5) [88–93]. For example, overexpression of the gene encoding the lectin-like chaperone calnexin resulted in 60–73 mg/L of the *Phanerochaete chrysosporium* manganese peroxidase in *A. niger* mutants, while the parental strain production reached only 14 mg/L [92]. However, a positive correlation between chaperone overexpression and protein secretion appears to depend specifically on the protein to be produced. For example, increased production of the chaperone protein BipA negatively affected production of the *P. chrysosporium* manganese peroxidase in *A. niger* [92], but improved production titers of the plant sweet protein thaumatin in *A. awamori* [93].

Table 5. Approaches for improving recombinant protein production through engineering unfolded-protein response (UPR) response.

Process	Modification	Performance	Improvement Factor	Reference
ER-stress and protein folding	Overexpression of *prpA* (multicopy integrated vector) in *A. niger* var. *awamori*	The level of chymosin by the control transformants was similar to the transformants with overexpression of the *prpA*	-	[90]
	Deletion of *prpA* in *A. niger* var. *awamori*	The production level of bovine prochymosin was lower than expected for randomly isolated transformants (3/19 transformants)	-	[90]
	Overexpression of *cypB* in *A. niger*	Twofold increase in glucoamylase production	2	[89]
	Insertion of multiple copies of *pdiA* in *A. awamori*	19 mg/L thaumatin compared to 5 mg/L in the parental strain	3.8	[91]
		Optimal bioreactor conditions: 150 mg/L thaumatin compared to 40 mg/L in the parental strain	3.8	
	Overexpression of *clxA* in *A.niger*	60–73 mg/L *P. chrysosporium* manganese peroxidase compared to 14 mg/L in the parental strain	4–5	[92]
	Overexpression of *bipA* in *A.niger*	*P. chrysosporium* manganese peroxidase was severely reduced-almost undetectable levels	-	[92]

Table 5. *Cont.*

Process	Modification	Performance	Improvement Factor	Reference
	Overexpression of *hacA* in *A. niger* var. *awamori*	13–34 mg/L chymosin compared to 12.5 mg/L in parental strain	1.3–2.8	[88]
		3.9–8.5 nkat/mL laccase compared to 0.9 nkat/mL in parental strain	3–7.6	
	Overexpression of *bipA* in *A. awamori*	20 mg/L thaumatin compared to 9 mg/L in the parental strain	2–2.5	[93]
	Constitutive expression of the active form of *hacA* cDNA in *A. oryzae*	2 mg/L neoculin compared to 1.5 mg/L in the control strain	1.5	[40]

- Glycosylation

As mentioned before, glycosylation affects protein activity and stability and in many cases it also influences protein folding and secretion efficiency [78,79]. *N*-glycosylation has been linked to an ER-quality control system of glycoprotein folding in filamentous fungi [94,95]. In fact, inhibiting the specific process in *A. nidulans* appeared to hinder secretion of α-galactosidase, resulting in the accumulation of the under-glycosylated protein to the cell wall [96]. Engineering *N*-glycosylation has been applied as a strategy to increase levels of heterologous protein secretion (Table 6) [97–99]. For example, improving a poorly used *N*-glycosylation site or adding a new one in a chymosin gene almost doubled the levels of secreted protein in *A. niger* strains [98].

Table 6. Approaches for improving recombinant protein production through engineering glycosylation.

Process	Modification	Performance	Improvement Factor	Reference
Glycosylation	Overexpression of *S. cerevisiae* DPM1 in *A. nidulans* strains impaired in DPMS activity	No significant increase in protein secretion observed Production of invertase and glucoamylase was higher but the proteins were trapped in the periplasmic space	28	[97]
	Bovine prochymosin synthetic gene with a single mutation (S335T)—This mutation resulted in a potentially better *N*-glycosylation site (NHT) in *A. niger*	207 IMCU/mL (0.9 g/L) chymosin compared to 90 IMCU/mL in the parental strain 90% of the chymosin molecules were glycosylated compared to 10% in the parental strain	3	[98]
	Bovine prochymosin synthetic gene with a N–S–T glycosylation site (TDNST) in the short peptide linker in *A. niger*	141 International Milk Clotting Unit/mL (0.6 g/L) chymosin compared to 90 IMCU/mL in the parental strain Same glycosylation pattern	1.5	[98]
	Deletion of *mnn9, mnn10, ochA* in *A. niger*	Δmnn9: 14.6% increase in *Tramete* laccase production	1.1	[99]
		Δmnn10: 12.7% increase	1.1	
		ΔochA: 7.2% increase	1.1	
		Δmnn9/ochA: 16.8% increase	1.2	

Strangely enough, it appears that glycosylation is not always essential for the secretion of proteins. For example, secretion of α-amylase was not affected at all when the antibiotic tunicamycin was used to block *N*-glycosylation in *A. oryzae* cultures [79]. In another attempt, overexpressing the yeast DPM1 gene, a key gene of *O*-glycosylation, improved the generation of native proteins in *O*-glycosylation-deficient *A. nidulans* strains, but the proteins produced were mostly localized in the periplasmic space [97]. In addition, deleting *algC*, a gene involved in early steps of *N*-glycosylation, resulted in an increase of the overall protein secretion in *A. niger* strains [100]. It is apparent that the role of glycosylation in protein production is not yet clearly understood and requires more research in order to be employed for improving heterologous protein production.

- Protein translocation

Following glycosylation and quality control in the ER, secreted proteins are packed in ER-derived vesicles (COPII-coated vesicles) and are transported to the Golgi apparatus. Proteins are further glycosylated there and are then guided to the plasma membrane in Golgi-derived vesicles. Additionally, other types of vesicles starting from Golgi carry ER-residents back to the ER or recycle important Golgi-enzymes and trafficking components (COPI-coated vesicles) [101] (Figure 1).

Regulating expression of key factors involved in vesicle trafficking (e.g., cargo receptors recruiting proteins into the vesicles) has been applied lately in order to assist protein transport through the secretory pathway and has improved production yields in aspergilli (Table 7) [61,102–104]. Indicatively, overproduction of the Rab GTPase RabD, a protein involved in cargo transport from the Golgi apparatus to the plasma membrane, increased the secretion yields of a fluorescent reporter protein (mRFP) in *A. nidulans* cultures by approximately 25% [61]. Similarly, deleting genes encoding the receptors AoVip36 and AoEmp47, which retain proteins in the ER, almost doubled the level of chymosin secreted in *A. oryzae* strains [104].

Table 7. Projects for improving recombinant protein production through engineering protein trafficking.

Process	Modification	Performance	Improvement Factor	Reference
Protein translocation	Deletion of *Aovip36* in *A. oryzae*	50 mg/L chymosin compared to 27 mg/L in the parental strain	1.9	[104]
		300% EGFP in culture supernatant compared to 100% in the parental strain The α-amylase activity (native protein) was reduced by approximately 30% compared with the activity in the control strain	3	
	Deletion of *AoEmp47* in *A. oryzae*	50 mg/L chymosin compared to 27 mg/L in the parental strain	1.9	[104]
		210% EGFP in culture supernatant compared to 100% in the parental strain No difference in the α-amylase activity (native protein)	2.1	
	Overexpression of *rabD* in *A. nidulans*	25% increase in mRFP secretion in submerged cultivations in shake flasks	1.3	[61]
		40% increase in RFP secretion in 2l bioreactor	1.4	

Table 7. *Cont.*

Process	Modification	Performance	Improvement Factor	Reference
	Deletion of *racA* in *A. niger*	Native GlaA secreted into the culture medium is four times more compared to its parental strain, when ensuring continuous high-level expression of *glaA*. Quantification was done by dot blot analysis using a monoclonal antibody (Arbitrary units)	4	[102]
	Overexpression of *arfA* in *A. niger*	Quantitative abundance of GlaA-dtomato reporter protein was 397.4 absolute fluorescence compared to 298.7 in control strain	1.3	[103]

- Protein degradation pathways—ERAD and Vacuole

Misfolded proteins that fail to be refolded properly through the UPR enter the ERAD pathway (Figure 1). They are transported from the ER to the cytoplasm, where they are ubiquitinated and degraded by the proteasome [105]. Disrupting key genes of the ERAD has been applied in order to study intracellular degradation of heterologous proteins and improve production yields in Aspergillus species (Table 8) [99,106,107]. Inactivation of *doaA*, which encodes a factor required for ubiquitin-mediated proteolysis, combined with induction of UPR-related genes (*sttC*) contributed to improved heterologous protein expression in *A. niger* [106].

Table 8. Approaches for improving recombinant protein production through engineering protein degradation pathways.

Process	Modification	Performance	Improvement Factor	Reference
Protein degradation pathways—ERAD and Vacuole	Deletion of *derA* and *derB* in *A. niger*	ΔderA: 80% decrease in *Tramete* laccase production	0.2	[99]
	-	ΔderB: 15.7% increase in *Tramete* laccase	1.15	
	Deletion of *doaA* and overexpression of *sttC* in *A. niger*	Higher GUS activity compared to parental strain (no quantitative data available)	-	[106]
	Disruption of *Aovps10* in *A. oryzae*	83.1 and 70.3 mg/L chymosin compared to 28.7 mg/L in parental strain	3–2.5	[108]
		22.6 and 24.6 mg/L human lysozyme compared to 11.1 mg/L in parental strain	2–2.2	
	Deletion of ERAD key genes (*derA*, *doaA*, *hrdC*, *mifA* and *mnsA*) in *A. niger*	ΔderA and ΔhrdC: 2-fold increase compared to parental strain (single-copy)	2	[107]
		ΔderA: 6-fold increase compared to parental strain (multi-copy)	6	
		Relative amount of intracellular GlaGus (β-glucuronidase levels) fusion protein detected in total protein extracts of strains with impaired ERAD and respective parental strain		

Table 8. *Cont.*

Process	Modification	Performance	Improvement Factor	Reference
Disruption of genes involved in autophagy in *A. oryzae*		ΔAoatg1: 60 mg/L chymosin	2.3	[109]
		ΔAoatg13: 37 mg/L chymosin	1.4	
		ΔAoatg4: 80 mg/L chymosin	3.1	
		ΔAoatg8: 66 mg/L chymosin	2.5	
		ΔAoatg15: 24 mg/L chymosin	1	
		Control: 26 mg/L chymosin	-	

During the last steps of secretion additional protein quality control mechanisms can target aberrant proteins to degradation. During autophagy proteins are guided through vesicle trafficking to vacuoles, where they get degraded (Figure 1). Disruption of the autophagic process has been proposed as a way to enhance production of recombinant heterologous proteins (Table 8) [108,109]. For example, deleting the vacuolar protein sorting receptor gene *AoVPS* in *A. oryzae* resulted in increased extracellular production levels of chymosin and human lysozyme in *A. oryzae* by 3- and 2-fold, respectively [108].

Collectively, these studies suggest that the fungal secretory pathway offers multiple engineering targets and opens up a new perspective on developing hypersecreting Aspergillus strains for industrial applications.

3.4.2. Carrier Proteins

A commonly applied approach for successful secretion of foreign proteins in Aspergillus cultures is the use of native, well-secreted carrier proteins. Fusion of a carrier protein to heterologously produced proteins appears to improve mRNA stability [75] and facilitate proper folding and translocation through the fungal secretory pathway [110]. Multiple proteins of industrial and pharmaceutical relevance have been produced and efficiently secreted following this strategy. Chymosin, a protease used as a milk clotting agent in cheese manufacturing, was one of the first examples of a heterologous protein to be produced in Aspergillus cultures using the natively secreted glucoamylase A (GlaA) as a carrier protein [111,112] or just the GlaA signal peptide [113]. Since then, the advantages of using entire or parts of carrier proteins were exploited further for the production of human interleukin-6 [114], antibodies [115,116], and other commercially relevant proteins [40,117] in several Aspergillus species improving production yields (Table 9).

As GlaA is the most abundant and highly secreted enzyme in most Aspergillus species, it is a popular carrier choice for heterologous protein production [113,114,116,117]. However, other naturally secreted proteins, or their signal peptides, have also been used successfully for this purpose, e.g., the *A. oryzae* α-amylase [40,118,119] and signal peptides of an *A. niger* endoxylanase [120] (Table 9).

Table 9. Approaches for improving recombinant protein production through the use of carrier proteins.

Process	Modification	Performance	Improvement Factor	Reference
Carriers	Prochymosin sequence fused to *A. niger* GlaA signal peptide in *A. nidulans*	146 µg/g dry weight chymosin compared to 93 µg/g dry weight in the control strain	1.56	[113]
	Prochymosin sequence fused to codons for the GlaA signal peptide, propeptide, and 11 amino acids of mature glucoamylase in *A. nidulans*	119 µg/g dry weight chymosin compared to 93 µg/g dry weight in the control strain	1.27	[113]

Table 9. *Cont.*

Process	Modification	Performance	Improvement Factor	Reference
	Prochymosin sequence fused to GlaA signal peptide and propeptide in *A. nidulans*	23 µg/g dry weight compared to 93 µg/g dry weight in the control strain	0.24	[113]
	Prochymosin sequence fused after the last codon of *A. awamori* GlaA in *A. awamori*	140 µg/mL secreted chymosin compared to 8 µg/mL in the control strain (prochymosin + GlaA signal peptide)	17.5	[111,112]
	hIL6 fused to 1-514 nt of *A. niger* GlaA in *A. niger*	15 mg/L hIL6 compared to less than 1 µg/L in the control strain (hIL6 fused to the GlaA signal peptide)	>15	[114]
	E. coli uidA (13-glucuronidase) and the *T. lanuginosa* lipase fused to *A. niger* var. *awamori* endoxylanase II secretion signals	798 Arbitrary units of glucuronidase activity/mg of total protein (the control did not carry uidA)	-	[120]
		47.5 Arbitrary units of lipase activity compared to 47 Arbitrary units when the native lipase signal is used	1	
	ScFv-LYS encoding fragments fused to *A. niger* propeptide + GlaA (514nt)	90 mg/L ScFv-LYS compared to 2–22 mg/L in the control strains (18 aa GlaA signal sequence + ScFv-LYS)	4–45	[115]
	Human antibodies (κ- and γ-chain of IgG1) fused to GlaA in *A. niger*	0.9 g/L of trastuzumab IgG1 and 0.2 g/L of Hu1D10	-	[116]
	Neoculin gene fused to α-amylase in *A. oryzae*	1.3 mg/L of NCL	-	[40]
	Bovine chymosin gene fused to α-amylase in *A. oryzae*	42 mg/L chymosin compared to 20 mg/L in strains with non-fused gene	2.1	[119]
	Hemicellulose degrading enzymes sequences fused to GlaA secretion peptide in *A. nidulans* strains	50–100 mg/L xylanase B, xylanase C, xylosidase D, arabinofuranosidase B, ferulic acid esterase and arabinase	-	[117]

Regardless of the carrier protein or the signal peptide used, the heterologous protein needs to be detached from the protein fusion after secretion in order to gain full activity. An alternative practice to downstream in vitro treatment with proteases is to incorporate a protease cleavage site (e.g., Lys-Arg) between the native and the foreign protein, which can be proteolytically cleaved by fungal endoproteases (e.g., KEXB endoprotease in *A. niger*) during secretion [114,121,122]. Optimization of the sequence upstream the KEXB cleavage site appeared to increase Trastuzumab light chain production in *A. niger* [122].

3.5. Proteases

Proteolytic degradation by extracellular fungal proteases is one of the main reasons why secreted yields of heterologous proteins fail to reach the gram-per-liter production level of native proteins in Aspergillus species. Several bioprocessing strategies, such as maintaining high pH during fermentation [123] or low temperatures during downstream processing, separating the product from the protease-containing medium and using protease inhibitors, are often used to decrease protein degradation. However, proteolysis still occurs, making production of foreign proteins inefficient [124].

An alternative and more efficient approach is the use of protease-deficient strains [124]. Conventional mutagenesis and genetic engineering were applied to several Aspergillus species in order to disrupt genes that encode extracellular proteases, e.g., aspergillopepsin A (pepA) [125,126] or protease regulatory genes [126,127] (Table 10). The specific mutants exhibit reduced extracellular proteolytic activity and often appear to be more efficient producers of heterologous proteins than the wild types [124]. Deletion of *pepA* in *A. awamori* resulted in an aspergillopepsin A-deficient mutant, with decreased proteolytic activity [125] and able to produce higher levels of bovine chymosin (~430 mg/L), when compared to the control strain (*A. awamori* strain GC12-ΔargB3, ΔpyrG5: ~180 mg/L) [128]. *A. niger* mutants lacking the transcription factor PrtT, which regulates expression of both aspergillopepsin A and B genes (*pepA* and *pepB*) [129], showed only 1–2% of the parental strain extracellular protease activity [126] and were used to produce highly stable heterologous cutinase with 1.7-fold increased activity [127].

Table 10. Approaches for improving recombinant protein production through disruption of protease genes.

Process	Modification	Performance	Improvement Factor	Reference
Proteases	Deletion of *pepA* in *A. awamori* strains	Decreased extracellular proteolytic activity compared to the wild type (immunoassay using antibodies specific for PepA, but absolute values for PepA concentration were not determined)	-	[125]
	Deletion of *pepA* in *A. awamori*	430 mg/L of chymosin compared to 180 mg/L in the parental strain	2.4	[128]
	Deletion of *pepA* in *A. niger* (AB1.18)	15–20% proteolytic activity compared to the parent strain AB4.1	-	[126]
	Mutation on *prtT* (UV irradiation) in *A. niger* (AB1.13)	1–2% proteolytic activity compared to the parent strain AB4.1	-	[126]
	Deletion of *prtR, pepA, cpI, tppA* in *A. oryzae*	ΔprtR/pepA/cpI: 24.23 mg/L of *Acremonium cellulolyticus* cellobiohydrolase	1.2	[133]
		ΔprtR/pepA/tppA: 21.30 mg/L	1.1	
		ΔprtR/cpI/tppA: 22.08 mg/L	1.1	
		ΔprtR/pepA/cpI/tppA: 19.93 mg/L compared to 19.54 mg/L in the control strains	1.02	
	Deletion of *alp* and *Npl* in *A. oryzae*	1041 U/g of *Candida antarctica* lipase B compared to 575 U/g in the parental strains	1.8	[132]
	Deletion of various proteases in *A. niger*	Δdpp4: 6% increase in *Tramete* laccase	1.1	[99]
		Δdpp5: 15.4% increase	1.2	
		ΔpepB: 8.6% increase	1.1	
		ΔpepD: 4.8% increase	1.0	
		ΔpepF: 5.3% increase	1.1	
		ΔpepAa: 0.5% increase	1.1	
		ΔpepAb: 13.4% increase	1.1	
		ΔpepAd: 2.7% increase	1.0	
		Δdpp4/dpp5: 26.6% increase	1.3	
	Disruption of *tppA* and *pepE* in *A. oryzae* strains	25.4 mg/L of human lysozyme compared to 15 mg/L in the parental strains	1.7	[118]
	Disruption of *tppA, pepE, nptB, dppIV* and *dppV* in *A. oryzae*	84.4 mg/L of chymosin compared to the 63.1 mg/L in the double protease gene disruptant (ΔtppA/pepE)	1.3	[130]
	Disruption of *tppA, pepE, nptB, dppIV,* and *dppV, alpA, pepA, AopepAa, AopepAd* and *cpI* in *A. oryzae*	109.4 mg/L of chymosin and 35.8 mg/L of human lysozyme compared to the quintuple protease gene disruptant (ΔtppA/pepE/nptB/dppIV/dppV; 84.4 mg/L and 26.5 mg/L, respectively)	1.3 and 1.35	[131]

Table 10. *Cont.*

Process	Modification	Performance	Improvement Factor	Reference
	Deletion of *prtT* in *A. niger*	36.3–36.7 U/mL of mL *G. cingulate* cutinase compared to 21.2–20.4 U/mL in the parental strain	1.7	[127]
		Stability: Cutinase activity retained at 80% over the entire 14-day incubation period, while the parental lost more than 50% of their initial activities after six days of incubation and retained negligible activity after 14 days	-	
	Deletion of *dppV* and *pepA* in *A. nidulans*	*P. sanguineus* laccase activity 0.5 U/mL compared to 0.04 U/mL in the control strain	12.5	[51]
	Deletion of *mnn9* and *pepA* in *A. nidulans*	*P. sanguineus* laccase activity 0.3 U/mL compared to 0.04 U/mL in the control strain	7.5	[51]

Research on Aspergillus protease repertoire and development of molecular tools for multiple gene targeting allowed the disruption of multiple protease-related genes in a single production host, a successful tactic to further decrease proteolytic degradation and therefore improve protein production titers (Table 10) [51,118,130,131]. Disruption of two protease genes in *A. oryzae* (*tppA* and *pepE*) resulted in production of 25.4 mg/L of human lysozyme (HLY), which represents a 63% increase in production yields compared to the control strain [118]. Subsequently, *A. oryzae* quintuple and decuple protease gene disruptants produced even higher HLY amounts (26.5 mg/L and 35.8 mg/L, respectively) [130,131]. Although multiple protease-related gene disruptions appear to be a time-consuming and tedious procedure [127], it is a strategy commonly used for the optimization of heterologous protein production in several Aspergillus systems of industrial interest [99,132,133].

3.6. Altering Fungal Morphology Using Genetic Engineering

Protein secretion in filamentous fungi has been shown to happen mostly at the tip of growing hypha, thus hyperbranched phenotypes are more desirable when developing a protein production platform. Additionally, combining a hyperbranched phenotype with shortened mycelia may result in reduced culture viscosity, which is beneficial for high density submerged fermentation [134,135].

Multiple studies have focused on the effect of morphology on protein production in aspergilli. Genetic engineering attempts [102,103] (Table 11) or variation of fermentation parameters (see also Section 4.2.: Fungal morphology and bioprocessing) have been employed to obtain hyperbranching strains that can secrete large amounts of proteins. In *A. niger* deleting the gene that encodes the Rho-GTPase RacA, which mediates actin polymerization and depolymerisation at the hyphal apex, generated a strain producing 20% more hyphal tips. Under continuous high-level expression, this hyperbranching strain produced four times more glucoamylase compared to its parental strain [102].

Table 11. Approaches for improving recombinant protein production through engineering genes involved in fungal morphology.

Process	Modification	Performance	Improvement Factor	Reference
Fungal morphology	Deletion of *racA* in *A. niger*	GlaA secreted into the culture medium is four times more compared to its parental strain, when ensuring continuous high-level expression of *glaA*. Quantification was done by dot blot analysis using a monoclonal antibody (Arbitrary units)	4	[102]
	Overexpression of *arfA* in *A. niger*	Quantitative abundance of GlaA-dtomato reporter protein was 397.4 absolute fluorescence compared to 298.7 in control strain	1.3	[103]

4. Fermentation Conditions for Improved Heterologous Production in *Aspergillus*

Development of most heterologous expression platforms begins with strain improvement, which hopefully results in obtaining strains able to produce large quantities of a specific protein. Once strain improvement is complete, the fermentation process for production of the desirable protein in large-scale has to be established [7,136]. Designing and setting up fungal fermentations is a complex process that has to be repeated every time a newly engineered strain is used or a new protein is to be produced. This process requires several optimization steps, starting from finding the optimal growth medium and fermentation parameters (temperature, pH, and oxygenation) to choosing the appropriate type of fermentation and the fungal morphology that favors high production yields of the specific protein [137–139].

4.1. Fermentation Conditions

Multiple strategies have been applied to optimize fermentation conditions for improving recombinant protein production in Aspergillus cultures (Table 12). Several studies have focused on the effect of growth medium and culture conditions on protein production [140–145]. MacKenzie et al. (1994) studied the effect that temperature and growth medium have on the production of hen eggwhite lysozyme (HEWL) in *A. niger* cultures. When a standard expression medium (1% *w/v* soluble starch and 50 mM sodium phosphate buffer) was used, 20–25 °C was the optimal temperature range to obtain lysozyme in previously observed levels (8–10 mg/L). In addition, as the HEWL gene was under the control of the *glaA* promoter, production of lysozyme was highly induced when soluble starch was used as carbon source. Using a richer medium with soy milk led to even higher yields of up to 30–60 mg/L lysozyme, but interestingly growth temperature was adjusted to 37 °C to achieve these levels [141]. In another attempt, the impact of the nitrogen source was evaluated [143]. Swift et al. (2000) showed that glucoamylase production was increased by 115% with the addition of casamino acids, yeast extract, peptone, and gelatin in cultures of a recombinant *A. niger* strain, compared to non-supplemented cultures [143].

Table 12. Approaches for improving recombinant protein production through bioprocessing modifications.

Process	Modification	Performance	Improvement Factor	Reference
Fermentation conditions	Effect of growth medium and temperature on hen egg white lysozyme (HEWL) production in *A. niger*	20–25 °C 8–10 mg/L HEWL while 30–37 °C 3–5 mg/L HEWL	Temperature: 2–2.6	[141]
		soluble starch: 8.0 mg/L HEWL	Carbon source: 1.7–2	
		maltose: 4.5 mg/L HEWL	-	
		glucose: 4.0 mg/L HEWL	-	
		xylose: 0.2 mg/L HEWL	-	
		soy milk medium: 30–60 mg/L HEWL	Rich medium: 3.8–7.5	
	Effect of organic nitrogen sources on recombinant glucoamylase production in *A. niger*	Unsupplemented: 44 mg glucoamylase/g biomass	-	[143]
		L-alanine: 32 mg glucoamylase/g biomass	0.7	
		L-methionine: 26 mg glucoamylase/g	0.6	
		casamino acids, yeast extract, peptone, and gelatin: 100 mg glucoamylase/g	2.2	
	Effect of agitation intensity on recombinant amyloglucosidase (AMG) production in *A. oryzae*	Titer at the end of the batch phase 525 rpm: 110 U/L AMG	-	[146]
		675 rpm: 230 U/L AMG	1.6	
		825 rpm: 370 U/L AMG	3.3	

Table 12. *Cont.*

Process	Modification	Performance	Improvement Factor	Reference
	Effects of bioprocess parameters—agitation intensity, initial glucose concentration, initial yeast extract concentration, and dissolved oxygen tension (DO)—on heterologous protein production in *A. oryzae*	Highest GFP yields were achieved under these conditions: agitation 400 rpm, glucose 25 g/L, yeast extract 0 g/dm3, DO 15%	-	[142]
	Effect of agitation intensity on recombinant glucose oxidase production in *A. niger*	200 rpm: 300 µkat/L of glucose oxidase	-	[144]
		500 rpm: 800 µkat/L of glucose oxidase	2.6	
		800 rpm: 600 µkat/L of glucose oxidase	1.3	
	Effect of temperature on *Pleurotus eryngii* versatile peroxidase production in *A. nidulans* and *A. niger*	-*A. nidulans* 31 °C: 24 U/L peroxidase activity	-	[145]
		28 °C: 80 U/L peroxidase activity	3.3	
		19 °C: 466 U/L peroxidase activity	19.4	
		-*A. niger* 28 °C: 107 U/L peroxidase activity	-	
		19 °C: 412 U/L peroxidase activity	3.8	
Fungal morphology	Effect of raising the viscosity of the medium by addition of polyvinylpyrrolidone-PVP (transition from aggregated mycelia (pellets) to dispersed mycelia) on hen egg white lysozyme (HEWL) in *A. niger*	Medium with no PVP: 110 mg/L fresh and 8 mg/g dry weight of HEWL	1.7	[147]
		Medium with PVP: 190 mg/L fresh and 14 mg/g dry weight of HEWL		
	Effect of addition of microparticles (linked to the formation of freely dispersed mycelium) on titers of native glucoamylase (GlaA) and recombinant fructofuranosidase (FF) produced in *A. niger*	No microparticles: 17 U/mL GlaA and 42 U/mL FF	3.5 GlaA 2–3.8 FF	[148]
		Talc microparticles: 61 U/mL GlaA and 92 U/mL FF FF production can reach up to 160 U/mL (10 g/L talc microparticles of size 6 mm)		
	Effect of addition of titanate microparticles (TiSiO$_4$, 8 mm) on titers of native glucoamylase (GlaA) and recombinant fructofuranosidase (FF) produced in *A. niger*	No microparticles: 19 U/mL GlaA and 40 U/mL FF	9.5 GlaA 3.7 FF	[149]
		Microparticles: 190 U/mL glucoamylase and 150 U/mL fructofuranosidase		
	Effect of growth type on hen egg white lysozyme (HEWL) production and protease activity in *A. niger*	Free suspension: 5.8 mg/g HEWL 95.3 U/g Protease activity	1.5	[140]
		Mycelial pellets: 5.0 mg/g HEWL 58.6 U/g Protease activity	1.2	
		Celite-560-immobilized cultures: 4.1 mg/g HEWL 56.3 U/g Protease activity	-	

Additional bioprocess parameters such as agitation intensity, initial nutrient concentration and dissolved oxygen levels were also studied with regard to heterologous protein production in different Aspergillus fermentation types [142,146].

4.2. Altering Fungal Morphology Using Bioprocessing

Filamentous fungi that grow in submerged cultures, including aspergilli, exhibit variable morphologies, which can affect the overall performance of the microorganism during fermentation. The commonly observed filamentous growth results in undesirable viscous cultures, which require high consumption of energy for agitation and usually complicate downstream processing. On the contrary, a pelleted morphology (growth in pellets) decreases viscosity of the culture, improves mixing, and facilitates downstream processing. However, when large pellets are formed in the culture, diffusion of oxygen and nutrients to the inner pellet core is hindered, resulting in reduced productivity [135].

Many studies have attempted to correlate high protein production yields to a specific type of fungal morphology. For *A. niger*, dispersed mycelial suspensions led to higher protein yields over the pelleted form [140] (Table 12). In fact, microparticles are often added into Aspergillus cultures in order to prevent formation of large pellets and to favor formation of disperse mycelium, and consequently recombinant protein production [144,147–149]. However, Gyamerah et al. (2002) showed that the free suspension culture presented a higher protease activity, compared to the cultures with immobilized fungal biomass (mycelial pellet or by entrapment in Celite beads), and this could be an additional limitation for recombinant protein production [140].

5. Conclusions and Future Perspectives

Filamentous fungi hold unlimited potential for industrial applications, from the development of meat-like products and biomaterials, to bioremediation and biofuel production. One of their best qualities, largely exploited by the industry, is their innate capacity for the secretion of enzymes, which facilitate downstream processing and product recovery. Moreover, their ability to produce complex proteins with post-translational modifications and the fact that they can be cultivated on inexpensive media makes them a promising alternative for production of eukaryotic proteins. Despite their undeniable potential though, filamentous fungi have not yet been exploited to the fullest in the industrial production of recombinant proteins.

Advances in the molecular toolkit available for genetic manipulation of several Aspergillus species opened up the path for developing them into production systems for recombinant proteins. Nevertheless, due to a number of factors described in the review, aspergilli have not yet met the expected production levels. Many studies that focused on engineering different steps of protein synthesis and secretion, or generating protease-deficient strains, have resulted in a significant increase of protein yields. Additionally, optimization of the fungal fermentation process has further improved protein production. However, there are aspects of the fungal physiology that limit protein production and remain unclear. Continuous data input from "omics" studies sheds light on the complex fungal mechanisms related to protein quality control and secretion stress, as well as their impact on protein productivity. The knowledge generated from these studies combined with advances in the field of synthetic biology will soon place Aspergillus, and possibly other filamentous fungi, in the race for the most efficient recombinant protein production system. Its potential as a large-scale production platform not only for recombinant proteins, but also for organic acids, bioactive compounds, enzymes, and peptides, as well as new perspectives related to the use of Aspergillus in waste treatment and bioremediation processes, prove that this fungus can provide sustainable solutions for multiple and diverse markets and industries.

Author Contributions: Conceptualization, F.N. and R.F.; writing—original draft preparation, F.N. and R.F.; writing—review—editing, F.N., U.H.M.; C.S. and R.F. supervision, C.S. and R.F.; project administration, C.S.; funding acquisition, C.S. and R.F. All authors have read and agreed to the published version of the manuscript.

Funding: FN's post-doctorate is supported through the project "PEPTIPORTE-Peptide production and robust export" funded by Région Hauts-de-France and Bpifrance (FRRI Hauts-de-France), grant number DOS0080512. Publication fee was partly funded by the Université de Picardie Jules Verne.

Conflicts of Interest: The authors declare no conflict of interest. The funders had no role in the design of the study; in the collection, analyses, or interpretation of data; in the writing of the manuscript; or in the decision to publish the results.

References

1. Puetz, J.; Wurm, F.M. Recombinant proteins for industrial versus pharmaceutical purposes: A review of process and pricing. *Processes* **2019**, *7*, 476. [CrossRef]
2. Choi, J.-M.; Han, S.-S.; Kim, H.-S. Industrial applications of enzyme biocatalysis: Current status and future aspects. *Biotechnol. Adv.* **2015**, *33*, 1443–1454. [CrossRef] [PubMed]
3. Baeshen, N.A.; Baeshen, M.N.; Sheikh, A.; Bora, R.S.; Ahmed, M.M.M.; Ramadan, H.A.I.; Saini, K.S.; Redwan, E.M. Cell factories for insulin production. *Microb. Cell Fact.* **2014**, *13*, 141. [CrossRef]
4. Khan, S.; Ullah, M.W.; Siddique, R.; Nabi, G.; Manan, S.; Yousaf, M.; Hou, H. Role of recombinant DNA technology to improve life. *Int. J. Genomics* **2016**, *2016*, 1–14. [CrossRef] [PubMed]
5. Walsh, G. Biopharmaceutical benchmarks 2018. *Nat. Biotechnol.* **2018**, *36*, 1136–1145. [CrossRef] [PubMed]
6. Ferreira, R.d.G.; Azzoni, A.R.; Freitas, S. Techno-economic analysis of the industrial production of a low-cost enzyme using *E. coli*: The case of recombinant β-glucosidase. *Biotechnol. Biofuels* **2018**, *11*, 81. [CrossRef]
7. Nevalainen, H.; Peterson, R. Making recombinant proteins in filamentous fungi—Are we expecting too much? *Front. Microbiol.* **2014**, *5*, 1–10. [CrossRef]
8. Demain, A.L.; Vaishnav, P. Production of recombinant proteins by microbes and higher organisms. *Biotechnol. Adv.* **2009**, *27*, 297–306. [CrossRef]
9. Baghban, R.; Farajnia, S.; Rajabibazl, M.; Ghasemi, Y.; Mafi, A.; Hoseinpoor, R.; Rahbarnia, L.; Aria, M. Yeast expression systems: Overview and recent advances. *Mol. Biotechnol.* **2019**, *61*, 365–384. [CrossRef]
10. Xie, Y.; Han, X.; Miao, Y. An effective recombinant protein expression and purification system in *Saccharomyces cerevisiae*. *Curr. Protoc. Mol. Biol.* **2018**, *123*, e62. [CrossRef]
11. Meyer, V.; Basenko, E.Y.; Benz, J.P.; Braus, G.H.; Caddick, M.X.; Csukai, M.; de Vries, R.P.; Endy, D.; Frisvad, J.C.; Gunde-Cimerman, N.; et al. Growing a circular economy with fungal biotechnology: A white paper. *Fungal Biol. Biotechnol.* **2020**, *7*, 5. [CrossRef] [PubMed]
12. Havlik, D.; Brandt, U.; Bohle, K.; Fleißner, A. Establishment of *Neurospora crassa* as a host for heterologous protein production using a human antibody fragment as a model product. *Microb. Cell Fact.* **2017**, *16*, 128. [CrossRef] [PubMed]
13. Landowski, C.P.; Mustalahti, E.; Wahl, R.; Croute, L.; Sivasiddarthan, D.; Westerholm-Parvinen, A.; Sommer, B.; Ostermeier, C.; Helk, B.; Saarinen, J.; et al. Enabling low cost biopharmaceuticals: High level interferon alpha-2b production in *Trichoderma reesei*. *Microb. Cell Fact.* **2016**, *15*, 104. [CrossRef]
14. Magaña-Ortíz, D.; Fernández, F.; Loske, A.M.; Gómez-Lim, M.A. Extracellular expression in *Aspergillus niger* of an antibody fused to *Leishmania* sp. antigens. *Curr. Microbiol.* **2018**, *75*, 40–48. [CrossRef]
15. Arnau, J.; Yaver, D.; Hjort, C.M. Strategies and challenges for the development of industrial enzymes using fungal cell factories. In *Grand Challenges in Fungal Biotechnology*; Springer: Cham, Switzerland, 2020; pp. 179–210, ISBN 9783030295417.
16. Sun, X.; Su, X. Harnessing the knowledge of protein secretion for enhanced protein production in filamentous fungi. *World J. Microbiol. Biotechnol.* **2019**, *35*, 54. [CrossRef] [PubMed]
17. Meyer, V. Genetic engineering of filamentous fungi—Progress, obstacles and future trends. *Biotechnol. Adv.* **2008**, *26*, 177–185. [CrossRef] [PubMed]
18. Gupta, S.K.; Shukla, P. Advanced technologies for improved expression of recombinant proteins in bacteria: Perspectives and applications. *Crit. Rev. Biotechnol.* **2016**, *36*, 1089–1098. [CrossRef]
19. Baeshen, M.N.; Al-Hejin, A.M.; Bora, R.S.; Ahmed, M.M.M.; Ramadan, H.A.I.; Saini, K.S.; Baeshen, N.A.; Redwan, E.M. Production of biopharmaceuticals in *E. coli*: Current scenario and future perspectives. *J. Microbiol. Biotechnol.* **2015**, *25*, 953–962. [CrossRef]
20. Contreras-Gómez, A.; Sánchez-Mirón, A.; García-Camacho, F.; Molina-Grima, E.; Chisti, Y. Protein production using the baculovirus-insect cell expression system. *Biotechnol. Prog.* **2014**, *30*, 1–18. [CrossRef]
21. Hunter, M.; Yuan, P.; Vavilala, D.; Fox, M. Optimization of protein expression in mammalian cells. *Curr. Protoc. Protein Sci.* **2019**, *95*, e77. [CrossRef]
22. Houdebine, L.-M. Production of pharmaceutical proteins by transgenic animals. *Comp. Immunol. Microbiol. Infect. Dis.* **2009**, *32*, 107–121. [CrossRef]
23. Łojewska, E.; Kowalczyk, T.; Olejniczak, S.; Sakowicz, T. Extraction and purification methods in downstream processing of plant-based recombinant proteins. *Protein Expr. Purif.* **2016**, *120*, 110–117. [CrossRef] [PubMed]

24. Yao, J.; Weng, Y.; Dickey, A.; Wang, K. Plants as factories for human pharmaceuticals: Applications and challenges. *Int. J. Mol. Sci.* **2015**, *16*, 28549–28565. [CrossRef]

25. Currie, J.N. The citric acid fermentation of *Aspergillus niger*. *J. Biol. Chem.* **1917**, *31*, 15–37.

26. Cairns, T.C.; Nai, C.; Meyer, V. How a fungus shapes biotechnology: 100 years of *Aspergillus niger* research. *Fungal Biol. Biotechnol.* **2018**, *5*, 13. [CrossRef]

27. Dai, Z.; Zhou, H.; Zhang, S.; Gu, H.; Yang, Q.; Zhang, W.; Dong, W.; Ma, J.; Fang, Y.; Jiang, M.; et al. Current advance in biological production of malic acid using wild type and metabolic engineered strains. *Bioresour. Technol.* **2018**, *258*, 345–353. [CrossRef] [PubMed]

28. Barrios-González, J.; Miranda, R.U. Biotechnological production and applications of statins. *Appl. Microbiol. Biotechnol.* **2010**, *85*, 869–883. [CrossRef]

29. Meyer, V.; Andersen, M.R.; Brakhage, A.A.; Braus, G.H.; Caddick, M.X.; Cairns, T.C.; de Vries, R.P.; Haarmann, T.; Hansen, K.; Hertz-Fowler, C.; et al. Current challenges of research on filamentous fungi in relation to human welfare and a sustainable bio-economy: A white paper. *Fungal Biol. Biotechnol.* **2016**, *3*, 6. [CrossRef] [PubMed]

30. Martins-Santana, L.; Nora, L.C.; Sanches-Medeiros, A.; Lovate, G.L.; Cassiano, M.H.A.; Silva-Rocha, R. Systems and synthetic biology approaches to engineer fungi for fine chemical production. *Front. Bioeng. Biotechnol.* **2018**, *6*, 117. [CrossRef]

31. Meyer, V.; Wu, B.; Ram, A.F.J. *Aspergillus* as a multi-purpose cell factory: Current status and perspectives. *Biotechnol. Lett.* **2011**, *33*, 469–476. [CrossRef]

32. Leynaud-Kieffer, L.M.C.; Curran, S.C.; Kim, I.; Magnuson, J.K.; Gladden, J.M.; Baker, S.E.; Simmons, B.A. A new approach to Cas9-based genome editing in *Aspergillus niger* that is precise, efficient and selectable. *PLoS ONE* **2019**, *14*, e0210243. [CrossRef] [PubMed]

33. Sarkari, P.; Marx, H.; Blumhoff, M.L.; Mattanovich, D.; Sauer, M.; Steiger, M.G. An efficient tool for metabolic pathway construction and gene integration for *Aspergillus niger*. *Bioresour. Technol.* **2017**, *245*, 1327–1333. [CrossRef] [PubMed]

34. Cairns, T.C.; Feurstein, C.; Zheng, X.; Zhang, L.H.; Zheng, P.; Sun, J.; Meyer, V. Functional exploration of co-expression networks identifies a nexus for modulating protein and citric acid titres in *Aspergillus niger* submerged culture. *Fungal Biol. Biotechnol.* **2019**, *6*, 18. [CrossRef]

35. Nødvig, C.S.; Nielsen, J.B.; Kogle, M.E.; Mortensen, U.H. A CRISPR-Cas9 system for genetic engineering of filamentous fungi. *PLoS ONE* **2015**, *10*, e0133085. [CrossRef] [PubMed]

36. Katayama, T.; Tanaka, Y.; Okabe, T.; Nakamura, H.; Fujii, W.; Kitamoto, K.; Maruyama, J.I. Development of a genome editing technique using the CRISPR/Cas9 system in the industrial filamentous fungus *Aspergillus oryzae*. *Biotechnol. Lett.* **2016**, *38*, 637–642. [CrossRef] [PubMed]

37. Fuller, K.K.; Chen, S.; Loros, J.J.; Dunlap, J.C. Development of the CRISPR/Cas9 system for targeted gene disruption in *Aspergillus fumigatus*. *Eukaryot. Cell* **2015**, *14*, 1073–1080. [CrossRef] [PubMed]

38. Dunn-Coleman, N.S.; Bloebaum, P.; Berka, R.M.; Bodie, E.; Robinson, N.; Armstrong, G.; Ward, M.; Przetak, M.; Carter, G.L.; LaCost, R.; et al. Commercial levels of chymosin production by *Aspergillus*. *Bio/Technology* **1991**, *9*, 976–981. [CrossRef]

39. Ward, P.P.; Lo, J.-Y.; Duke, M.; May, G.S.; Headon, D.R.; Conneely, O.M. Production of biologically active recombinant human lactoferrin in *Aspergillus Oryzae*. *Nat. Biotechnol.* **1992**, *10*, 784–789. [CrossRef] [PubMed]

40. Nakajima, K.I.; Asakura, T.; Maruyama, J.I.; Morita, Y.; Oike, H.; Shimizu-Ibuka, A.; Misaka, T.; Sorimachi, H.; Arai, S.; Kitamoto, K.; et al. Extracellular production of neoculin, a sweet-tasting heterodimeric protein with taste-modifying activity, by *Aspergillus oryzae*. *Appl. Environ. Microbiol.* **2006**, *72*, 3716–3723. [CrossRef] [PubMed]

41. Van Den Hondel, C.A.M.J.J.; Punt, P.J.; Van Gorcom, R.F.M. *Heterologous Gene Expression in Filamentous Fungi*; Academic Press, Inc.: Cambridge, MA, USA, 1991.

42. Fleißner, A.; Dersch, P. Expression and export: Recombinant protein production systems for *Aspergillus*. *Appl. Microbiol. Biotechnol.* **2010**, *87*, 1255–1270. [CrossRef] [PubMed]

43. Rendsvig, J.K.H.; Workman, C.T.; Hoof, J.B. Bidirectional histone-gene promoters in *Aspergillus*: Characterization and application for multi-gene expression. *Fungal Biol. Biotechnol.* **2019**, *6*, 24. [CrossRef] [PubMed]

44. Qiu, R.; Zhu, X.; Liu, L.; Tang, G. Detection of a protein, AngCP, which binds specifically to the three upstream regions of glaA gene in *A. niger* T21. *Sci. China Ser. C* **2002**, *45*, 527. [CrossRef] [PubMed]

45. Liu, L.; Liu, J.; Qiu, R.X.; Zhu, X.G.; Dong, Z.Y.; Tang, G.M. Improving heterologous gene expression in *Aspergillus niger* by introducing multiple copies of protein-binding sequence containing CCAAT to the promoter. *Lett. Appl. Microbiol.* **2003**, *36*, 358–361. [CrossRef]

46. Moralejo, F.-J.; Cardoza, R.-E.; Gutierrez, S.; Martin, J.F. Thaumatin production in *Aspergillus awamori* by use of expression cassettes with strong fungal promoters and high gene dosage. *Appl. Environ. Microbiol.* **1999**, *65*, 1168–1174. [CrossRef]

47. Pachlinger, R.; Mitterbauer, R.; Adam, G.; Strauss, J. Metabolically independent and accurately adjustable *Aspergillus* sp. expression system. *Appl. Environ. Microbiol.* **2005**, *71*, 672–678. [CrossRef] [PubMed]

48. Bando, H.; Hisada, H.; Ishida, H.; Hata, Y.; Katakura, Y.; Kondo, A. Isolation of a novel promoter for efficient protein expression by *Aspergillus oryzae* in solid-state culture. *Appl. Microbiol. Biotechnol.* **2011**, *92*, 561–569. [CrossRef] [PubMed]

49. Gressler, M.; Hortschansky, P.; Geib, E.; Brock, M. A new high-performance heterologous fungal expression system based on regulatory elements from the *Aspergillus terreus* terrein gene cluster. *Front. Microbiol.* **2015**, *6*. [CrossRef]

50. Jun, W.; Yuk, S.L. Polynucleotide Fragment, Expression Vector Comprising Same, *Aspergillus niger* Genetic Engineering Strain and Application of *Aspergillus niger* Genetic Engineering. CN Patent No. CN107304431A, 31 October 2017.

51. Li, W.; Yu, J.; Li, Z.; Yin, W.B. Rational design for fungal laccase production in the model host *Aspergillus nidulans*. *Sci. China Life Sci.* **2019**, *62*, 84–94. [CrossRef]

52. Yin, W.; Li, W.; Ma, Z. Construction and Application of Heterologous Expression System of *Aspergillus nidulans*. CN Patent No. CN108795970A, 13 November 2018.

53. Li, M.; Lu, F.; Chen, Y. Fungus Promoter and Application Thereof 2019. CN Patent No. CN110331144A, 15 October 2019.

54. Gladden, J.M.; Campen, S.A.; Zhang, J.; Magnuson, J.K.; Baker, S.E.; Simmons, B.A. Promoter Useful for High Expression of a Heterologous Gene of Interest in *Aspergillus niger*. U.S. Patent No. US20,190,169,584A1, 6 June 2019.

55. Meyer, V.; Wanka, F.; van Gent, J.; Arentshorst, M.; van den Hondel, C.A.M.J.J.; Ram, A.F.J. Fungal gene expression on demand: An inducible, tunable, and metabolism-independent expression system for *Aspergillus niger*. *Appl. Environ. Microbiol.* **2011**, *77*, 2975–2983. [CrossRef]

56. Vogt, K.; Bhabhra, R.; Rhodes, J.C.; Askew, D.S. Doxycycline-regulated gene expression in the opportunistic fungal pathogen *Aspergillus fumigatus*. *BMC Microbiol.* **2005**, *5*, 1–11. [CrossRef] [PubMed]

57. Rantasalo, A.; Landowski, C.P.; Kuivanen, J.; Korppoo, A.; Reuter, L.; Koivistoinen, O.; Valkonen, M.; Penttilä, M.; Jäntti, J.; Mojzita, D. A universal gene expression system for fungi. *Nucleic Acids Res.* **2018**, *46*, e111. [CrossRef] [PubMed]

58. Verdoes, J.C.; Punt, P.J.; Schrickx, J.M.; van Verseveld, H.W.; Stouthamer, A.H.; van den Hondel, C.A.M.J.J. Glucoamylase overexpression in *Aspergillus niger*: Molecular genetic analysis of strains containing multiple copies of the glaA gene. *Transgenic Res.* **1993**, *2*, 84–92. [CrossRef] [PubMed]

59. Wallis, G.L.F.; Swift, R.J.; Hemming, F.W.; Trinci, A.P.J.; Peberdy, J.F. Glucoamylase overexpression and secretion in *Aspergillus niger*: Analysis of glycosylation. *Biochim. Biophys. Acta* **1999**, *1472*, 576–586. [CrossRef]

60. Tada, S.; Iimura, Y.; Gomi, K.; Takahashi, K.; Hara, S.; Yoshizawa, K. Cloning and nucleotide sequence of the genomic taka-amylase a gene of *Aspergillus oryzae*. *Agric. Biol. Chem.* **1989**, *53*, 593–599. [CrossRef]

61. Schalén, M.; Anyaogu, D.C.; Hoof, J.B.; Workman, M. Effect of secretory pathway gene overexpression on secretion of a fluorescent reporter protein in *Aspergillus nidulans*. *Fungal Biol. Biotechnol.* **2016**, *3*, 1–14. [CrossRef]

62. Lubertozzi, D.; Keasling, J.D. Marker and promoter effects on heterologous expression in *Aspergillus nidulans*. *Appl. Microbiol. Biotechnol.* **2006**, *72*, 1014–1023. [CrossRef]

63. Verdoes, J.C.; Punt, P.J.; van den Hondel, C.A.M.J.J. Molecular genetic strain improvement for the overproduction of fungal proteins by filamentous fungi. *Appl. Microbiol. Biotechnol.* **1995**, *43*, 195–205. [CrossRef]

64. Gouka, R.J.; Hessing, J.G.M.; Stam, H.; Musters, W.; van den Hondel, C.A.M.J.J. A novel strategy for the isolation of defined pyrG mutants and the development of a site-specific integration system for *Aspergillus awamori*. *Curr. Genet.* **1995**, *27*, 536–540. [CrossRef]

65. Qin, L.; Jiang, X.; Dong, Z.; Huang, J.; Chen, X. Identification of two integration sites in favor of transgene expression in *Trichoderma reesei*. *Biotechnol. Biofuels* **2018**, *11*, 142. [CrossRef]

66. Hansen, B.G.; Salomonsen, B.; Nielsen, M.T.; Nielsen, J.B.; Hansen, N.B.; Nielsen, K.F.; Regueira, T.B.; Nielsen, J.; Patil, K.R.; Mortensen, U.H. Versatile enzyme expression and characterization system for *Aspergillus nidulans*, with the *Penicillium brevicompactum* polyketide synthase gene from the mycophenolic acid gene cluster as a test case. *Appl. Environ. Microbiol.* **2011**, *77*, 3044–3051. [CrossRef]

67. Mauro, V.P. Codon optimization in the production of recombinant biotherapeutics: Potential risks and considerations. *BioDrugs* **2018**, *32*, 69–81. [CrossRef] [PubMed]

68. Gustafsson, C.; Minshull, J.; Govindarajan, S.; Ness, J.; Villalobos, A.; Welch, M. Engineering genes for predictable protein expression. *Protein Expr. Purif.* **2012**, *83*, 37–46. [CrossRef] [PubMed]

69. Tanaka, M.; Tokuoka, M.; Gomi, K. Effects of codon optimization on the mRNA levels of heterologous genes in filamentous fungi. *Appl. Microbiol. Biotechnol.* **2014**, *98*, 3859–3867. [CrossRef] [PubMed]

70. Tokuoka, M.; Tanaka, M.; Ono, K.; Takagi, S.; Shintani, T.; Gomi, K. Codon optimization increases steady-state mRNA levels in *Aspergillus oryzae* heterologous gene expression. *Appl. Environ. Microbiol.* **2008**, *74*, 6538–6546. [CrossRef]

71. Koda, A.; Bogaki, T.; Minetoki, T.; Hirotsune, M. High expression of a synthetic gene encoding potato α-glucan phosphorylase in *Aspergillus niger*. *J. Biosci. Bioeng.* **2005**, *100*, 531–537. [CrossRef]

72. Gouka, R.J.; Punt, P.J.; Hessing, J.G.M.; Van Den Hondel, C.A.M.J.J. Analysis of heterologous protein production in defined recombinant *Aspergillus awamori* strains. *Appl. Environ. Microbiol.* **1996**, *62*, 1951–1957. [CrossRef]

73. Tanaka, M.; Tokuoka, M.; Shintani, T.; Gomi, K. Transcripts of a heterologous gene encoding mite allergen Der f 7 are stabilized by codon optimization in *Aspergillus oryzae*. *Appl. Microbiol. Biotechnol.* **2012**, *96*, 1275–1282. [CrossRef]

74. Yu, C.-H.; Dang, Y.; Zhou, Z.; Wu, C.; Zhao, F.; Sachs, M.S.; Liu, Y. Codon usage influences the local rate of translation elongation to regulate co-translational protein folding. *Mol. Cell* **2015**, *59*, 744–754. [CrossRef]

75. Gouka, R.J.; Punt, P.J.; Van Den Hondel, C.A.M.J.J. Glucoamylase gene fusions alleviate limitations for protein production in *Aspergillus awamori* at the transcriptional and (post)translational levels. *Appl. Environ. Microbiol.* **1997**, *63*, 488–497. [CrossRef]

76. Carraway, K.L.; Hull, S.R. O-glycosylation pathway for mucin-type glycoproteins. *BioEssays* **1989**, *10*, 117–121. [CrossRef]

77. Bause, E. Structural requirements of N-glycosylation of proteins. Studies with proline peptides as conformational probes. *Biochem. J.* **1983**, *209*, 331–336. [CrossRef] [PubMed]

78. Deshpande, N.; Wilkins, M.R.; Packer, N.; Nevalainen, H. Protein glycosylation pathways in filamentous fungi. *Glycobiology* **2008**, *18*, 626–637. [CrossRef] [PubMed]

79. Eriksen, S.H.; Jensen, B.; Olsen, J. Effect of N-linked glycosylation on secretion, activity, and stability of α-amylase from *Aspergillus oryzae*. *Curr. Microbiol.* **1998**, *37*, 117–122. [CrossRef] [PubMed]

80. Maras, M.; van Die, I.; Contreras, R.; van den Hondel, C.A. Filamentous fungi as production organisms for glycoproteins of bio-medical interest. *Glycoconj. J.* **1999**, *16*, 99–107. [CrossRef] [PubMed]

81. Kainz, E.; Gallmetzer, A.; Hatzl, C.; Nett, J.H.; Li, H.; Schinko, T.; Pachlinger, R.; Berger, H.; Reyes-Dominguez, Y.; Bernreiter, A.; et al. N-Glycan modification in *Aspergillus* species. *Appl. Environ. Microbiol.* **2008**, *74*, 1076–1086. [CrossRef]

82. Kasajima, Y.; Yamaguchi, M.; Hirai, N.; Ohmachi, T.; Yoshida, T. In vivo expression of UDP-N-acetylglucosamine: Alpha-3-D-mannoside beta-1,2-N-acetylglucosaminyltransferase I (GnT-1) in *Aspergillus oryzae* and effects on the sugar chain of alpha-amylase. *Biosci. Biotechnol. Biochem.* **2006**, *70*, 2662–2668. [CrossRef]

83. Ichishima, E.; Taya, N.; Ikeguchi, M.; Chiba, Y.; Nakamura, M.; Kawabata, C.; Inoue, T.; Takahashi, K.; Minetoki, T.; Ozeki, K.; et al. Molecular and enzymic properties of recombinant 1, 2-alpha-mannosidase from *Aspergillus saitoi* overexpressed in *Aspergillus oryzae* cells. *Biochem. J.* **1999**, *339*, 589–597. [CrossRef]

84. Kalsner, I.; Hintz, W.; Reid, L.S.; Schachter, H. Insertion into *Aspergillus nidulans* of functional UDP-GlcNAc: Alpha 3-D- mannoside beta-1,2-N-acetylglucosaminyl-transferase I, the enzyme catalysing the first committed step from oligomannose to hybrid and complex N-glycans. *Glycoconj. J.* **1995**, *12*, 360–370. [CrossRef]

85. Guillemette, T.; van Peij, N.N.M.E.; Goosen, T.; Lanthaler, K.; Robson, G.D.; van den Hondel, C.A.M.J.J.; Stam, H.; Archer, D.B. Genomic analysis of the secretion stress response in the enzyme-producing cell factory *Aspergillus niger*. *BMC Genom.* **2007**, *8*, 158. [CrossRef]

86. Kwon, M.J.; Jørgensen, T.R.; Nitsche, B.M.; Arentshorst, M.; Park, J.; Ram, A.F.; Meyer, V. The transcriptomic fingerprint of glucoamylase over-expression in *Aspergillus niger*. *BMC Genom.* **2012**, *13*, 701. [CrossRef]

87. Geysens, S.; Whyteside, G.; Archer, D.B. Genomics of protein folding in the endoplasmic reticulum, secretion stress and glycosylation in the aspergilli. *Fungal Genet. Biol.* **2009**, *46*, S121–S140. [CrossRef]

88. Valkonen, M.; Ward, M.; Wang, H.; Penttilä, M.; Saloheimo, M. Improvement of foreign-protein production in *Aspergillus niger* var. awamori by constitutive induction of the unfolded-protein response. *Appl. Environ. Microbiol.* **2003**, *69*, 6979–6986. [CrossRef] [PubMed]

89. Derkx, P.M.F.; Madrid, S.M. Peptidyl prolyl cis-trans isomerases 2000. World Patent WO0018934, 6 April 2000.

90. Wang, H.; Ward, M. Molecular characterization of a PDI-related gene prpA in *Aspergillus niger* var. awamori. *Curr. Genet.* **2000**, *37*, 57–64. [CrossRef]

91. Moralejo, F.; Watson, A.; Jeenes, D.; Archer, D.; Martín, J. A defined level of protein disulfide isomerase expression is required for optimal secretion of thaumatin by *Aspergillus awamori*. *Mol. Genet. Genom.* **2001**, *266*, 246–253. [CrossRef] [PubMed]

92. Conesa, A.; Jeenes, D.; Archer, D.B.; van den Hondel, C.A.; Punt, P.J. Calnexin overexpression increases manganese peroxidase production in *Aspergillus niger*. *Appl. Environ. Microbiol.* **2002**, *68*, 846–851. [CrossRef] [PubMed]

93. Lombraña, M.; Moralejo, F.J.; Pinto, R.; Martín, J.F. Modulation of *Aspergillus awamori* thaumatin secretion by modification of bipA gene expression. *Appl. Environ. Microbiol.* **2004**, *70*, 5145–5152. [CrossRef]

94. Banerjee, S.; Vishwanath, P.; Cui, J.; Kelleher, D.J.; Gilmore, R.; Robbins, P.W.; Samuelson, J. The evolution of N-glycan-dependent endoplasmic reticulum quality control factors for glycoprotein folding and degradation. *Proc. Natl. Acad. Sci. USA* **2007**, *104*, 11676–11681. [CrossRef]

95. Zhang, L.; Feng, D.; Fang, W.; Ouyang, H.; Luo, Y.; Du, T.; Jin, C. Comparative proteomic analysis of an *Aspergillus fumigatus* mutant deficient in glucosidase I (AfCwh41). *Microbiology* **2009**, *155*, 2157–2167. [CrossRef]

96. Ríos, S.; Fernández-Monistrol, I.; Laborda, F. Effect of tunicamycin on α-galactosidase secretion by *Aspergillus nidulans* and the importance of N-glycosylation. *FEMS Microbiol. Lett.* **1994**, *120*, 169–175. [CrossRef]

97. Perlińska-lenart, U.; Kurzątkowski, W.; Janas, P.; Palamarczyk, G.; Kruszewska, J.S. Protein production and secretion in an *Aspergillus nidulans* mutant impaired in glycosylation. *Acta Biochim. Pol.* **2005**, *52*, 195–205. [CrossRef]

98. Van den Brink, H.J.M.; Petersen, S.G.; Rahbek-Nielsen, H.; Hellmuth, K.; Harboe, M. Increased production of chymosin by glycosylation. *J. Biotechnol.* **2006**, *125*, 304–310. [CrossRef] [PubMed]

99. Wang, H. Gene Inactivated Mutants with Altered Protein Production. World Patent No. WO2006110677, 19 October 2006.

100. Dai, Z.; Aryal, U.K.; Shukla, A.; Qian, W.J.; Smith, R.D.; Magnuson, J.K.; Adney, W.S.; Beckham, G.T.; Brunecky, R.; Himmel, M.E.; et al. Impact of alg3 gene deletion on growth, development, pigment production, protein secretion, and functions of recombinant *Trichoderma reesei* cellobiohydrolases in *Aspergillus niger*. *Fungal Genet. Biol.* **2013**, *61*, 120–132. [CrossRef] [PubMed]

101. Pantazopoulou, A. The Golgi apparatus: Insights from filamentous fungi. *Mycologia* **2016**, *108*, 603–622. [CrossRef] [PubMed]

102. Fiedler, M.R.M.; Barthel, L.; Kubisch, C.; Nai, C.; Meyer, V. Construction of an improved *Aspergillus niger* platform for enhanced glucoamylase secretion. *Microb. Cell Fact.* **2018**, *17*, 1–12. [CrossRef]

103. Fiedler, M.R.M.; Cairns, T.C.; Koch, O.; Kubisch, C.; Meyer, V. Conditional expression of the small GTPase ArfA impacts secretion, morphology, growth, and actin ring position in *Aspergillus niger*. *Front. Microbiol.* **2018**, *9*, 1–17. [CrossRef]

104. Hoang, H.-D.; Maruyama, J.; Kitamoto, K. Modulating endoplasmic reticulum-golgi cargo receptors for improving secretion of carrier-fused heterologous proteins in the filamentous fungus *Aspergillus oryzae*. *Appl. Environ. Microbiol.* **2015**, *81*, 533–543. [CrossRef]

105. Bonifacino, J.S.; Weissman, A.M. Ubiquitin and the control of protein fate in the secretory and endocytic pathways. *Annu. Rev. Cell Dev. Biol.* **1998**, *14*, 19–57. [CrossRef]

106. Jacobs, D.I.; Olsthoorn, M.M.A.; Maillet, I.; Akeroyd, M.; Breestraat, S.; Donkers, S.; van der Hoeven, R.A.M.; van den Hondel, C.A.M.J.J.; Kooistra, R.; Lapointe, T. Effective lead selection for improved protein production in *Aspergillus niger* based on integrated genomics. *Fungal Genet. Biol.* **2009**, *46*, S141–S152. [CrossRef]

107. Carvalho, N.D.S.P.; Arentshorst, M.; Kooistra, R.; Stam, H.; Sagt, C.M.; Van Den Hondel, C.A.M.J.J.; Ram, A.F.J. Effects of a defective ERAD pathway on growth and heterologous protein production in *Aspergillus niger*. *Appl. Microbiol. Biotechnol.* **2011**, *89*, 357–373. [CrossRef]

108. Yoon, J.; Aishan, T.; Maruyama, J.; Kitamoto, K. Enhanced production and secretion of heterologous proteins by the filamentous fungus *Aspergillus oryzae* via disruption of vacuolar protein sorting receptor gene Aovps10. *Appl. Environ. Microbiol.* **2010**, *76*, 5718–5727. [CrossRef]

109. Yoon, J.; Kikuma, T.; Maruyama, J.; Kitamoto, K. Enhanced production of bovine chymosin by autophagy deficiency in the filamentous fungus *Aspergillus oryzae*. *PLoS ONE* **2013**, *8*, e62512. [CrossRef] [PubMed]

110. Zhang, N.; An, Z. Heterologous protein expression in yeasts and filamentous fungi. In *Manual of Industrial Microbiology and Biotechnology*; Baltz, R.H., Demain, A.L., Davies, J.E., Bull, A.T., Junker, B., Katz, L., Lynd, L.R., Masurekar, P., Reeves, C.D., Zhao, H., Eds.; American Society of Microbiology Press: Washington, DC, USA, 2010; pp. 145–156, ISBN 9781555815127.

111. Lawlis, V.B. DNA Sequences, Vectors, and Fusion Polypeptides to Increase Secretion of Desired Polypeptides from Filamentous Fungi. World Patent No. WO9015860, 27 December 1990.

112. Ward, M.; Wilson, L.J.; Kodama, K.H.; Rey, M.W.; Berka, R.M. Improved production of chymosin in *Aspergillus* by expression as a glucoamylase-chymosin fusion. *Nat. Biotechnol.* **1990**, *8*, 435–440. [CrossRef] [PubMed]

113. Cullen, D.; Gray, G.L.; Wilson, L.J.; Hayenga, K.J.; Lamsa, M.H.; Rey, M.W.; Norton, S.; Berka, R.M. Controlled expression and secretion of bovine chymosin in *Aspergillus nidulans*. *Nat. Biotechnol.* **1987**, *5*, 369–376. [CrossRef]

114. Broekhuijsen, M.P.; Mattern, I.E.; Contreras, R.; Kinghorn, J.R.; van den Hondel, C.A.M.J.J. Secretion of heterologous proteins by *Aspergillus niger*: Production of active human interleukin-6 in a protease-deficient mutant by KEX2-like processing of a glucoamylase-hIL6 fusion protein. *J. Biotechnol.* **1993**, *31*, 135–145. [CrossRef]

115. Frenken, L.G.J.; Van Gorcom, R.F.M.; Hessing, J.G.M.; Van Den Hondel, C.A.; Musters, W.; Verbakel, J.M.A.; Verrips, C.T. Process for Producing Fusion Proteins Comprising Scfv Fragments by a Transformed Mould. World Patent No. WO9429457, 22 December 1994.

116. Ward, M.; Lin, C.; Victoria, D.C.; Fox, B.P.; Fox, J.A.; Wong, D.L.; Meerman, H.J.; Pucci, J.P.; Fong, R.B.; Heng, M.H.; et al. Characterization of humanized antibodies secreted by *Aspergillus niger*. *Appl. Environ. Microbiol.* **2004**, *70*, 2567–2576. [CrossRef]

117. Segato, F.; Damásio, A.R.L.; Gonçalves, T.A.; de Lucas, R.C.; Squina, F.M.; Decker, S.R.; Prade, R.A. High-yield secretion of multiple client proteins in *Aspergillus*. *Enzyme Microb. Technol.* **2012**, *51*, 100–106. [CrossRef] [PubMed]

118. Jin, F.J.; Watanabe, T.; Juvvadi, P.R.; Maruyama, J.; Arioka, M.; Kitamoto, K. Double disruption of the proteinase genes, *tppA* and *pepE*, increases the production level of human lysozyme by *Aspergillus oryzae*. *Appl. Microbiol. Biotechnol.* **2007**, *76*, 1059–1068. [CrossRef]

119. Ohno, A.; Maruyama, J.; Nemoto, T.; Arioka, M.; Kitamoto, K. A carrier fusion significantly induces unfolded protein response in heterologous protein production by *Aspergillus oryzae*. *Appl. Microbiol. Biotechnol.* **2011**, *92*, 1197–1206. [CrossRef]

120. Gouka, R.J.; Van Den Hondel, C.A.; Musters, W.; Stam, H.; Verbakel, J.M.A. Process for Producing/Secreting a Protein by a Transformed Mould Using Expression/Secretion Regulating Regions Derived from an Aspergillus Endoxylanase II Gene. World Patent No. WO9312237, 24 June 1993.

121. Jalving, R.; Van De Vondervoort, P.J.I.; Visser, J.; Schaap, P.J. Characterization of the Kexin-Like Maturase of *Aspergillus niger*. *Appl. Environ. Microbiol.* **2000**, *66*, 363–368. [CrossRef]

122. Wang, H.; Ward, M. Kex2 Cleavage Regions of Recombinant Fusion Proteins. U.S. Patent No. US8,936,917B2, 20 January 2015.

123. O'Donnell, D.; Wang, L.; Xu, J.; Ridgway, D.; Gu, T.; Moo-Young, M. Enhanced heterologous protein production in *Aspergillus niger* through pH control of extracellular protease activity. *Biochem. Eng. J.* **2001**, *8*, 187–193. [CrossRef]

124. Van Den Hombergh, J.P.T.W.; Van De Vondervoort, P.J.I.; Fraissinet-Tachet, L.; Visser, J. *Aspergillus* as a host for heterologous protein production: The problem of proteases. *Trends Biotechnol.* **1997**, *15*, 256–263. [CrossRef]

125. Berka, R.M.; Ward, M.; Wilson, L.J.; Hayenga, K.J.; Kodama, K.H.; Carlomagne, L.P.; Thompson, S.A. Molecular cloning and deletion of the gene encoding aspergillopepsin A from *Aspergillus awamori*. *Gene* **1990**, *86*, 153–162. [CrossRef]

126. Mattern, I.E.; van Noort, J.M.; van den Berg, P.; Archer, D.B.; Roberts, I.N.; van den Hondep, C.A. Isolation and characterization of mutants of Aspergillus niger deficient in extracellular proteases. *Mol. Gen. Genet.* **1992**, 332–336. [CrossRef]

127. Kamaruddin, N.; Storms, R.; Mahadi, N.M.; Illias, R.M.; Bakar, F.D.A.; Murad, A.M.A. Reduction of extracellular proteases increased activity and stability of heterologous protein in *Aspergillus niger*. *Arab. J. Sci. Eng.* **2018**, *43*, 3327–3338. [CrossRef]

128. Berka, R.M.; Hayenga, K.; Lawlis, V.B.; Ward, M. Aspartic Proteinase Deficient Filamentous Fungi. World Patent No. WO199000192, 11 January 1990.

129. Punt, P.J.; Schuren, F.H.J.; Lehmbeck, J.; Christensen, T.; Hjort, C.; van den Hondel, C.A.M.J.J. Characterization of the *Aspergillus niger* prtT, a unique regulator of extracellular protease encoding genes. *Fungal Genet. Biol.* **2008**, *45*, 1591–1599. [CrossRef]

130. Yoon, J.; Kimura, S.; Maruyama, J.; Kitamoto, K. Construction of quintuple protease gene disruptant for heterologous protein production in *Aspergillus oryzae*. *Appl. Microbiol. Biotechnol.* **2009**, *82*, 691–701. [CrossRef]

131. Yoon, J.; Maruyama, J.; Kitamoto, K. Disruption of ten protease genes in the filamentous fungus *Aspergillus oryzae* highly improves production of heterologous proteins. *Appl. Microbiol. Biotechnol.* **2011**, *89*, 747–759. [CrossRef] [PubMed]

132. Lehmbeck, J. Novel Host Cells and Methods of Producing Proteins. World Patent No. WO9812300, 26 March 1998.

133. Shinkawa, S.; Mitsuzawa, S.; Tanaka, M.; Imai, T. *Aspergillus* Mutant Strain and Transformant Thereof. U.S. Patent No. US9,567,563, 11 August 2016.

134. Taheri-Talesh, N.; Horio, T.; Araujo-Bazán, L.; Dou, X.; Espeso, E.A.; Peñalva, M.A.; Osmani, S.A.; Oakley, B.R. The tip growth apparatus of *Aspergillus nidulans*. *Mol. Biol. Cell* **2008**, *19*, 1439–1449. [CrossRef]

135. Grimm, L.H.; Kelly, S.; Krull, R.; Hempel, D.C. Morphology and productivity of filamentous fungi. *Appl. Microbiol. Biotechnol.* **2005**, *69*, 375–384. [CrossRef]

136. Sisniega, H.; Río, J.-L.; Amaya, M.-J.; Faus, I. Strategies for large-scale production of recombinant proteins in filamentous fungi. In *Microbial Processes and Products*; Humana Press: Totowa, NJ, USA, 2005; Volume 18, pp. 225–237.

137. El-Enshasy, H.A. Filamentous fungal cultures—Process characteristics, products, and applications. In *Bioprocessing for Value-Added Products from Renewable Resources*; Elsevier: Amsterdam, The Netherlands, 2007; pp. 225–261.

138. Workman, M.; Andersen, M.R.; Thykaer, J. Integrated approaches for assessment of cellular performance in industrially relevant filamentous fungi. *Ind. Biotechnol.* **2013**, *9*, 337–344. [CrossRef]

139. Wang, L.; Ridgway, D.; Gu, T.; Moo-Young, M. Bioprocessing strategies to improve heterologous protein production in filamentous fungal fermentations. *Biotechnol. Adv.* **2005**, *23*, 115–129. [CrossRef] [PubMed]

140. Gyamerah, M.; Merichetti, G.; Adedayo, O.; Scharer, J.; Moo-Young, M. Bioprocessing strategies for improving hen egg-white lysozyme (HEWL) production by recombinant *Aspergillus niger* HEWL WT-13-16. *Appl. Microbiol. Biotechnol.* **2002**, *60*, 403–407. [CrossRef] [PubMed]

141. MacKenzie, D.A.; Gendron, L.C.G.; Jeenes, D.J.; Archer, D.B. Physiological optimization of secreted protein production by *Aspergillus niger*. *Enzyme Microb. Technol.* **1994**, *16*, 276–280. [CrossRef]

142. Wang, L.; Ridgway, D.; Gu, T.; Moo-Young, M. Effects of process parameters on heterologous protein production in *Aspergillus niger* fermentation. *J. Chem. Technol. Biotechnol.* **2003**, *78*, 1259–1266. [CrossRef]

143. Swift, R.J.; Karandikar, A.; Griffen, A.M.; Punt, P.J.; van den Hondel, C.A.M.J.J.; Robson, G.D.; Trinci, A.P.J.; Wiebe, M.G. The Effect of organic nitrogen sources on recombinant glucoamylase production by *Aspergillus niger* in chemostat culture. *Fungal Genet. Biol.* **2000**, *31*, 125–133. [CrossRef]

144. El-Enshasy, H.; Kleine, J.; Rinas, U. Agitation effects on morphology and protein productive fractions of filamentous and pelleted growth forms of recombinant *Aspergillus niger*. *Process Biochem.* **2006**, *41*, 2103–2112. [CrossRef]

145. Eibes, G.M.; Lú-Chau, T.A.; Ruiz-Dueñas, F.J.; Feijoo, G.; Martínez, M.J.; Martínez, A.T.; Lema, J.M. Effect of culture temperature on the heterologous expression of *Pleurotus eryngii* versatile peroxidase in *Aspergillus* hosts. *Bioprocess Biosyst. Eng.* **2009**, *32*, 129–134. [CrossRef]

146. Amanullah, A.; Christensen, L.H.; Hansen, K.; Nienow, A.W.; Thomas, C.R. Dependence of morphology on agitation intensity in fed-batch cultures of *Aspergillus oryzae* and its implications for recombinant protein production. *Biotechnol. Bioeng.* **2002**, *77*, 815–826. [CrossRef]

147. Archer, D.B.; MacKenzie, D.A.; Ridout, M.J. Heterologous protein secretion by *Aspergillus niger* growing in submerged culture as dispersed or aggregated mycelia. *Appl. Microbiol. Biotechnol.* **1995**, *44*, 157–160. [CrossRef]

148. Driouch, H.; Sommer, B.; Wittmann, C. Morphology engineering of *Aspergillus niger* for improved enzyme production. *Biotechnol. Bioeng.* **2010**, *105*, 1058–1068. [CrossRef]

149. Driouch, H.; Hänsch, R.; Wucherpfennig, T.; Krull, R.; Wittmann, C. Improved enzyme production by bio-pellets of *Aspergillus niger*: Targeted morphology engineering using titanate microparticles. *Biotechnol. Bioeng.* **2012**, *109*, 462–471. [CrossRef] [PubMed]

Erratum

Erratum: Ntana, F., et al. Aspergillus: A Powerful Protein Production Platform. *Catalysts* 2020, *10*, 1064

Fani Ntana [1], Uffe Hasbro Mortensen [2], Catherine Sarazin [1,*] and Rainer Figge [1]

[1] Unité de Génie Enzymatique et Cellulaire, UMR 7025 CNRS/UPJV/UTC, Université de Picardie Jules Verne, 80039 Amiens, France; fani.ntana@u-picardie.fr (F.N.); rainer.figge@u-picardie.fr (R.F.)

[2] Department of Biotechnology and Biomedicine, Technical University of Denmark, Søltofts Plads, Building 223, 2800 Kongens Lyngby, Denmark; um@bio.dtu.dk

* Correspondence: catherine.sarazin@u-picardie.fr; Tel.: +33-3-22-82-75-95

Received: 24 November 2020; Accepted: 25 November 2020; Published: 30 November 2020

The author wishes to make the following erratum to this paper [1]: Update due to some reporting errors in Tables 2, 8, 10 and 12.

Due to typographical errors concerning reference [47] and [51,52], replace:

Table 2. Approaches for improving recombinant protein production through promoter engineering.

Process	Modification	Performance	Improvement Factor	Reference
Promoters	Use of several promoters (P) in *A. awamori*	PB2 from *Acremonium chrysogenum*: 0.25–2 mg/L thaumatin	-	[46]
		PpcbC from *Penicillium chrysogenum*: 0.25–2 mg/L thaumatin		
		PgdhA from *A. awamori*: 1–9 mg/L thaumatin		
		PgpdA from *A. nidulans*: 0.75–11 mg/L thaumatin		
	Insertion of multiple copies of an activator protein-binding site from the *cis*-regulatory region of *A. niger glaA* to the new promoter in *A. niger*	396.0 ± 51.5 mg/L of *Vitreoscilla* hemoglobin compared to 19.7 ± 4.8 mg/L from the strain with 1 copy	20	[45]
	Use of hybrid promoters (combination of a human hERa-activated promoter (pERE), *S. cerevisiae* URA3 promoter and *A. nidulans nirA* promoter) in *A. nidulans*	pERE-URA-nirA + *lacZ*: 25 U of β-galactosidase activity/mg of protein	-	[47]
		pERE-URA-RS (random stuffer-link) + *lacZ*: 100 U of β-galactosidase activity/mg of protein	4	
		pERE-RS-nirA + *lacZ*: 1400 U of β-galactosidase activity/mg of protein [1 pM inducer (DES)]	56	
	Use of a hemolysin-like protein promoter (Phyl) for heterologous production in *A. oryzae*	Reporter gene: Endoglucanase Cel B Pamy: 24.1 ± 5.5 U/mL, Phyl: 57.9 ± 17.4 U/mL	2.4	[48]
		Reporter gene: *Trichoderma* endoglucanase I Pamy: 7.7 ± 3.9 U/mL, Phyl: 27.8 ± 1.3 U/mL	3.6	
		Reporter gene: *Trichoderma* endoglucanase III Pamy:4.0 ± 0.6 U/mL, hyl:31.7 ± 3.3 U/mL	7.9	
	Regulatory elements (TerR and PterA) from *A. terreus* terrain gene cluster for *E. coli lacZ* expression in *A. niger*	Promoter activity ~5000 mU/mg when TerR under PgpdA (No activity when TerR under the native promoter)	-	[49]
		Promoter activity ~10,000 mU/mg (when TerR under PgpdA in 2 copies)	2	
		Promoter activity ~15,000 mU/mg (when TerR under PamyB)	3	

Table 2. *Cont.*

Process	Modification		Performance	Improvement Factor	Reference
	A. niger α-glucosyltransferase produced under the *A. niger* pyruvate kinase promoter		2000 U/mL total activity of α-glucosyltransferase compared to 600 U/mL in the wild type	3.3	[50]
	Overexpression of the transcription factor RsmA, while the aflR promoter was inserted in front of the *pslcc* in *A. nidulans*		60,000 U/mL of *Pycnoporus sanguineus* laccase compared to 4000 U/mL in the control strain	15	[51,52]
	A novel promoter from *Talaromyces emersonii* (Pglucan1200) for expressing *glaA* in *A. niger*		6000 U/mL of GlaA, enzyme activity increased by about 25% compared to 5000 U/mL in the strain with the PglaA	1.2	[53]
	The constitutive promoter of *ecm33* (Pecm33) from *A. niger* in *A. niger*	Maltose:	Pecm33 activity induced by 1.7 compared to PglaA activity that induced by 2.7	-	[54]
		Glucose:	Pecm33 activity induced by 1.1 compared to PglaA activity that induced by 1.8		
		Xylose:	Pecm33 activity induced by 2 compared to PglaA activity that induced by 1.3 Increased Pecm33 activity at 37 °C		

with

Table 2. Approaches for improving recombinant protein production through promoter engineering.

Process	Modification	Performance	Improvement Factor	Reference
Promoters	Use of several promoters (P) in *A. awamori*	PB2 from *Acremonium chrysogenum*: 0.25–2 mg/L thaumatin	-	[46]
		PpcbC from *Penicillium chrysogenum*: 0.25–2 mg/L thaumatin		
		PgdhA from *A. awamori*: 1–9 mg/L thaumatin		
		PgpdA from *A. nidulans*: 0.75–11 mg/L thaumatin		
	Insertion of multiple copies of an activator protein-binding site from the *cis*-regulatory region of *A. niger glaA* to the new promoter in *A. niger*	396.0 ± 51.5 mg/L of *Vitreoscilla* hemoglobin compared to 19.7 ± 4.8 mg/L from the strain with 1 copy	20	[45]
	Use of hybrid promoters (combination of a human hERa-activated promoter (pERE), *S. cerevisiae* URA3 promoter and *A. nidulans nirA* promoter) in *A. nidulans*	pERE-RS-nirA+ *lacZ*: 25 U of β-galactosidase activity/mg of protein	-	[47]
		pERE-URA-nirA+ *lacZ*: 100 U of β-galactosidase activity/mg of protein	4	
		pERE-URA-RS + *lacZ*: 1400 U of β-galactosidase activity/mg of protein [1 pM inducer (DES)]	56	
	Use of a hemolysin-like protein promoter (Phyl) for heterologous production in *A. oryzae*	Reporter gene: Endoglucanase Cel B Pamy: 24.1 ± 5.5 U/mL, Phyl: 57.9 ± 17.4 U/mL	2.4	[48]
		Reporter gene: *Trichoderma* endoglucanase I Pamy: 7.7 ± 3.9 U/mL, Phyl: 27.8 ± 1.3 U/mL	3.6	
		Reporter gene: *Trichoderma* endoglucanase III Pamy:4.0 ± 0.6 U/mL, hyl:31.7 ± 3.3 U/mL	7.9	
	Regulatory elements (TerR and PterA) from *A. terreus* terrain gene cluster for *E. coli lacZ* expression in *A. niger*	Promoter activity ~5000 mU/mg when TerR under PgpdA (No activity when TerR under the native promoter)	-	[49]
		Promoter activity ~10,000 mU/mg (when TerR under PgpdA in 2 copies)	2	
		Promoter activity ~15,000 mU/mg (when TerR under PamyB)	3	

Table 2. *Cont.*

Process	Modification	Performance	Improvement Factor	Reference
	A. niger α-glucosyltransferase produced under the *A. niger* pyruvate kinase promoter	2000 U/mL total activity of α-glucosyltransferase compared to 600 U/mL in the wild type	3.3	[50]
	Overexpression of the transcription factor RsmA, while the aflR promoter was inserted in front of the *pslcc* in *A. nidulans*	0.06 U/mL of *Pycnoporus sanguineus* laccase compared to 0.004 U/mL in the control strain	15	[51,52]
	A novel promoter from *Talaromyces emersonii* (Pglucan1200) for expressing *glaA* in *A. niger*	6000 U/mL of GlaA compared to 5000 U/mL in the strain with the PglaA	1.2	[53]
	The constitutive promoter of *ecm33* (Pecm33) from *A. niger* in *A. niger* Maltose:	Pecm33 activity induced by 1.7 compared to PglaA activity that induced by 2.7	-	[54]
	Glucose:	Pecm33 activity induced by 1.1 compared to PglaA activity that induced by 1.8		
	Xylose:	Pecm33 activity induced by 2 compared to PglaA activity that induced by 1.3 Increased Pecm33 activity at 37 °C		

Due to a typographical error concerning reference [109], replace:

Table 8. Approaches for improving recombinant protein production through engineering protein degradation pathways.

Process	Modification	Performance	Improvement Factor	Reference
Protein degradation pathways— ERAD and Vacuole	Deletion of *derA* and *derB* in *A. niger*	ΔderA: 80% decrease in *Tramete* laccase production	0.2	[99]
	-	ΔderB: 15.7% increase in *Tramete* laccase	1.15	
	Deletion of *doaA* and overexpression of *sttC* in *A. niger*	Higher GUS activity compared to parental strain (no quantitative data available)	-	[106]
	Disruption of *Aovps10* in *A. oryzae*	83.1 and 70.3 mg/L chymosin compared to 28.7 mg/L in parental strain	3–2.5	[108]
		22.6 and 24.6 mg/L human lysozyme compared to 11.1 mg/L in parental strain	2–2.2	
	Deletion of ERAD key genes (*derA*, *doaA*, *hrdC*, *mifA* and *mnsA*) in *A. niger*	ΔderA and ΔhrdC: 2-fold increase compared to parental strain (single-copy)	2	[107]
		ΔderA: 6-fold increase compared to parental strain (multi-copy) Relative amount of intracellular GlaGus (β-glucuronidase levels) fusion protein detected in total protein extracts of strains with impaired ERAD and respective parental strain	6	

Table 8. *Cont.*

Process	Modification	Performance	Improvement Factor	Reference
	Disruption of genes involved in autophagy in *A. oryzae*	ΔAoatg1: 60 mg/L chymosin	2.3	[109]
		ΔAoatg13: 37 mg/L chymosin	1.4	
		ΔAoatg4: 80 mg/L chymosin	3.1	
		ΔAoatg8: 66 mg/L chymosin	2.5	
		ΔAoatg15: Not detectable	-	
		Control: 26 mg/L chymosin	-	

with

Table 8. Approaches for improving recombinant protein production through engineering protein degradation pathways.

Process	Modification	Performance	Improvement Factor	Reference
Protein degradation pathways—ERAD and Vacuole	Deletion of *derA* and *derB* in *A. niger*	ΔderA: 80% decrease in *Tramete* laccase production	0.2	[99]
	-	ΔderB: 15.7% increase in *Tramete* laccase	1.15	
	Deletion of *doaA* and overexpression of *sttC* in *A. niger*	Higher GUS activity compared to parental strain (no quantitative data available)	-	[106]
	Disruption of *Aovps10* in *A. oryzae*	83.1 and 70.3 mg/L chymosin compared to 28.7 mg/L in parental strain	3–2.5	[108]
		22.6 and 24.6 mg/L human lysozyme compared to 11.1 mg/L in parental strain	2–2.2	
	Deletion of ERAD key genes (*derA*, *doaA*, *hrdC*, *mifA* and *mnsA*) in *A. niger*	ΔderA and ΔhrdC: 2-fold increase compared to parental strain (single-copy)	2	[107]
		ΔderA: 6-fold increase compared to parental strain (multi-copy) Relative amount of intracellular GlaGus (β-glucuronidase levels) fusion protein detected in total protein extracts of strains with impaired ERAD and respective parental strain	6	
	Disruption of genes involved in autophagy in *A. oryzae*	ΔAoatg1: 60 mg/L chymosin	2.3	[109]
		ΔAoatg13: 37 mg/L chymosin	1.4	
		ΔAoatg4: 80 mg/L chymosin	3.1	
		ΔAoatg8: 66 mg/L chymosin	2.5	
		ΔAoatg15: 24 mg/L chymosin	1	
		Control: 26 mg/L chymosin	-	

Due to typographical errors concerning reference [126] and [51], replace:

Table 10. Approaches for improving recombinant protein production through disruption of protease genes.

Process	Modification	Performance	Improvement Factor	Reference
Proteases	Deletion of *pepA* in *A. awamori* strains	Decreased extracellular proteolytic activity compared to the wild type (immunoassay using antibodies specific for PepA, but absolute values for PepA concentration were not determined)	-	[125]
	Deletion of *pepA* in *A. awamori*	430 mg/L of chymosin compared to 180 mg/L in the parental strain	2.4	[128]
	Deletion of *pepA* in *A. niger* (AB1.1)	15–20% proteolytic activity compared to the parent strain AB4.1	-	[126]
	Mutation on *prtT* (UV irradiation) in *A. niger* (AB1.13)	1–2% proteolytic activity compared to the parent strain AB4.1	-	[126]
	Deletion of *prtR*, *pepA*, *cpI*, *tppA* in *A. oryzae*	ΔprtR/pepA/cpI: 24.23 mg/L of *Acremonium cellulolyticus* cellobiohydrolase	1.2	[133]
		ΔprtR/pepA/tppA: 21.30 mg/L	1.1	
		ΔprtR/cpI/tppA: 22.08 mg/L	1.1	
		ΔprtR/pepA/cpI/tppA: 19.93 mg/L compared to 19.54 mg/L in the control strains	1.02	
	Deletion of *alp* and *Npl* in *A. oryzae*	1041 U/g of *Candida antarctica* lipase B compared to 575 U/g in the parental strains	1.8	[132]
	Deletion of various proteases in *A. niger*	Δdpp4: 6% increase in *Tramete* laccase	1.1	[99]
		Δdpp5: 15.4% increase	1.2	
		ΔpepB: 8.6% increase	1.1	
		ΔpepD: 4.8% increase	1.0	
		ΔpepF: 5.3% increase	1.1	
		ΔpepAa: 0.5% increase	1.1	
		ΔpepAb: 13.4% increase	1.1	
		ΔpepAd: 2.7% increase	1.0	
		Δdpp4/dpp5: 26.6% increase	1.3	
	Disruption of *tppA* and *pepE* in *A. oryzae* strains	25.4 mg/L of human lysozyme compared to 15 mg/L in the parental strains	1.7	[118]
	Disruption of *tppA*, *pepE*, *nptB*, *dppIV* and *dppV* in *A. oryzae*	84.4 mg/L of chymosin compared to the 63.1 mg/L in the double protease gene disruptant (ΔtppA/pepE)	1.3	[130]
	Disruption of *tppA*, *pepE*, *nptB*, *dppIV*, and *dppV*, *alpA*, *pepA*, *AopepAa*, *AopepAd* and *cpI* in *A. oryzae*	109.4 mg/L of chymosin and 35.8 mg/L of human lysozyme compared to the quintuple protease gene disruptant (ΔtppA/pepE/nptB/dppIV/dppV; 84.4 mg/L and 26.5 mg/L, respectively)	1.3 and 1.35	[131]
	Deletion of *prtT* in *A. niger*	36.3–36.7 U/mL of mL *G. cingulate* cutinase compared to 21.2–20.4 U/mL in the parental strain	1.7	[127]
		Stability: Cutinase activity retained at 80% over the entire 14-day incubation period, while the parental lost more than 50% of their initial activities after six days of incubation and retained negligible activity after 14 days	-	
	Deletion of *dppV* and *pepA* in *A. nidulans*	*P. sanguineus* laccase activity 500,000 U/mL compared to 40,000 U/mL in the control strain	12.5	[51]
	Deletion of *mnn9* and *pepA* in *A. nidulans*	*P. sanguineus* laccase activity 300,000 U/mL compared to 40,000 U/mL in the control strain	7.5	[51]

with

Table 10. Approaches for improving recombinant protein production through disruption of protease genes.

Process	Modification	Performance	Improvement Factor	Reference
Proteases	Deletion of *pepA* in *A. awamori* strains	Decreased extracellular proteolytic activity compared to the wild type (immunoassay using antibodies specific for PepA, but absolute values for PepA concentration were not determined)	-	[125]
	Deletion of *pepA* in *A. awamori*	430 mg/L of chymosin compared to 180 mg/L in the parental strain	2.4	[128]
	Deletion of *pepA* in *A. niger* (AB1.18)	15–20% proteolytic activity compared to the parent strain AB4.1	-	[126]
	Mutation on *prtT* (UV irradiation) in *A. niger* (AB1.13)	1–2% proteolytic activity compared to the parent strain AB4.1	-	[126]
	Deletion of *prtR, pepA, cpI, tppA* in *A. oryzae*	ΔprtR/pepA/cpI: 24.23 mg/L of *Acremonium cellulolyticus* cellobiohydrolase	1.2	[133]
		ΔprtR/pepA/tppA: 21.30 mg/L	1.1	
		ΔprtR/cpI/tppA: 22.08 mg/L	1.1	
		ΔprtR/pepA/cpI/tppA: 19.93 mg/L compared to 19.54 mg/L in the control strains	1.02	
	Deletion of *alp* and *Npl* in *A. oryzae*	1041 U/g of *Candida antarctica* lipase B compared to 575 U/g in the parental strains	1.8	[132]
	Deletion of various proteases in *A. niger*	Δdpp4: 6% increase in *Tramete* laccase	1.1	[99]
		Δdpp5: 15.4% increase	1.2	
		ΔpepB: 8.6% increase	1.1	
		ΔpepD: 4.8% increase	1.0	
		ΔpepF: 5.3% increase	1.1	
		ΔpepAa: 0.5% increase	1.1	
		ΔpepAb: 13.4% increase	1.1	
		ΔpepAd: 2.7% increase	1.0	
		Δdpp4/dpp5: 26.6% increase	1.3	
	Disruption of *tppA* and *pepE* in *A. oryzae* strains	25.4 mg/L of human lysozyme compared to 15 mg/L in the parental strains	1.7	[118]
	Disruption of *tppA, pepE, nptB, dppIV* and *dppV* in *A. oryzae*	84.4 mg/L of chymosin compared to the 63.1 mg/L in the double protease gene disruptant (ΔtppA/pepE)	1.3	[130]
	Disruption of *tppA, pepE, nptB, dppIV,* and *dppV, alpA, pepA, AopepAa, AopepAd* and *cpI* in *A. oryzae*	109.4 mg/L of chymosin and 35.8 mg/L of human lysozyme compared to the quintuple protease gene disruptant (ΔtppA/pepE/nptB/dppIV/dppV; 84.4 mg/L and 26.5 mg/L, respectively)	1.3 and 1.35	[131]
	Deletion of *prtT* in *A. niger*	36.3–36.7 U/mL of mL *G. cingulate* cutinase compared to 21.2–20.4 U/mL in the parental strain	1.7	[127]
		Stability: Cutinase activity retained at 80% over the entire 14-day incubation period, while the parental lost more than 50% of their initial activities after six days of incubation and retained negligible activity after 14 days	-	
	Deletion of *dppV* and *pepA* in *A. nidulans*	*P. sanguineus* laccase activity 0.5 U/mL compared to 0.04 U/mL in the control strain	12.5	[51]
	Deletion of *mnn9* and *pepA* in *A. nidulans*	*P. sanguineus* laccase activity 0.3 U/mL compared to 0.04 U/mL in the control strain	7.5	[51]

Due to a typographical error concerning reference [144], replace:

Table 12. Approaches for improving recombinant protein production through bioprocessing modifications.

Process	Modification	Performance	Improvement Factor	Reference
Fermentation conditions	Effect of growth medium and temperature on hen egg white lysozyme (HEWL) production in *A. niger*	20–25 °C 8–10 mg/L HEWL while 30–37 °C 3–5 mg/L HEWL	Temperature: 2–2.6	[141]
		soluble starch: 8.0 mg/L HEWL	Carbon source: 1.7–2	
		maltose: 4.5 mg/L HEWL	-	
		glucose: 4.0 mg/L HEWL	-	
		xylose:0.2 mg/L HEWL	-	
		soy milk medium: 30–60 mg/L HEWL	Rich medium: 3.8–7.5	
	Effect of organic nitrogen sources on recombinant glucoamylase production in *A. niger*	Unsupplemented: 44 mg glucoamylase/g biomass	-	[143]
		L-alanine: 32 mg glucoamylase/g biomass	0.7	
		L-methionine: 26 mg glucoamylase/g	0.6	
		casamino acids, yeast extract, peptone, and gelatin: 100 mg glucoamylase/g	2.2	
	Effect of agitation intensity on recombinant amyloglucosidase (AMG) production in *A. oryzae*	Titer at the end of the batch phase 525 rpm: 110 U/L AMG	-	[146]
		675 rpm: 230 U/L AMG	1.6	
		825 rpm: 370 U/L AMG	3.3	
	Effects of bioprocess parameters—agitation intensity, initial glucose concentration, initial yeast extract concentration, and dissolved oxygen tension (DO)—on heterologous protein production in *A. oryzae*	Highest GFP yields were achieved under these conditions: agitation 400 rpm, glucose 25 g/L, yeast extract 0 g/dm^3, DO 15%	-	[142]
	Effect of agitation intensity on recombinant glucose oxidase production in *A. niger*	200 rpm: 300 mkat/L of glucose oxidase	-	[144]
		500 rpm: 800 mkat/L of glucose oxidase	2.6	
		800 rpm: 600 mkat/L of glucose oxidase	1.3	
	Effect of temperature on *Pleurotus eryngii* versatile peroxidase production in *A. nidulans* and *A. niger*	-*A. nidulans* 31 °C: 24 U/L peroxidase activity	-	[145]
		28 °C: 80 U/L peroxidase activity	3.3	
		19 °C: 466 U/L peroxidase activity	19.4	
		-*A. niger* 28 °C: 107 U/L peroxidase activity	-	
		19 °C: 412 U/L peroxidase activity	3.8	
Fungal morphology	Effect of raising the viscosity of the medium by addition of polyvinylpyrrolidone-PVP (transition from aggregated mycelia (pellets) to dispersed mycelia) on hen egg white lysozyme (HEWL) in *A. niger*	Medium with no PVP: 110 mg/L fresh and 8 mg/g dry weight of HEWL	1.7	[147]
		Medium with PVP: 190 mg/L fresh and 14 mg/g dry weight of HEWL		

Table 12. *Cont.*

Process	Modification	Performance	Improvement Factor	Reference
	Effect of addition of microparticles (linked to the formation of freely dispersed mycelium) on titers of native glucoamylase (GlaA) and recombinant fructofuranosidase (FF) produced in *A. niger*	No microparticles: 17 U/mL GlaA and 42 U/mL FF		
		Talc microparticles: 61 U/mL GlaA and 92 U/mL FF. FF production can reach up to 160 U/mL (10 g/L talc microparticles of size 6 mm)	3.5 GlaA 2–3.8 FF	[148]
	Effect of addition of titanate microparticles (TiSiO$_4$, 8 mm) on titers of native glucoamylase (GlaA) and recombinant fructofuranosidase (FF) produced in *A. niger*	No microparticles: 19 U/mL GlaA and 40 U/mL FF		
		Microparticles: 190 U/mL glucoamylase and 150 U/mL fructofuranosidase	9.5 GlaA 3.7 FF	[149]
	Effect of growth type on hen egg white lysozyme (HEWL) production and protease activity in *A. niger*	Free suspension: 5.8 mg/g HEWL 95.3 U/g Protease activity	1.5	
		Mycelial pellets: 5.0 mg/g HEWL 58.6 U/g Protease activity	1.2	[140]
		Celite-560-immobilized cultures: 4.1 mg/g HEWL 56.3 U/g Protease activity	-	

with

Table 12. Approaches for improving recombinant protein production through bioprocessing modifications.

Process	Modification	Performance	Improvement Factor	Reference
Fermentation conditions	Effect of growth medium and temperature on hen egg white lysozyme (HEWL) production in *A. niger*	20–25 °C 8–10 mg/L HEWL while 30–37 °C 3–5 mg/L HEWL	Temperature: 2–2.6	
		soluble starch: 8.0 mg/L HEWL	Carbon source: 1.7–2	
		maltose: 4.5 mg/L HEWL	-	[141]
		glucose: 4.0 mg/L HEWL	-	
		xylose:0.2 mg/L HEWL	-	
		soy milk medium: 30–60 mg/L HEWL	Rich medium: 3.8–7.5	
	Effect of organic nitrogen sources on recombinant glucoamylase production in *A. niger*	Unsupplemented: 44 mg glucoamylase/g biomass	-	
		L-alanine: 32 mg glucoamylase/g biomass	0.7	
		L-methionine: 26 mg glucoamylase/g	0.6	[143]
		casamino acids, yeast extract, peptone, and gelatin: 100 mg glucoamylase/g	2.2	

Table 12. *Cont.*

Process	Modification	Performance	Improvement Factor	Reference
	Effect of agitation intensity on recombinant amyloglucosidase (AMG) production in *A. oryzae*	Titer at the end of the batch phase 525 rpm: 110 U/L AMG	-	[146]
		675 rpm: 230 U/L AMG	1.6	
		825 rpm: 370 U/L AMG	3.3	
	Effects of bioprocess parameters—agitation intensity, initial glucose concentration, initial yeast extract concentration, and dissolved oxygen tension (DO)—on heterologous protein production in *A. oryzae*	Highest GFP yields were achieved under these conditions: agitation 400 rpm, glucose 25 g/L, yeast extract 0 g/dm^3, DO 15%	-	[142]
	Effect of agitation intensity on recombinant glucose oxidase production in *A. niger*	200 rpm: 300 µkat/L of glucose oxidase	-	[144]
		500 rpm: 800 µkat/L of glucose oxidase	2.6	
		800 rpm: 600 µkat/L of glucose oxidase	1.3	
	Effect of temperature on *Pleurotus eryngii* versatile peroxidase production in *A. nidulans* and *A. niger*	-*A. nidulans* 31 °C: 24 U/L peroxidase activity	-	[145]
		28 °C: 80 U/L peroxidase activity	3.3	
		19 °C: 466 U/L peroxidase activity	19.4	
		-*A. niger* 28 °C: 107 U/L peroxidase activity	-	
		19 °C: 412 U/L peroxidase activity	3.8	
Fungal morphology	Effect of raising the viscosity of the medium by addition of polyvinylpyrrolidone-PVP (transition from aggregated mycelia (pellets) to dispersed mycelia) on hen egg white lysozyme (HEWL) in *A. niger*	Medium with no PVP: 110 mg/L fresh and 8 mg/g dry weight of HEWL	1.7	[147]
		Medium with PVP: 190 mg/L fresh and 14 mg/g dry weight of HEWL		
	Effect of addition of microparticles (linked to the formation of freely dispersed mycelium) on titers of native glucoamylase (GlaA) and recombinant fructofuranosidase (FF) produced in *A. niger*	No microparticles: 17 U/mL GlaA and 42 U/mL FF	3.5 GlaA 2–3.8 FF	[148]
		Talc microparticles: 61 U/mL GlaA and 92 U/mL FF FF production can reach up to 160 U/mL (10 g/L talc microparticles of size 6 mm)		
	Effect of addition of titanate microparticles (TiSiO$_4$, 8 mm) on titers of native glucoamylase (GlaA) and recombinant fructofuranosidase (FF) produced in *A. niger*	No microparticles: 19 U/mL GlaA and 40 U/mL FF	9.5 GlaA 3.7 FF	[149]
		Microparticles: 190 U/mL glucoamylase and 150 U/mL fructofuranosidase		
	Effect of growth type on hen egg white lysozyme (HEWL) production and protease activity in *A. niger*	Free suspension: 5.8 mg/g HEWL 95.3 U/g Protease activity	1.5	[140]
		Mycelial pellets: 5.0 mg/g HEWL 58.6 U/g Protease activity	1.2	
		Celite-560-immobilized cultures: 4.1 mg/g HEWL 56.3 U/g Protease activity	-	

This update does not change any of the scientific results of the paper. The authors would like to apologize for any inconvenience caused to the readers by these changes. The manuscript will be updated and the original will remain online on the article webpage: https://www.mdpi.com/2073-4344/10/9/1064.

Reference

1. Ntana, F.; Mortensen, U.H.; Sarazin, C.; Figge, R. Aspergillus: A Powerful Protein Production Platform. *Catalysts* **2020**, *10*, 1064. [CrossRef]

Publisher's Note: MDPI stays neutral with regard to jurisdictional claims in published maps and institutional affiliations.

MDPI
St. Alban-Anlage 66
4052 Basel
Switzerland
Tel. +41 61 683 77 34
Fax +41 61 302 89 18
www.mdpi.com

Catalysts Editorial Office
E-mail: catalysts@mdpi.com
www.mdpi.com/journal/catalysts

www.ingramcontent.com/pod-product-compliance
Lightning Source LLC
LaVergne TN
LVHW070858170726
843515LV00005B/676